Lecture Notes in Physics

Springer
Berlin
Heidelberg
New York
Barcelona
Budapest
Hong Kong
London
Milan
Paris
Santa Clara
Singapore
Tokyo

The Editorial Policy for Proceedings

The series Lecture Notes in Physics reports new developments in physical research and teaching – quickly, informally, and at a high level. The proceedings to be considered for publication in this series should be limited to only a few areas of research, and these should be closely related to each other. The contributions should be of a high standard and should avoid lengthy redraftings of papers already published or about to be published elsewhere. As a whole, the proceedings should aim for a balanced presentation of the theme of the conference including a description of the techniques used and enough motivation for a broad readership. It should not be assumed that the published proceedings must reflect the conference in its entirety. (A listing or abstracts of papers presented at the meeting but not included in the proceedings could be added as an appendix.)

When applying for publication in the series Lecture Notes in Physics the volume's editor(s) should submit sufficient material to enable the series editors and their referees to make a fairly accurate evaluation (e.g. a complete list of speakers and titles of papers to be presented and abstracts). If, based on this information, the proceedings are (tentatively) accepted, the volume's editor(s), whose name(s) will appear on the title pages, should select the papers suitable for publication and have them refereed (as for a journal) when appropriate. As a rule discussions will not be accepted. The series editors and Springer-Verlag will normally not interfere with the detailed editing except in fairly obvious cases or on technical matters.

Final acceptance is expressed by the series editor in charge, in consultation with Springer-Verlag only after receiving the complete manuscript. It might help to send a copy of the authors' manuscripts in advance to the editor in charge to discuss possible revisions with him. As a general rule, the series editor will confirm his tentative acceptance if the final manuscript corresponds to the original concept discussed, if the quality of the contribution meets the requirements of the series, and if the final size of the manuscript does not greatly exceed the number of pages originally agreed upon. The manuscript should be forwarded to Springer-Verlag shortly after the meeting. In cases of extreme delay (more than six months after the conference) the series editors will check once more the timeliness of the papers. Therefore, the volume's editor(s) should establish strict deadlines, or collect the articles during the conference and have them revised on the spot. If a delay is unavoidable, one should encourage the authors to update their contributions if appropriate. The editors of proceedings are strongly advised to inform contributors about these points at an early stage.

The final manuscript should contain a table of contents and an informative introduction accessible also to readers not particularly familiar with the topic of the conference. The contributions should be in English. The volume's editor(s) should check the contributions for the correct use of language. At Springer-Verlag only the prefaces will be checked by a copy-editor for language and style. Grave linguistic or technical shortcomings may lead to the rejection of contributions by the series editors. A conference report should not exceed a total of 500 pages. Keeping the size within this bound should be achieved by a stricter selection of articles and not by imposing an upper limit to the length of the individual papers. Editors receive jointly 30 complimentary copies of their book. They are entitled to purchase further copies of their book at a reduced rate. As a rule no reprints of individual contributions can be supplied. No royalty is paid on Lecture Notes in Physics volumes. Commitment to publish is made by letter of interest rather than by signing a formal contract. Springer-Verlag secures the copyright for each volume.

The Production Process

The books are hardbound, and the publisher will select quality paper appropriate to the needs of the author(s). Publication time is about ten weeks. More than twenty years of experience guarantee authors the best possible service. To reach the goal of rapid publication at a low price the technique of photographic reproduction from a camera-ready manuscript was chosen. This process shifts the main responsibility for the technical quality considerably from the publisher to the authors. We therefore urge all authors and editors of proceedings to observe very carefully the essentials for the preparation of camera-ready manuscripts, which we will supply on request. This applies especially to the quality of figures and halftones submitted for publication. In addition, it might be useful to look at some of the volumes already published. As a special service, we offer free of charge LaTeX and TeX macro packages to format the text according to Springer-Verlag's quality requirements. We strongly recommend that you make use of this offer, since the result will be a book of considerably improved technical quality. To avoid mistakes and time-consuming correspondence during the production period the conference editors should request special instructions from the publisher well before the beginning of the conference. Manuscripts not meeting the technical standard of the series will have to be returned for improvement.

For further information please contact Springer-Verlag, Physics Editorial Department II, Tiergartenstrasse 17, D-69121 Heidelberg, Germany

Claudio Chiuderi   Giorgio Einaudi (Eds.)

Plasma Astrophysics

Lectures Held at the
Astrophysics School VII
Organized by the European Astrophysics Doctoral Network
(EADN) in San Miniato, Italy, 4–14 October 1994

Springer

Editors

Claudio Chiuderi
Dipartimento di Astronomia
Università di Firenze
I-50125 Firenze, Italy

Giorgio Einaudi
Dipartimento di Fisica
Università di Pisa
I-56100 Pisa, Italy

Cataloging-in-Publication Data applied for.

Die Deutsche Bibliothek - CIP-Einheitsaufnahme

Plasma astrophysics : lectures held at the Astrophysics School
VII in San Miniato, Italy, 3 - 14 October 1994 / Claudio
Chiuderi ; Giorgio Einaudi (ed.). Organized by the European
Astrophysics Doctoral Network (EADN). - Berlin ; Heidelberg
; New York ; Barcelona ; Budapest ; Hong Kong ; London ;
Milan ; Paris ; Santa Clara ; Singapore ; Tokyo : Springer, 1996
  (Lecture notes in physics ; 468)
  ISBN 3-540-61014-6
NE: Chiuderi, Claudio [Hrsg.]; Astrophysics School <7, 1994, San
  Miniato>; European Astrophysics Doctoral Network; GT

ISBN 3-540-61014-6 Springer-Verlag Berlin Heidelberg New York

Typesetting: Camera-ready by the authors
SPIN: 10520044   55/3142-543210 - Printed on acid-free paper

Preface

The European Astrophysics Doctoral Network (EADN) is a federation of 32 astrophysics institutes in 17 western European countries supported by the European Union through the ERASMUS and "Human Capital and Mobility" programs, as well as by national funds. The aim of the EADN is to stimulate the mobility of graduate students in astrophysics who are preparing their doctoral theses within Europe and, more generally, to promote international cooperation between young European researchers at a very early stage in their careers.

Among the initiatives of the EADN, a particularly important one is the organization of Predoctoral Astrophysics Schools for graduate students at the beginning of their doctoral research in astrophysics. Schools are held on an annual basis, last two weeks and provide the students with an excellent opportunity to contact the international scientific community. During their stay at the school they are exposed to the major fields of astrophysics and usually have a good opportunity of ample discussions among themselves and with the lecturers who are always well-established authorities in their own fields.

The seventh course of the Astrophysics Predoctoral School took place in the period 4–14 October, 1994, in San Miniato, a little medieval town in central Italy. It was attended by 32 students from 15 different countries. A special grant from the EU made it possible to support the participation of 5 extra students from eastern European countries. The general subject of the school was the astrophysics of plasmas, a field of increasing importance in all different branches of modern astrophysics. As usual in the EADN schools, the first week covered the more theoretical aspects of the subject, whereas the second one concentrated on more observational and instrumental topics. Eight lecturers delivered almost 60 hours of lectures, which made the schedule a very charged one for the students. On top of that, a series of informal seminars were held by the students themselves: each speaker presented his or her own research to the colleagues so that everyone was aware of the interests of the others. In spite of the unusual hour and of the heavy workload the seminars were prepared with enthusiasm and followed with interest.

This book collects the lecture notes of seven of the eight courses given at the school. We have been forced to leave out the notes of one of the courses to avoid further delay of the already late publication of this volume.

The school was supported mainly through funds from the European Union. Additional contributions came from Italian and Swedish national funds and from the Department of Astronomy and Space Science of the University of Florence. 28 out of 37 students received some support from us, although at different levels. Most of the other students were supported directly by their national authorities or universities. The school benefited also from the generosity of the local bank, Cassa di Risparmio di S. Miniato, who allowed us to use, free of charge, the facilities of the Centro di Studi *"I Cappuccini"*, and provided every student with free copies of the transparencies of all lectures.

The appreciation that many students expressed on their leaving has been our best reward and has confirmed to us the validity of this approach of intensive learning at the Predoctoral Summer Schools.

Florence, January 1996   Claudio Chiuderi and Giorgio Einaudi

Contents

Part I. Basic Theory

Part II. Observational and Instrumental Methods

Resistive and Collisionless Magnetic Reconnection

Dieter Biskamp

Max-Planck-Institut für Plasmaphysik, EURATOM Association,
D-85748 Garching, Germany

Abstract: A review of the present status of the theory of magnetic reconnection is given. First the basic concepts are introduced, the MHD equations, the generalized Ohm's law and the conservation laws. In strongly collisional plasmas reconnection proceeds via resistive current sheets, the theory of which is presented in some detail. At weak collisionality the reconnection dynamics is dominated by nondissipative effects, which in low-β plasmas are mainly the Hall term and electron inertia. The equations of electron magnetohydrodynamics are introduced, which allow a rather complete analysis of the reconnection configuration, yielding reconnection rates independent of the smallness parameters of the system. Including the coupling to the ions the reconnection rate is found to depend only on ion inertia. Hence collisionless reconnection can be very rapid proceeding essentially on the Alfvén time scale.

1 Introduction

Magnetic reconnection is probably the most important phenomenon in the dynamics of electrically conducting fluids, primarily in plasmas. The basic picture is that of two magnetic field lines or thin flux tubes, more properly speaking, which are moved around by the fluid owing to the property of ideal magnetic flux conservation, until they "touch" each other at some point, where by the effect of small but finite resistivity (or some other flux conservation breaking or nonideal process) they are cut and reconnected in a different way. This localized process may change the global field line connectivity permitting large-scale fluid motions tapping some reservoir of free energy which would otherwise be inaccessible. It has become clear that almost all rapid dynamic processes in magnetized plasmas involve reconnection. The main objective of reconnection theory is therefore to describe *fast* reconnection in order to explain the observations of explosive processes such as disruptions in tokamak plasmas or flares in the solar atmosphere, where nonideal effects seem to be very weak.

In this paper I try to give an overview of the present status of the theory of fast reconnection. In section 2 the basic concepts are introduced. I discuss the conservation of magnetic flux and of magnetic helicity and give criteria how to decide whether a process actually involves reconnection. Ohm's law is given in general form and the various nonideal effects are discussed which dominate the reconnection process under the widely different plasma conditions occuring in nature and the laboratory.

Section 3 deals with the most commonly used and rather well understood case of quasi-stationary resistive reconnection, which takes place in current sheets. I first discuss the qualitative properties of a current sheet, usually called a Sweet-Parker sheet. For the central region of a resistive sheet a simple power expansion yields an interesting result concerning the magnetic structure at the neutral point. The edge regions of a current sheet show a complex structure. I will then discuss the scaling laws for stationary driven resistive reconnection. As an example of a selfconsistent current sheet reconnection process the coalescence of two flux bundles is presented. The results on resistive reconnection are in obvious contrast to the behavior predicted by Petschek's reconnection theory, which is shown to be valid only as a phenomenological model.

In section 4 collisionless reconnection processes are discussed, which dominate the dynamics in high-temperature low-density plasmas. I first consider the magnitude and basic properties of the three nondissipative terms in Ohm's law, in particular the different types of dispersive waves these terms introduce. In the low-β case, on which I concentrate, the Hall term is the dominant effect, which generates the whistler mode. On small distances, where the electrons decouple from the ions, the dynamics is described by the equations of electron magnetohydrodynamics. It will be shown that in this framework reconnection rates depend only on the macroscopic configuration being independent of the smallness parameters of the theory. Introducing the coupling to the ions does not change the reconnection mechanism, and reconnection rates are found to be fast depending primarily on the Alfvén time scale.

Section 5 gives the conclusions, where I also add some comments on kinetic effects. Though a number of references are given at the end, the list is far from complete. For more details and further references I refer to a recent review of resistive reconnection theory (Biskamp, 1994). Since the results on collisionless reconnection are rather recent, the literature is still scarce.

2 Basic Concepts

2.1 The MHD Equations

The basic fluid description of a plasma is given by the MHD equations for the mass density ρ, mass velocity $\boldsymbol{v}$ and isotropic pressure p, which are coupled to the magnetic field $\boldsymbol{B}$ generated by the current density $\boldsymbol{j} = (c/4\pi)\nabla \times \boldsymbol{B}$,

$$\partial_t \rho + \nabla \cdot \boldsymbol{v}\rho = 0 \quad , \tag{1}$$

$$\rho\left(\partial_t \boldsymbol{v} + \boldsymbol{v} \cdot \nabla \boldsymbol{v}\right) = -\nabla p + \frac{1}{c}\boldsymbol{j} \times \boldsymbol{B} + \boldsymbol{F}_\pi \quad , \tag{2}$$

$$\partial_t \boldsymbol{B} = -c\nabla \times \boldsymbol{E} \quad , \tag{3}$$

where $\boldsymbol{E}$ is determined by Ohm's law (which is in fact the equation of motion of the electron fluid):

$$\boldsymbol{E} + \frac{1}{c}\boldsymbol{v} \times \boldsymbol{B} = \boldsymbol{R} \quad . \tag{4}$$

The viscous force $\boldsymbol{F}_\pi$ is usually approximated by an isotropic diffusion term $\boldsymbol{F}_\pi = \rho\nu\nabla^2\boldsymbol{v}$ and $\boldsymbol{R}$ represents the small nonideal effects in Ohm's law, which allow reconnection to occur and which will be discussed in more detail in section 2.4. In resistive MHD $\boldsymbol{R} = \eta\boldsymbol{j}$. The pressure p is determined by the equation of state of the plasma. In the absence of heat conduction the change of p is adiabatic:

$$\partial_t p + \boldsymbol{v} \cdot \nabla p = -\gamma p \nabla \cdot \boldsymbol{v} \quad . \tag{5}$$

In the incompressible limit, when plasma velocities are small compared with the phase velocities of the compressible waves, one has $\nabla \cdot \boldsymbol{v} = 0$. In this case the density of a fluid element is constant and ρ can be assumed homogeneous $\rho = \rho_0$ without much loss of generality. Then the pressure can be eliminated from eq.(2) by applying the curl operator, and the MHD equations reduce to two equations for the vorticity $\boldsymbol{\omega}$ and $\boldsymbol{B}$. With the normalizations

$$\boldsymbol{x} \to \frac{\boldsymbol{x}}{L}, \quad \boldsymbol{v} \to \frac{\boldsymbol{v}}{v_A}, \quad \rho \to \frac{\rho}{\rho_0} = 1, \quad \boldsymbol{B} \to \frac{\boldsymbol{B}}{B_0} \quad ,$$

where L is a typical spatial scale, B_0 a typical field intensity, $\rho_0 = nm_i$ a typical mass density, and $v_A = B_0/\sqrt{4\pi\rho_0}$ the Alfvén velocity, the equations are written in nondimensional form :

$$\partial_t \boldsymbol{\omega} - \nabla \times (\boldsymbol{v} \times \boldsymbol{\omega}) - \nabla \times (\boldsymbol{j} \times \boldsymbol{B}) = \nu\nabla^2\boldsymbol{\omega} \quad , \tag{6}$$

$$\partial_t \boldsymbol{B} - \nabla \times (\boldsymbol{v} \times \boldsymbol{B}) = \eta\nabla^2 B \quad , \tag{7}$$

$$\nabla \cdot \boldsymbol{v} = \nabla \cdot \boldsymbol{B} = 0, \quad \boldsymbol{\omega} = \nabla \times \boldsymbol{v}, \quad \boldsymbol{j} = \nabla \times \boldsymbol{B} \quad .$$

Here $\eta(= \eta c^2/4\pi v_A L)$ is the normalized resistivity and $\nu(= \nu/v_A L)$ the normalized kinematic viscosity. The quantity $S = \eta^{-1}$ is called the Lundquist number, and $Pr_m = \nu/\eta$ the magnetic Prandtl number.

A further important quantity is the magnetic Reynolds number $R_m = vL/(c^2\eta/4\pi)$ with v some average fluid velocity. R_m characterizes the ratio of the convective and the diffusion terms in eq.(7) and is related to the Lundquist number by $R_m = Sv/v_A$.

Since in reconnection theory consideration is often limited to 2D systems, I also give eqs. (6), (7) in this case. With z as ignorable coordinate, one conveniently introduces the magnetic flux function ψ and the stream function φ by

$$\boldsymbol{B} = \hat{\boldsymbol{z}} \times \nabla\psi + \hat{\boldsymbol{z}}B_z, \quad \boldsymbol{v} = \hat{\boldsymbol{z}} \times \nabla\varphi + \hat{\boldsymbol{z}}v_z,$$

following the equations

$$\partial_t \omega + \boldsymbol{v} \cdot \nabla \omega - \boldsymbol{B} \cdot \nabla j = \nu \nabla^2 \omega \quad , \tag{8}$$

$$\partial_t \psi + \boldsymbol{v} \cdot \nabla \psi = \eta \nabla^2 \psi \quad , \tag{9}$$

$$\omega = \nabla^2 \varphi, \quad j = \nabla^2 \psi \quad .$$

Knowing ψ and φ the remaining quantities B_z, v_z and p can be computed a posteriori.

2.2 Magnetic Helicity

The individuality of magnetic field lines in a plasma, which is only destroyed by rare reconnection events, is based on the property of flux conservation in the ideal MHD limit. Consider the time derivative of the magnetic flux through the surface $F(t)$ bounded by the closed curve $l(t)$ moving with the plasma:

$$\frac{d}{dt} \int_F \boldsymbol{B} \cdot d\boldsymbol{F} = \int_F \partial_t \boldsymbol{B} \cdot d\boldsymbol{F} + \oint_l \boldsymbol{B} \cdot (\boldsymbol{v} \times d\boldsymbol{l}) = 0 \quad , \tag{10}$$

which can easily be verified by inserting Faraday's law (7) for $\eta = 0$ and applying Stoke's theorem. Sweeping the boundary curve l along $\boldsymbol{B}$ defines a flux tube and in the limit of vanishing tube diameter a field line.

A measure of the connectivity of the set of flux tubes forming a magnetic configuration is given by the magnetic helicity

$$H = \int_V \boldsymbol{A} \cdot \boldsymbol{B} dV, \quad \boldsymbol{B} = \nabla \times \boldsymbol{A} \quad . \tag{11}$$

(The form is reminiscent of the hydrodynamic helicity $\int \boldsymbol{v} \cdot \boldsymbol{\omega} dV$.) H is an ideal MHD invariant, as can readily be shown using the gauge $\boldsymbol{E} = -\frac{1}{c} \partial_t \boldsymbol{A}$ $(= -\frac{1}{c} \boldsymbol{v} \times \boldsymbol{B})$

$$\begin{aligned} \frac{dH}{dt} &= \int \left(\boldsymbol{B} \cdot \partial_t \boldsymbol{A} + \boldsymbol{A} \cdot \partial_t \boldsymbol{B} \right) dV \\ &= \int \boldsymbol{A} \cdot \nabla \times (\boldsymbol{v} \times \boldsymbol{B}) \, dV \\ &= \oint \left(\boldsymbol{A} \cdot \boldsymbol{v} \boldsymbol{B} - \boldsymbol{A} \cdot \boldsymbol{B} \boldsymbol{v} \right) \cdot d\boldsymbol{F} \quad , \end{aligned} \tag{12}$$

which vanishes for $v_n = 0, B_n = 0$. The latter condition is also necessary for gauge invariance of H. Making a gauge transformation $\boldsymbol{A}' = \boldsymbol{A} + \nabla \chi$, we have

$$H' - H = \int \boldsymbol{B} \cdot \nabla \chi dV = \oint \chi \boldsymbol{B} \cdot d\boldsymbol{F} \quad . \tag{13}$$

While in laboratory plasma devices these boundary conditions are often satisfied by a surrounding metallic vessel, astrophysical systems are usually open, such that the helicity (11) is not defined. One can, however, give a generalized helicity-like expression (Finn and Antonsen, 1985)

$$\bar{H} = \int (\boldsymbol{A} + \boldsymbol{A}) \cdot (\boldsymbol{B} - \boldsymbol{B}_0) dV \quad , \tag{14}$$

where $\boldsymbol{B}_0 = \nabla \times \boldsymbol{A}_0$ is a static field with the same asymptotic behavior $lim_{|x| \to \infty}(\boldsymbol{B} - \boldsymbol{B}_0) = 0$, for instance the actual field taken at some time $t_0, \boldsymbol{B}_0 = \boldsymbol{B}(t_0)$. It is easy to show that $\bar{H}$ is gauge invariant and conserved.

Let us come back to the main property of the magnetic helicity, namely to characteristize the topological complexity of a configuration. We first realize that H is constant for a closed flux tube of volume $V_\epsilon(t)$, $H_\epsilon = \int_{V_\epsilon(t)} \boldsymbol{A} \cdot \boldsymbol{B} dV$,

$$\frac{dH_\epsilon}{dt} = \int \partial_t (\boldsymbol{A} \cdot \boldsymbol{B}) \, dV + \oint \boldsymbol{A} \cdot \boldsymbol{B} \boldsymbol{v} \cdot d\boldsymbol{F} = 0 \quad , \tag{15}$$

using eq. (12). Gauge invariance is automatically satisfied since by definition $B_n = 0$. H_ϵ is a topological invariant measuring the linkages, knots and twists of an ensemble of flux tubes, i.e. a magnetic configuration, where $\boldsymbol{B}$ is confined to a number of thin tubes. Their radius should be small enough so that the internal field distribution can be neglected, but not the internal twist of the field lines around the tube axis corresponding to finite parallel current density. (Because of its internal structure a flux tube of infinitesimal diameter representing a field line in a conducting fluid, differs from a strictly one-dimensional field line, for which H_ϵ would not be defined). For an isolated flux tube H_ϵ depends on the knottedness N and the total twist T, the latter consisting of the internal twist w of the field lines in the tube around the tube axis and the external twist or torsion τ of the tube axis (Moffat and Ricca, 1992):

$$H_\varepsilon = (N + w + \tau) \phi^2 \quad , \tag{16}$$

where ϕ is in the flux in the tube. Let us consider some examples. An unknotted $(N = 0)$ untwisted $(T = 0)$ tube has vanishing helicity,

$$H_\varepsilon = \int_{V_\epsilon} \boldsymbol{A} \cdot \boldsymbol{B} dV = \phi \oint \boldsymbol{A} \cdot d\boldsymbol{l} = 0 \quad , \tag{17}$$

where the integration path l can be taken along the tube axis, and H_ε vanishes, since $\oint \boldsymbol{A} \cdot d\boldsymbol{l} = \int_F \boldsymbol{B} \cdot d\boldsymbol{F}$ and the flux across the surface F bounded by l is zero. If the tube has an internal twist $w = n$, i.e. the field lines wind around the tube axis n times, the integration path, too, encloses the tube n times, hence $\oint \boldsymbol{A} \cdot d\boldsymbol{l} = n\phi$ and

$$H_\epsilon = n\phi^2 \quad . \tag{18}$$

(In principle n can be any real number). If a tube with internal twist $w = 1$ (Fig. 1a) is contorted by 180° to a figure eight (Fig. 1b), the field in the tube is no longer twisted, w being replaced τ, such that $H_\epsilon = w + \tau = \phi^2$ is not changed. The characteristic feature of the configuration in Fig. 1b is the crossing, which may have two signatures, $\diagup\!\!\!\diagdown$ or $\diagdown\!\!\!\diagup$, the latter giving a negative contribution to H_ϵ. It can be shown (Pfister and Gekelman, 1990; Moffat and Ricca, 1992) that the helicity of a knotted, plane, untwisted $(w = 0)$ tube is given by the sum of its crossings

$$H_\epsilon = (N + \tau)\,\phi^2 = (N_+ - N_-)\,\phi^2 \quad , \tag{19}$$

an example being the right-handed trefoil knot (Fig. 2), which has $H_\epsilon = 3\phi^2$. The property (19) is also valid for an arbitrary set of interlinked, knotted tubes. Consider for instance two interlinked tubes (Fig. 3). Relation (19) immediately yields $H_\epsilon = 2\phi_1\phi_2$, which is identical with the result of a direct calculation:

$$H_\epsilon = \int_{V_1} \boldsymbol{A} \cdot \boldsymbol{B} dV + \int_{V_2} \boldsymbol{A} \cdot \boldsymbol{B} dV$$
$$= \phi_1 \int_{l_2} \boldsymbol{A} \cdot d\boldsymbol{l} + \phi_2 \int_{l_1} \boldsymbol{A} \cdot d\boldsymbol{l} = 2\phi_1\phi_2 \quad . \tag{20}$$

It is instructive to visualize the connection between linking and twist using the ribbon model (Pfister and Gekelman, 1990). Take a ribbon, twist it by 360° and join the ends. Now cut the ribbon along its axis into two narrower ribbons with the "fluxes" ϕ_1, ϕ_2, where $\phi = \phi_1 + \phi_2$ is the "flux" of the original ribbon. The result are two twisted interlinked ribbons. The helicity is not changed,

$$H_\epsilon = 2\phi_1\phi_2 + \phi_1^2 + \phi_2^2 = \phi^2 \quad , \tag{21}$$

using eqs. (18) and (20).

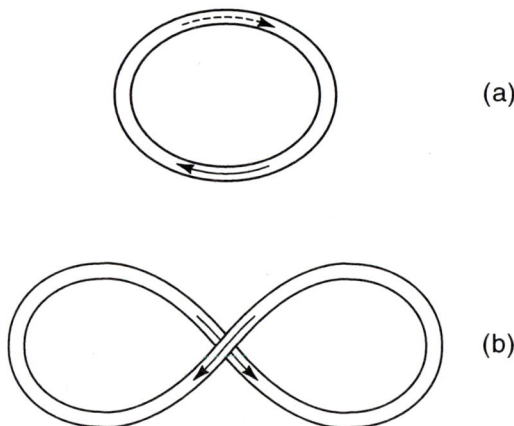

Fig. 1. An internally twisted flux tube (a) has the same helicity as a externally contorted one (b)

An important property is that the helicity is not changed by reconnection, i.e. by cutting and reconnecting two flux tubes taking into account the internal field line distribution of the tubes. As a simple instructive example consider the crossed configuration Fig. 1b. Reconnection implies to connect the field lines on the lower side of the upper tube branch with those on the upper side of the

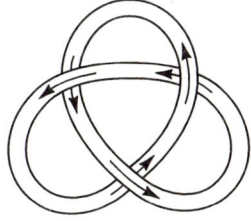

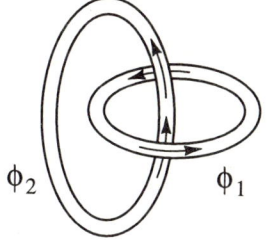

Fig. 2. Right-handed trefoil knot   **Fig. 3.** Two linked flux tubes

lower tube branch, which yields two closed tube with a twist $w = 1/2$ each. Hence the corresponding change of tube topology does not change the helicity. This property does not seem to be generally known. In particular in Taylor's magnetic relaxation theory (Taylor, 1986) it is assumed that during a dynamic, probably turbulent phase reconnection changes the helicity H_ϵ of the individual flux tubes, while only the total helicity H is conserved. Since, however, the individuality of the flux tubes is changed by reconnection, the conservation of H_ϵ is only of formal significance and does not change the result of Taylor's theory, the relaxation to a linear force-free state.

2.3 Criteria for Reconnection

An important point in understanding the physics of MHD processes is to know, whether some, usually geometrically complicated, dynamical process involves reconnection. In symmetric configurations (plane, axisymmetric, or helical) the behavior of the magnetic field is described by a flux function ψ (cf. eq. (9)), the component of the vector potential in the direction of the ignorable coordinate. However intricate the dynamics may be, field lines remain located on magnetic surfaces $\psi = const$. Hence simple inspection of the change of $\psi(x, y, t)$ allows to decide, whether the field line topology has changed implying reconnection. This is illustrated in Fig. 4, which gives a 2D model of the geomagnetic tail. While in (a) all field lines connect to the earth, the left hand boundary, field lines in the plasmoid, the hatched region in (b), obviously do not. Hence the transition from (a) to (b) involves reconnection. Regions of different field line topology are separated by separatrices, joining in X-type neutral points, or Y-points in the presence of extended current sheets, which will be discussed in section 3.

In general nonsymmetric systems conditions are more complicated. Figure 5 gives the generalization of Fig. 4b, a tail model of finite extent in the z direction with a finite component B_z (Schindler et al., 1988). Though all field lines connect to the earth, reconnection has taken place in the formation of the plasmoid, as indicated in Fig. 5b. Hence it is not the change of the global field line topology, which indicates reconnection, but the *local* field line connectivity. In complicated

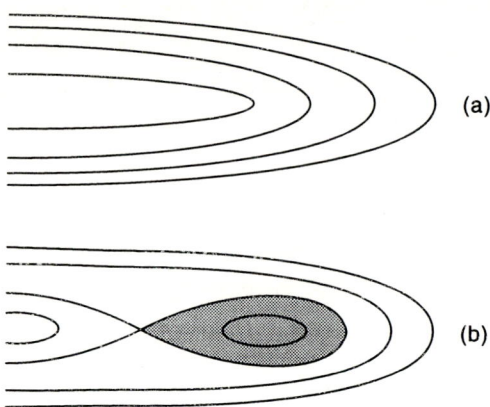

(a)

(b)

Fig. 4. 2D magnetotail models. Reconnection is required to go from state (a) to state (b)

systems this idea is often of little use. Instead one should go back to eq. (4), which shows that reconnection occurs if $E_\parallel$ is finite. Consider the time derivative of the total helicity of a system. (In the case of an open system we have to use the generalized expression (14)). With the gauge $\boldsymbol{E} = -\frac{1}{c}\partial_t \boldsymbol{A}$ and $\boldsymbol{E} \to 0$ asymptotically we obtain

$$\frac{dH}{dt} = \frac{d\overline{H}}{dt} = -2c \int \boldsymbol{E} \cdot \boldsymbol{B} dV \quad . \tag{22}$$

Hence if H (or $\bar{H}$) changes, $E_\parallel$ is finite and reconnection occurs in the system. This does not contradict the result of section 2.2, that reconnection of flux tubes does not change their helicity. What is meant there, is a change of order unity, which does not preclude a change of order η as implied in eq. (22).

2.4 Generalized Ohm's Law

I now discuss more in detail the different nonideal effects in Ohm's law represented by $\boldsymbol{R}$ in eq. (4). From the electron equation of motion

$$m_e n \left(\partial_t \boldsymbol{v}_e + \boldsymbol{v}_e \cdot \nabla \boldsymbol{v}_e\right) = -en \left(\boldsymbol{E} + \frac{\boldsymbol{v}_e}{c} \times \boldsymbol{B}\right) - \nabla p_e - m_e n \nu_{ei} \left(\boldsymbol{v}_e - \boldsymbol{v}_i\right) + \boldsymbol{F}_{\pi e} \quad , \tag{23}$$

we obtain the general form of $\boldsymbol{R}$

$$\begin{aligned}
\boldsymbol{E} + \frac{\boldsymbol{v}}{c} \times \boldsymbol{B} &= \boldsymbol{R} \\
&= \eta \boldsymbol{j} + \mu_e \frac{m_e}{e} \nabla^2 \boldsymbol{v}_e \\
&\quad + \frac{1}{nec} \boldsymbol{j} \times \boldsymbol{B} - \frac{1}{ne} \nabla p_e - \frac{m_e}{e} \left(\partial_t \boldsymbol{v}_e + \boldsymbol{v}_e \cdot \nabla \boldsymbol{v}_e\right) \quad ,
\end{aligned} \tag{24}$$

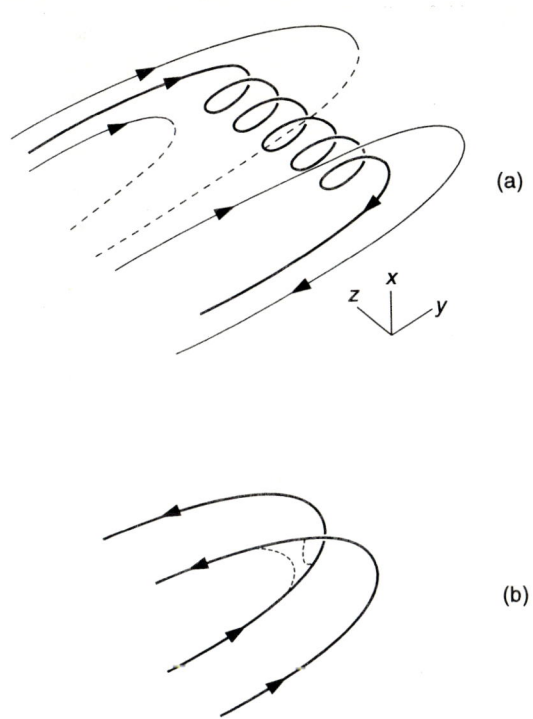

Fig. 5. 3D magnetotail model, corresponding to the state in Fig. 4(b)

where $v = v_i$ and $j = ne\,(v_i - v_e)$. The first two terms on the right represent the dissipative effects, resistivity and electron viscosity (again assuming for simplicity isotropic viscosity), while the remaining effects are nondisspative or collisionless, the Hall term, electron pressure and electron inertia. The magnitude of the individual terms becomes clear when writing eq. (24) in nondimensional form. With the normalization $t \to t v_A / L,\, x \to x/L,\, B \to B/B_0,\, E \to Ec/B_0 v_A$ (the latter follows from Faraday's law (3)), $v_e \to v_e 4\pi ne L / cB_0$, and $p_e \to p_e / p_0$ one obtains

$$\boldsymbol{R} = \eta \boldsymbol{j} + \mu_e \nabla^2 \boldsymbol{v}_e$$
$$+\, d_i \left[\boldsymbol{j} \times \boldsymbol{B} - \beta \nabla p_e - d_e^2 \left(\partial_t \boldsymbol{v}_e + \boldsymbol{v}_e \nabla \boldsymbol{v}_e \right) \right] \tag{25}$$

where $\eta = S^{-1}$, $\mu_e = \mu_e d_e^2 / v_A L$, $d_e = c/\omega_{pe} L$, $d_i = c/\omega_{pi} L$, $\beta = 4\pi p_0 / B_0^2$. For $v_e \sim j_{\parallel}/en$ the relative magnitude of the dissipative terms is $L^2 \nu_{ei}/\mu_e$, where L is a typical current gradient scale length in the dissipation region (see section 3). For collisional electron viscosity $\mu_e \sim \rho_e^2 \nu_{ee} \sim \rho_e^2 \nu_{ei}$ the ratio is ρ_e^2 / L^2, which is negligibly small in most cases of interest. However, μ_e may be strongly enhanced, for instance by the effect of small-scale field line stochasticity (Kaw et al., 1979). The relative magnitude of the nondissipative terms will be discussed in section 4.1.

3 Quasi-Stationary Resistive Reconnection

Resistivity has been the prototypical flux conservation breaking effect used in most of the previous theoretical work on reconnection.

For $R_m \gg 1$ fast reconnection can only arise, if the spatial scales involved in the reconnection process are much shorter than the global scale L. Hence the process must be localized, and the preferred loci are X-type neutral points in the magnetic configuration. Figure 6 illustrates the behavior in the vicinity of an X-point, plasma flowing into the reconnection region from above and below and leaving sidewise. Since fluid velocities even in fast reconnection are small compared with compressional wave velocities, flows in reconnection theory are usually assumed incompressible. Figure 6 also shows, that the configuration is likely to be flattened into a sheet. In fact resistive reconnection occurs in current sheets.

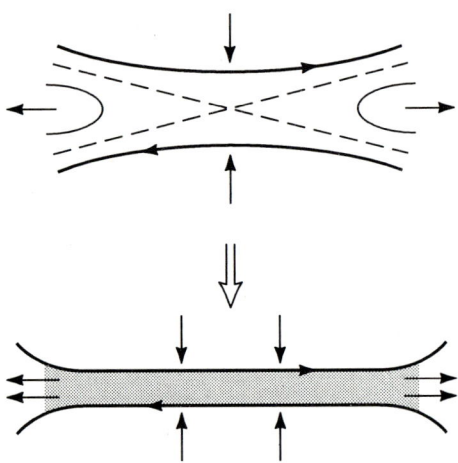

Fig. 6. Current sheet generation at an X-point

In general only a small component of $\boldsymbol{B}$, the poloidal field $\boldsymbol{B}_\perp$, vanishes at an X-point, while the system varies only slowly along the main field. Hence one usually restricts consideration to 2D geometry. True magnetic neutral points, where the total magnetic field vanishes, called magnetic nulls, may also occur in a configuration. On theoretical grounds such points have been attributed a particular significance for reconnection (Bulanov et al., 1986; Lau and Finn,

1990). Recent nonlinear 3D computations, however, indicate that nulls do not play an important role (Politano et al.,1995).

3.1 Basic Properties of Current Sheets

A dynamical current sheet, usually called a Sweet-Parker sheet, is a quasi one-dimensional stationary configuration as illustrated in Fig. 7, where the plasma inflow balances resistive diffusion. Assuming incompressibility a current sheet is characterized by six quantities, the magnetic field immediately outside the sheet called the upstream field B_0 (the downstream field at the sheet edges is small), the upstream flow u_0 into the sheet perpendicular to the field, the downstream flow v_0 along the sheet, the width δ and the length Δ, and the resistivity η. These quantities are connected by three relations derived from the continuity equation, Ohm's law and the equation of motion, assuming stationarity. Integrating $\nabla \cdot \rho v = \rho_0 \nabla \cdot v = 0$ over the domain indicated in Fig. 8 one obtains

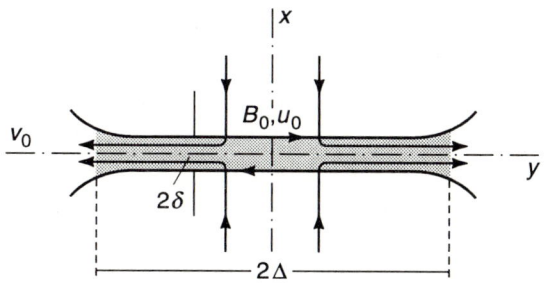

Fig. 7. Resistive (Sweet-Parker) current sheet

$$u_0 \Delta = v_0 \delta \tag{26}$$

ignoring profile effects. The z component of Ohm's law (4) with $\boldsymbol{R} = \eta \boldsymbol{j}$ and $E_z = E = const$ gives

$$u_0 B_0 = \eta j \simeq \eta \frac{B_0}{\delta} \quad . \tag{27}$$

Since in general $u_0 \ll B_0$, the inertia term is neglible in the force balance across the sheet, $\partial_x(p + B^2/2) = 0$, hence

$$B_0^2/2 = p_m - p_0 \quad , \tag{28}$$

where p_0 is the upstream pressure and p_m the pressure maximum in the sheet center. Along the sheet, where B_x and hence the magnetic force essentially vanish, the pressure gradient accelerates the fluid, $v_y \partial_y v_y = -\partial_y p$. Integration between sheet center and edge yields

$$v_0^2/2 = p_m - p_0 \tag{29}$$

assuming that at the edge the pressure has dropped to its upstream value. Combining (28) and (29), we obtain the important relation that the downstream velocity equals the upstream Alfvén velocity

$$v_0 = B_0 = v_A \quad . \tag{30}$$

Equations (26) and (27) can be used to express two of the remaining five quantities by the other three, for instance

$$\frac{u_0}{v_A} = \left(\frac{\eta}{B_0 \Delta}\right)^{1/2} \equiv S_0^{-1/2} \tag{31}$$

$$\frac{\delta}{\Delta} = S_0^{-1/2} \quad . \tag{32}$$

Equation (31) shows that the reconnection rate $\partial_t \psi = E = u_0 B_0 \sim \eta^{1/2}$ for macroscopic sheet length $\Delta \sim L$, which is far too slow to explain the time scales observed for instance in a solar flare. For sufficiently fast reconnection current sheets of much smaller dimensions are required, as is for instance assumed in Petschek's theory (erroneously, see section 3.6), or develop in MHD turbulence (Biskamp and Welter, 1989).

There has been some discussion about the force acting along the sheet. It is occasionally (and erroneously) claimed that the magnetic force ("sling shot effect") is the dominant one with the argument that in the vorticity form (8) of the equation of motion only the magnetic force appears. This puzzle can, however, easily be solved. Integrating this equation over the domain indicated in Fig. 8 and using Gauss's theorem one obtains

$$\oint v_n \omega \, dl = \oint B_n j \, dl \quad . \tag{33}$$

Since ω vanishes exactly on the paths 3 and 4 and effectively on 1, only 2 contributes to the kinetic term

$$\oint v_n \omega \, dl \simeq \int_2 v_y \partial_x v_y \, dx = v_0^2/2 \quad , \tag{34}$$

while in the magnetic term only 3 and 4 contribute, the main contribution coming from 4.

$$\oint B_n j \, dl \simeq \int_4 B_y \partial_x B_y = B_0^2/2 \quad , \tag{35}$$

which gives exactly the result (30). The contribution from path 3 is negligible, since B_x is small, as is shown explicitly in section 3.2.

It should be emphasized that current sheets are formed under quite general conditions. The field need not be antiparallel, vanishing in the neutral plane, but only a component, say B_y, has to change sign. The location of current sheets in a general sheared magnetic configuration depends on the structure of the flows as excited for instance by an instability. A strong axial field $B_z \gg B_y$ provides also

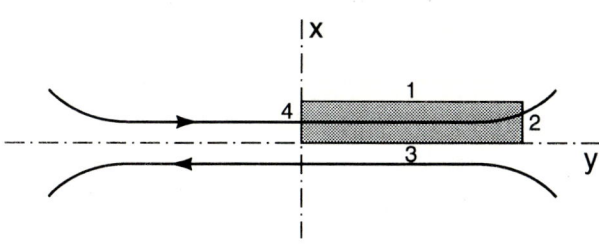

Fig. 8. Integration domain in eq. (33)

a justification of the incompressibility assumption, which might appear doubtful in the sheet, where the velocity becomes large reaching the Alfvén speed of the reconnected field component B_y. If the latter is small compared with the axial component, the total Alfvén velocity is much greater than the flow speed, which guarantees that the perpendicular flow is a nearly incompressible $\mathbf{E} \times \mathbf{B}$ convection.

I should point out that a Sweet-Parker current sheet is also a vorticity sheet, as is clear from the flow pattern illustrated in Fig. 7. The vorticity distribution has a quadrupole structure (instead of the monopole structure of the current density), because ω vanishes on the symmetry lines $x = 0$ and $y = 0$. Since $u_0 \sim \eta^{1/2} B_0$, the magnitude of the vorticity is, however, small $\omega \sim \eta^{1/2} j$, hence for $\nu \sim \eta$ in eqs. (8), (9) viscous dissipation $\epsilon_\nu = \nu \int \omega^2 dV$ is smaller than Ohmic dissipation $\epsilon_\eta = \eta \int j^2 dV$.

3.2 Stationary Solution at the Neutral Point

One can show by a simple analysis (Cowley, 1975) that the magnetic field in the vicinity of the neutral point in the sheet center differs basically from a normal X-point. For a symmetric configuration, where the stagnation point of the flow coincides with the magnetic neutral point at $(x, y) = (0, 0)$, we assume ψ to be even and φ odd:

$$\psi = \sum_{m,n \geq 0} \psi_{2m,2n} \frac{x^{2m} y^{2n}}{(2m)!(2n)!}, \quad \varphi = \sum_{m,n \geq 0} \varphi_{2m+1,2n+1} \frac{x^{2m+1} y^{2n+1}}{(2m+1)!(2n+1)!} \tag{36}$$

with $\psi_{mn} = \partial_x^m \partial_y^n \psi|_{x,y=0}$ etc. In steady state eqs. (8), (9) read

$$\partial_x \varphi \partial_y \psi - \partial_y \varphi \partial_x \psi = \eta \left(\partial_x^2 \psi + \partial_y^2 \psi \right) - E \quad , \tag{37}$$

$$\partial_x \varphi \partial_y \omega - \partial_y \varphi \partial_x \omega - \partial_x \psi \partial_y j + \partial_y \psi \partial_x j = \nu \left(\partial_x^2 \omega + \partial_y^2 \omega \right) \quad , \tag{38}$$

where $E = \partial_t \psi = const.$ Inserting expressions (36) gives to lowest order

$$\eta (\psi_{20} + \psi_{02}) = E \quad . \tag{39}$$

Differentiating eq. (37) twice with respect to x and with respect to y at the origin, one obtains

$$2\varphi_{11}\psi_{20} + \eta\left(\psi_{40} + \psi_{22}\right) = 0 \quad , \tag{40}$$

$$2\varphi_{11}\psi_{02} - \eta\left(\psi_{22} + \psi_{04}\right) = 0 \quad . \tag{41}$$

Differentiating eq. (38) once with respect to both x and y at the origin gives

$$-\psi_{20}\left(\psi_{22} + \psi_{04}\right) + \psi_{02}\left(\psi_{40} + \psi_{22}\right) = \nu\left(\varphi_{51} + 2\varphi_{33} + \varphi_{15}\right) \quad , \tag{42}$$

which becomes by use of (40), (41)

$$-\frac{4}{\eta}\phi_{11}\psi_{20}\psi_{02} = \nu\left(\phi_{51} + 2\phi_{33} + \phi_{15}\right) \quad . \tag{43}$$

First consider the case $\nu = 0$. Assuming streamlines to form hyperbolae, i.e. $\varphi_{11} \neq 0$, either ψ_{20} or ψ_{02} (not both because of eq.(39)) must vanish. This implies, that field lines do not form hyperbolae, in particular the separatrix branches do not intersect at a finite angle, but osculate. While in a true X-point configuration the downstream field increases linearly $B_x = \psi_{02}y$, it is cubic in the osculating case $B_x \propto y^3$, whereas the velocity remains linear $v_y = \varphi_{11}y$. This behavior indicates an inherent tendency toward current sheet formation at an X-point. For finite viscosity ν the argument is no longer strictly valid. Numerical simulations, however, show that the inviscid behavior remains essentially true as long as $\nu \lesssim \eta$.

3.3 Current Sheet Edge Region

In a seminal paper Syrovatskii (Syrovatskii, 1971) presents a simple theory of current sheets as branch cuts of a complex potential function, which form at singular X-points, i.e. X-point, where the electric field is finite. He derives an expression of the sheet current density integrated across the sheet width δ

$$J(y) = 2\left(\frac{I}{2\pi} + \frac{\Delta^2}{2} - y^2\right) \Big/ \sqrt{\Delta^2 - y^2} \quad , \tag{44}$$

where I is the total sheet current and Δ the length of the sheet. Expression (44) shows the interesting feature, that $J(y)$ reverses sign, being positive, i.e. in the direction of the total current, in the central part $|y| < y_0$, and negative in the outer parts $|y| > y_0$, where $y_0^2 = (I/2\pi) + \Delta^2/2$. At the sheet edges $|y| = \Delta$ the current density becomes in general singular.

Resistive MHD simulations essentially confirm Syrovatskii's prediction (44). In fact the current density in the sheet reverses sign and exhibits a quasi-singular structure in the edge regions of the sheet (see Biskamp, 1994, Figs. 14, 15). The current density changes sign at the position, where the velocity reaches its maximum along the sheet, $v = v_A$. The fluid is subsequently decelerated in the negative current density part and finally completely blocked and turned around and accelerated again along smaller secondary current sheets, formed on both sides of the quasi-singular edges of the primary sheet. For still smaller η tertiary sheets become visible, leading to a sequence of higher order sheets in the limit $\eta \to 0$. Such behavior, however, rests on the assumptions of spatial symmetry and stationarity. For nonsymmetric, nonstationary conditions a turbulent state consisting of an irregular ensemble of micro-current sheets is generated (Biskamp and Welter, 1989).

3.4 Scaling Laws for Driven Resistive Reconnection

The concept of driven or forced reconnection plays an important role in reconnection theory. While originally the term refers to open externally forced systems in contrast to closed systems, where reconnection occurs spontaneously as an internal process, the concept can be applied much more generally. Assuming that reconnection is localized in space, one may restrict consideration to a small region L around this position instead of the entire system of size $\Lambda \gg L$. On the other hand, L should be large compared with the scales of the reconnecting structures, for instance $L \gg \Delta$ in the case of a single current sheet of length Δ, so that these are not affected by the artificial boundaries of the subsystem L. This procedure allows to simplify the geometry substantially. It also allows to assume stationarity even for a nonstationary global system. Since the coupling to the latter occurs by the boundary conditions imposed on the subsystem, which vary on the global time scale Λ/v_A, while the subsystem varies on the time scale L/v_A, we may consider the subsystem in a relaxed steady state (if such a state exists).

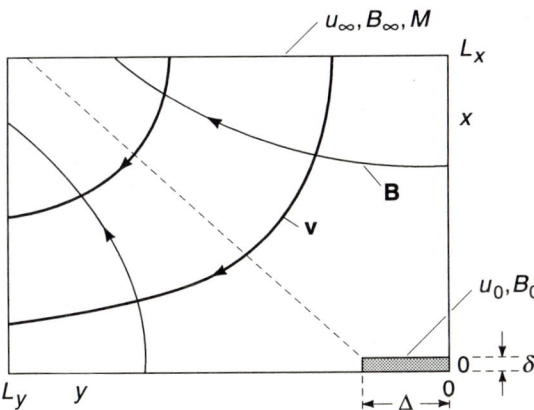

Fig. 9. Computational domain in forced reconnection studies

In this sense the subsystem constitutes a stationary driven reconnection configuration. By assuming up-down and right-left symmetry only a quadrant must be considered, for instance the upper left quadrant shown in Fig. 9, where the main parameters are indicated. The plasma along with its frozen-in magnetic field is injected from above and after reconnection ejected to the left. While u_∞, B_∞ are given, the internal parameters u_0, B_0 in front of the current layer and the dimensions δ, Δ are determined by the reconnection process. Before discussing the general scaling laws of these quantities, let us consider the case of weak driving $E = \partial_t \psi = u_\infty \lesssim \eta$ (with the normalization $B_\infty = 1$). To lowest order in E the stationary equations (8), (9) are

$$\boldsymbol{B} \cdot \nabla j = 0 \quad , \tag{45}$$

$$\boldsymbol{v} \cdot \nabla \psi = -E + \eta j \quad . \tag{46}$$

If j vanishes asymptotically, eq. (45) implies $j = 0$ everywhere for an X-point configuration with open field lines. There is a simple similarity solution:

$$\psi = \tfrac{1}{2} \left(x^2 - y^2 \right) \quad , \tag{47}$$

$$\phi = \tfrac{1}{2} E \ln \left| \frac{x+y}{x-y} \right| \quad . \tag{48}$$

Since, however, $\omega = 4Exy/(x^2-y^2)^2$ becomes singular on the separatrix $x = \pm y$, the inertial term $\boldsymbol{v} \cdot \nabla \omega$, though formally of order E^2, cannot be neglected, hence j cannot vanish. For $E \lesssim \eta$ the deviation from the solution (47), (48) remains small when regularized by finite viscosity. For larger driving E, however, the configuration is substantially changed by the formation of a current sheet at the X-point.

In this case the configuration can only be computed numerically. A series of simulation runs (Biskamp, 1986) performed for different values of E and η, and identical boundary profile functions, gives the following scaling laws of the parameters of the current sheet

$$B_0 \sim E^2/\eta \quad , \tag{49}$$

$$u_0 = E/B_0 \sim \eta/E \quad , \tag{50}$$

$$\Delta \sim E^4/\eta^2 \quad , \tag{51}$$

$$\delta \sim E\eta^0 \quad . \tag{52}$$

Hence increasing E or decreasing η leads to a rapid stretching of the sheet length Δ as illustrated in Fig. 10. The width δ does not decrease but increases with E contrary to exspectation, the reason being the increase (pile-up) of the magnetic field B_0 in front of the layer and the corresponding decrease of u_0, since $u_0 B_0 = E$.

The scaling laws (49)-(52) reflect an important property of the Sweet-Parker sheet. Consider the average inertia force along the sheet. Since the velocity increases about linearly $v_y \sim B_0 y/\Delta$, we have

$$\overline{v_y \partial_y v_y} \simeq B_0^2/2\Delta \simeq E/2\delta \quad ,$$

using mass conservation $u_0 \Delta = B_0 \delta$ and Ohm's law $u_0 B_0 = E$. Hence the scaling (52) implies that the inertia force is invariant to changes of E and η, in particular remains finite for $\eta \to 0$.

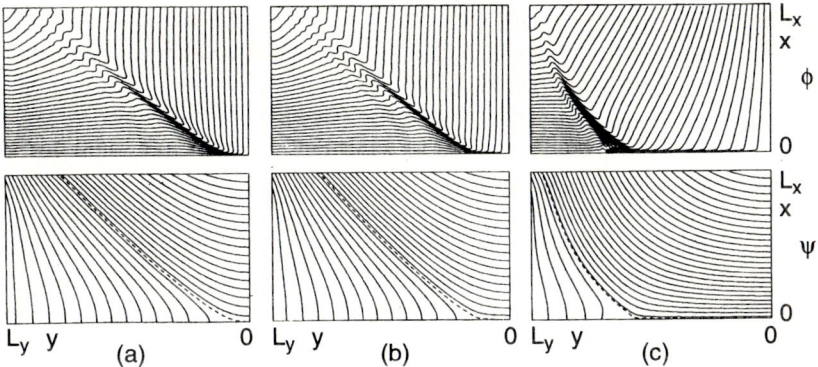

Fig. 10. Contours of ϕ and ψ of driven reconnection states with identical boundary conditions but different η : (a) $\eta = \eta_0$, (b) $\eta = \eta_0/2$, (c) $\eta = \eta_0/4$

When the sheet length Δ reaches the system size L as in Fig. 10c, Δ cannot increase further, and the scaling laws (49)-(52) are no longer valid. Instead we have $\Delta = L$. From $u_0 B_0 = E = \eta B_0/\delta$ one obtains $u_0 = \eta/\delta$, $B_0 = E\delta/\eta$. Using mass conservation we find

$$B_0 \sim \left(E^2 L/\eta\right)^{1/3} \quad , \tag{53}$$

$$\delta \sim \left(\eta^2 L/E\right)^{1/3} \quad . \tag{54}$$

3.5 Magnetic Island Coalescence

The most instructive paradigm of fast magnetic reconnection is the coalescence of magnetic islands (e.g. Biskamp, 1993), or more generally of two flux bundles. Consider for instance the following corrugated sheet pinch equilibrium

$$\psi(x,y) = B_\infty a \ln\left(\cosh\frac{y}{a} + \varepsilon\cos\frac{x}{a}\right) \quad , \tag{55}$$

which consists of a periodic chain of magnetic islands around the midplane $y = 0$ imbedded in a strong antiparallel field $B_x \to \pm B_\infty$ for $y \to \pm\infty$. The island width w is given by the relation $\cosh(w/2a) = 1 + 2\varepsilon$, in particular $w/a \simeq 4\sqrt{\varepsilon}$ for $w \ll a$. (The equilibrium corresponds to a finite amplitude tearing mode of wavenumber $k_x a = 1$, the marginally stable mode in the noncorrugated ($\varepsilon = 0$) sheet pinch.) The configuration (55) is ideally unstable to pairwise coalescence of islands for any island width. The nonlinear process, which has been investigated by numerical simulations (Fig. 11), can be divided into (a) the ideal MHD phase, where the islands are freely accelerated toward each other by the force between the currents in the islands, which leads to flux compression and current sheet formation; (b) the quasi-stationary reconnection phase. For intermediate values of the resistivity $\eta = S^{-1} = 10^{-4} - 10^{-2}$ a selfsimilar behavior is observed, with the scaling laws of the upstream quantities u_0, B_0 taken just in front of the sheet

$$u_0 \sim \eta^{1/3}, \quad B_0 \sim \eta^{-1/3} \tag{56}$$

and geometric properties of the sheet

$$\Delta \sim w \sim \eta^0, \quad \delta \sim \eta^{2/3} \quad , \tag{57}$$

which corresponds to the scaling (53), (54). Hence the reconnection rate E is independent of η, $E = u_0 B_0 \sim \eta^0$.

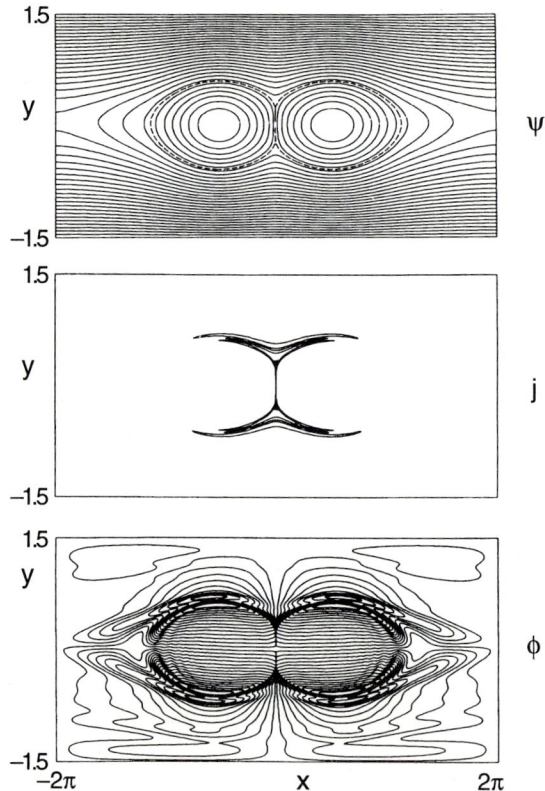

Fig. 11. Coalescence of magnetic islands illustrated by contours of ψ, j, ϕ

Obviously the scaling law (56) is only valid within a certain η range, since B_0 cannot exceed the maximum value B_m, which would be obtained in the ideal case $\eta = 0$, when the motion is reversed because of the repelling force by the compressed magnetic field. For smaller values of η, roughly $\eta < 10^{-4}$, one recovers Sweet-Parker scaling

$$\overline{u}_0 \sim \eta^{1/2}, \quad \overline{B}_0 \simeq B_m \sim \eta^0 \quad , \tag{58}$$

as found in Fig. 12 from the derivative of $\Delta x(t)$ in the quasi-stationary reconnection phase $t > 100$. The bars on $\overline{u}_0, \overline{B}_0$ indicate a time average, since the motion of the islands toward the sheet is modulated by a sloshing due to the inertia effect, the kinetic energy of the fluid in the island gained in the ideal acceleration phase.

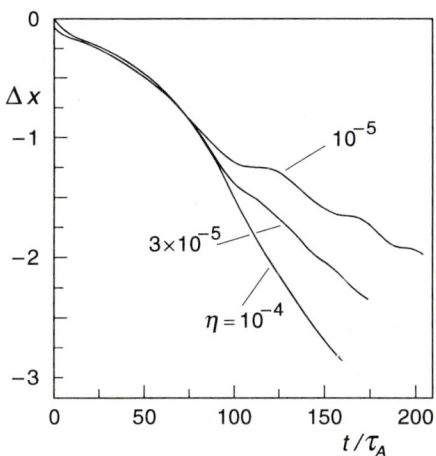

Fig. 12. Island coalescence. $\Delta x =$ change of O-point position, $\Delta x = -\pi$ corresponding to complete coalescence

3.6 Petschek's Quasi-Ideal Reconnection Model

For two decades Petschek's slow-shock model (Petschek, 1964) was the the generally accepted theory of magnetic reconnection. In recent years it has, however, been realized, that the model is not valid in the usually adopted framework of resistive MHD in the limit of small resistivity. Petschek's configuration, illustrated in Fig. 13, is characterized by two pairs of slow shocks standing back to back against the upstream flow, which they deviate by roughly 90° into the downstream cone α between the shocks. Current density and vorticity are concentrated in the shocks and the central current sheet. The shocks derive their properties from the slow magnetosonic wave, a slow compressible mode, which survives with finite phase velocity in the incompressible limit, $\omega^2 = k_\parallel^2 v_A^2 = k^2 B_n^2$, where B_n is the component normal to the wavefront. Hence for a given speed the flow can always be supersonic with respect to this mode, if the angle between wavefront and magnetic field is sufficiently small.

Petschek's configuration is a solution of the ideal MHD equations valid outside the singularities. The jump conditions at the shocks determine the position

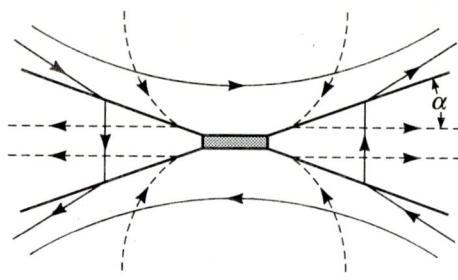

Fig. 13. Petschek's reconnection model

of the latter, i.e. the downstream angle α in terms of the upstream velocity. The reconnection rate is essentially independent of the resistivity, which was a most welcome feature of the model. The solution must, however, be matched to the resistive solution in the current sheet, which is very difficult and has previously been essentially ignored. Petschek *assumes* a current sheet of dimensions $\Delta \sim \delta \sim \eta$ adjusting automatically to the external ideal solution. This is, however, not true. Numerical solutions of the full resistive problem show that Δ does not shrink with η but becomes macroscopically large, following the scaling laws (49)-(52), which changes the external ideal configuration fundamentally and leads to a slow Sweet-Parker type reconnection rate $E = O(\eta^{1/2})$. Petschek's model may, however, be applicable, if the conditions in the reconnection region are alleviated. This is achieved by a locally enhanced effective resistivity $\eta_{eff} = O\,(1)$ due to turbulence generated for instance by the high current density in the sheet. Macroscopic current sheet formation can also be avoided, if collisionless reconnection processes are dominant as discussed in the following section.

4 Collisionless Reconnection

Since at high magnetic Reynolds numbers resistive current sheet reconnection becomes rather slow $E \sim \eta^{1/2} \sim R_m^{-1}$, the question arises, what processes can lead to faster reconnection in order to account for the observed time scales. For dense plasmas, where resistive MHD remains valid, R_m is large mainly because of large spatial scales L. Reconnection is increased by the instability of the current sheet, which occurs above a (somewhat loosely defined) threshold R_c (Biskamp, 1986). For $R_m \gtrsim R_c$ this instability leads to the formation of plasmoids, which are ejected along the sheet, while in the limit $R_m \gg R_c$ the configuration around the reconnection region becomes fully MHD turbulent, which gives rise to reconnection rates roughly independent of the Reynolds number (Biskamp and Welter, 1989).

4.1 Collisionless Reconnection Mechanisms

In many hot and dilute plasmas, however, both in the laboratory as well as in space, collisions are sufficiently infrequent such that the dissipative effects are no longer dominant in Ohm's law (24). This happens, if resistive scales $\delta \sim \eta^{1/2}$ are smaller than the length scales characterizing the collisionless terms in $\boldsymbol{R}$, i.e. d_i and $\rho_s (= d_i \beta^{1/2})$, the ion Larmor radius at the electron temperature. From eq. (25) we see that in the usual case of $\beta < 1$, the Hall term is formally the most important collisionless effect. One should, however, note that the Hall term by itself cannot give rise to reconnection, since it simply means that the field is frozen to the electron fluid. Reconnection requires finite $E_\parallel$. As will be seen later, even the collisionless effects in $E_\parallel$, the pressure and the inertia terms, cannot prevent the system from becoming singular, and hence some dissipation, however small, is required.

In the low-β limit the pressure term can be neglected. Inserting $\boldsymbol{E}$ into Faraday's law and linearizing the latter together with the equation of motion (6) about a homogeneous magnetic field gives the dispersion relation

$$\omega^2 = k_\parallel^2 v_A^2 \frac{1 + k^2 d_i^2}{\left(1 + k^2 d_e^2\right)^2} \quad . \tag{59}$$

Hence for spatial scales smaller than the ion inertia length the mode properties are changed from nondispersive Alfvén waves to dispersive whistler waves, which for $k d_i > 1$ follow the dispersion relation written in the usual form

$$\omega^2 = \Omega_e^2 d_e^4 k_\parallel^2 k^2 / \left(1 + d_e^2 k^2\right)^2 \quad , \tag{60}$$

since $v_A d_i = \Omega_i d_i^2 = \Omega_e d_e^2$. For $k d_e < 1$ the mode is dominated by the Hall term and the group velocity increases with k, $\partial \omega / \partial k \propto k$, while for $k d_e > 1$, where electron inertia dominates, the group velocity becomes very small. These properties determine the reconnection behavior in the low-β quasi-collisionless regime discussed in section 4.2

In the high-β regime the pressure effect dominates in Ohm's law. In this case the dispersion relation of linear modes is (including electron inertia)

$$\omega^2 = k_\parallel^2 v_A^2 \frac{1 + k^2 \rho_s^2}{1 + k^2 d_e^2} \quad . \tag{61}$$

Again, the mode becomes dispersive for sufficiently small scales. If $\rho_s > d_e$ the group velocity increases with k for $\rho_s^{-1} < k < d_e^{-1}$. This mode is called kinetic Alfvén wave. Since collisionless reconnection in the pressure-dominated regime, which is determined by the properties of the kinetic Alfvén wave, has recently been studied in several papers (Aydemir, 1992; Kleva et al., 1995), I shall restrict the discussion to the low-β, whistler-dominated case presenting some new, yet unpublished results.

4.2 Electron Magnetohydrodynamics and Collisionless Reconnection

To explore the effect of the whistler mode on the reconnection process, we first restrict consideration to scales $< d_i$, where the ion dynamics can be neglected. The system is determined by the electron velocity $\boldsymbol{v}_e = -\boldsymbol{j}/ne$ and the selfconsistent magnetic field following Faraday's law (3) with $\boldsymbol{E}$ inserted from Ohm's law,

$$\partial_t \left(\boldsymbol{B} - d_e^2 \nabla^2 \boldsymbol{B}\right) - \nabla \times \left[\boldsymbol{v}_e \times \left(\boldsymbol{B} - d_e^2 \nabla^2 \boldsymbol{B}\right)\right] = -\mu_e \nabla^{(4)} \boldsymbol{B} \quad , \tag{62}$$

$$\boldsymbol{v}_e = -\alpha \nabla \times \boldsymbol{B} \quad , \tag{63}$$

where $\alpha = c/4\pi ne$ is the Hall parameter. Linearization of eq. (62) gives the whistler dispersion relation (60). Equations (62), (63) are called electron magnetohydrodynamics (EMHD) (Kingsep et al., 1990; Bulanov et al., 1992; Drake et al., 1994). Since neglecting the ion dynamics, $m_i \rightarrow \infty$, corresponds to Alfvén time $t_A = L/v_A \rightarrow \infty$, it is convenient to use the whistler time $t_W = L^2/\alpha B_0 = L^2/\Omega_e d_e^2$ as time unit, which formally amounts to $\alpha \rightarrow 1$ in eq. (63). On the right side electron viscosity is chosen as the most important dissipative effect, since resistivity (= electron friction) can in general not prevent the formation of singular gradients.

Whistler modes are destabilized by a strong gradient of the current density (Mikhailowskii, 1974), where the most unstable modes are perpendicular to $\boldsymbol{j}$, i.e. to $\boldsymbol{B}$ in a low-β plasma. This suggests to study whistler-related processes in the plane perpendicular to the main field $\boldsymbol{B}_0$. Choosing $\boldsymbol{B}_0 = B_0 \hat{\boldsymbol{z}}$ the 2D version of eq. (62) can be written in terms of the flux function ψ for the poloidal field $\boldsymbol{B}_\perp = \hat{\boldsymbol{z}} \times \nabla \psi$ and the axial field fluctuation $\delta B_z = B_z - B_0$, which we denote by φ_e, since it plays the role of a stream function of the electron velocity, $\boldsymbol{v}_e = \hat{\boldsymbol{z}} \times \nabla \varphi_e$:

$$\partial_t \left(\psi - d_e^2 j\right) + \boldsymbol{v}_e \cdot \nabla \left(\psi - d_e^2 j\right) = -\mu_e \nabla^2 j \quad , \tag{64}$$

$$\partial_t \left(\varphi_e - d_e^2 \omega_e\right) + \boldsymbol{v}_e \cdot \nabla \left(\varphi_e - d_e^2 \omega_e\right) + \boldsymbol{B}_\perp \cdot \nabla j = -\mu_e \nabla^2 \omega_e \quad , \tag{65}$$

$$j = \nabla^2 \psi, \quad \omega_e = \nabla^2 \varphi_e \quad .$$

In the limit $\mu_e \rightarrow 0$ eqs. (63), (64) conserve the energy

$$W = \tfrac{1}{2} \int \left[\varphi_e^2 + B_\perp^2 + d_e^2 \left(v_e^2 + j^2\right)\right] dV \quad , \tag{66}$$

which consists of the magnetic energy and the kinetic energy of the electron fluid.

Let us consider the problem of stationary reconnection in the framework of the EMHD equations (64), (65). We start by noting that for $d_e = 0$ the stationary equations

$$E + \boldsymbol{v}_e \cdot \nabla \psi = 0 \quad ,$$

$$\boldsymbol{B}_\perp \cdot \nabla j = 0$$

have the similarity solution (47), (48) for ψ and φ_e, valid for any reconnection rate E (in contrast to the MHD case, where this solution is only valid for sufficiently small E, Biskamp, 1986). The upstream flow converges toward the X-point and the downstream flow diverges away from it. Finite dissipation is only needed to smooth the flow singularity on the separatrix $x = \pm y$.

In the vicinity of the X-point the similarity solution (48) for φ_e breaks down because of the effect of electron inertia. The magnetic flux is, however, not affected, so that eq. (47) remains valid also in the inertia-dominated region. The invariance of ψ originates from the ideal conservation of the quantity $F = \psi - d_e^2 \nabla^2 \psi$, which is essentially the z component of the canonical momentum. Integration using the appropriate Green's function gives

$$\psi = \int G \left(\frac{x - x'}{d_e}, \frac{y - y'}{d_e} \right) F(x', y') \, d^2 x' \quad .$$

This form indicates that ψ varies only on scales $> d_e$, i.e. does not depend on the local current distribution inside this region. The fact that ψ is known allows an analysis of the entire reconnection configuration (Biskamp and Drake, 1995). The main results are

(a) The inertia-dominated region, where the flow deviates from the similarity solution, has the scale

$$l \sim (d_e^2 E)^{1/3} \quad .$$

(b) Embedded in this region is a sheet-like substructure, where vorticity and current density are maximal. The length Δ of this sheet is

$$\Delta \sim l \quad ,$$

which shrinks to zero for $d_e \to 0$. Hence in contrast to resistive MHD there is no macroscopic current sheet. The width δ of the sheet is determined by dissipative effects

$$\delta \sim \mu_e^{1/2} \quad .$$

(c) The outflow velocity v_0 from the sheet equals the current density j_m in the sheet

$$v_0 \sim j_m \quad ,$$

which differs basically from the MHD result eq. (30). The total current in the sheet becomes negligible in the limit of vanishing dissipation, which is consistent with the persistence of the X-point solution for ψ.

(d) For sufficiently large sheet aspect ratio Δ/δ the electron flow along the sheet becomes Kelvin-Helmholtz unstable generating whistler turbulence which gives rise to a finite energy dissipation rate in the limit $\mu_e \to 0$.

(e) The current sheet may adjust to any reconnection rate E. Hence reconnection in a given global configuration is independent of the smallness parameters d_e, μ_e.

The analytical results (a) - (e) are essentially confirmed by numerical simulations using eqs. (64), (65), where we consider the coalescence of two flux bundles. Figure 14 illustrates the evolution showing ψ and φ_e at two times for a case with

$d_e = 0.03$. The conspicuous feature is that the flux surfaces appear to be pulled into the reconnection region rather than pushed against it as in the MHD case (cf. Figs. 10, 11). This behavior is due to the flow pattern, exhibiting streamlines converging toward the X-point corresponding to acceleration, which agrees with the similarity solution (48). In the MHD case by contrast the flow is rather uniform across the flux bundle, even decelerating in front of the current sheet as shown in Fig. 11c. Figure 15 gives blowups of the central region showing the onset of instability. The turbulence is clearly generated by the collimated flow in the sheet edge region and not by the current density.

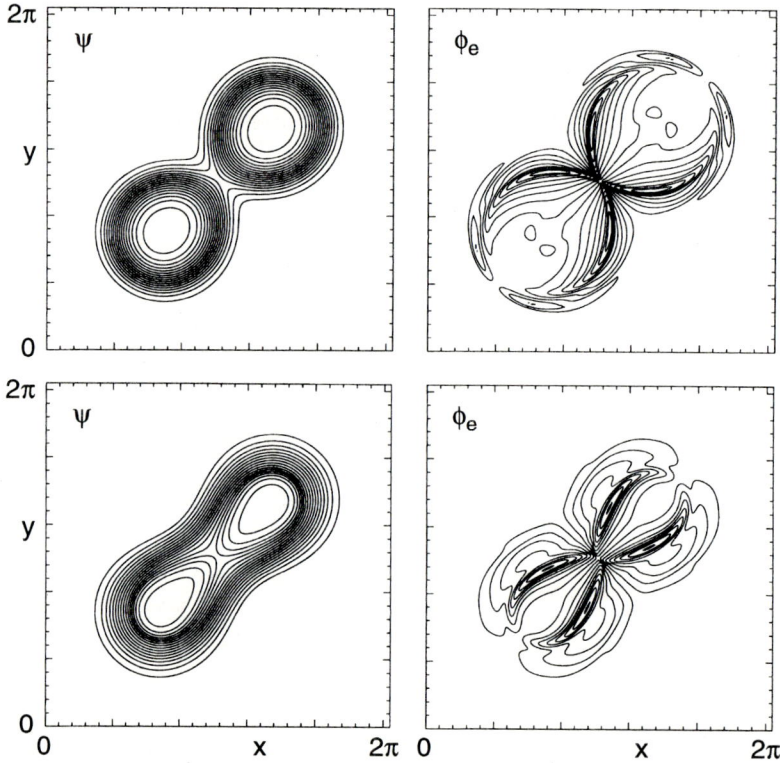

Fig. 14. Coalescence of magnetic flux bundles : EMHD simulations. Contours of ψ and φ_e at two different times

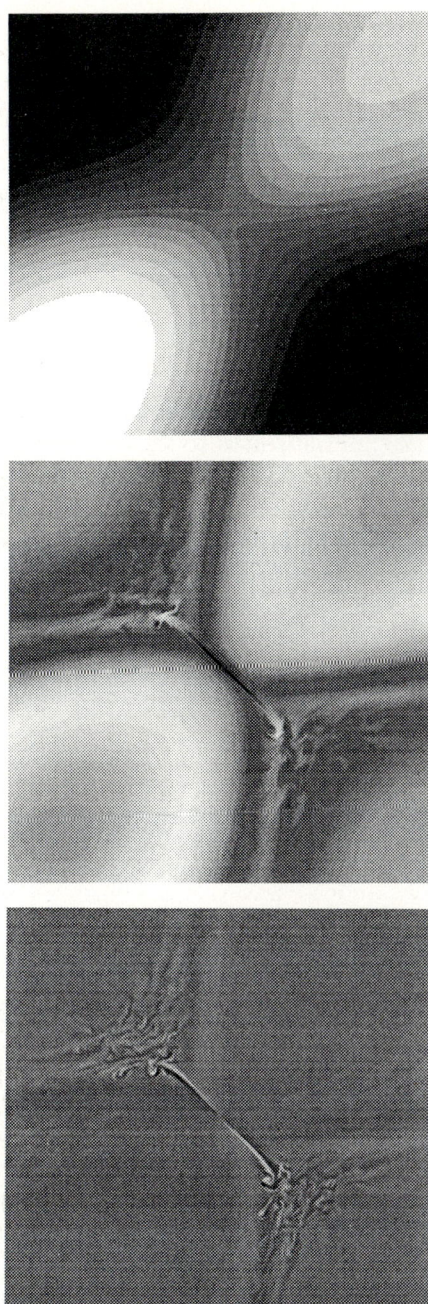

Fig. 15. Gray scale plots of ψ, j, ω_e(from above) in the reconnection region showing the generation of turbulence in the downstream cone

4.3 Coupling to Finite Mass Ions

The approximation of infinitely massive ions, on which EMHD is based, is only valid for magnetic structures $L < d_i$. In most applications, however, the global scales are much larger, so that the coupling to the ions cannot be neglected. On scales $L > d_i$ ions and electrons in fact move together, which justifies neglecting the Hall term in the MHD approximation. We therefore introduce the ion stream function $\varphi_i, v_i = \hat{z} \times \nabla\varphi_i, \omega_i = \nabla^2\varphi_i$. Adding the ion and the electron fluid equations gives

$$d_i^2 \left(\partial_t\omega_i + v_i \cdot \nabla\omega_i\right) + d_e^2 \left(\partial_t\omega_e + v_e \cdot \nabla\omega_e\right) - B_\perp \cdot \nabla j$$
$$= \mu_i\nabla^2\omega_i + \mu_e\nabla^2\omega_e \quad , \tag{67}$$

maintaining the normalization to the whistler time t_W introduced in section 4.2. (Compared with the normalization to the Alfvén time used in MHD theory the difference is a factor $t_A/t_W = d_i$). Neglecting the out-of-plane ion motion (essentially the motion parallel to the main magnetic field), the out-of-plane current density j is unchanged and hence eq. (64) remains valid. The perpendicular current density is, however, modified,

$$j_\perp = -\hat{z} \times \nabla\left(\varphi_e - \varphi_i\right) = -\hat{z} \times \nabla\delta B_z \quad ,$$

whence the fluctuation of the axial field is

$$\delta B_z = \varphi_e - \varphi_i \quad . \tag{68}$$

Substituting δB_z for φ_e in eq. (65) and noting that the electron inertia term contains only the electron flow, eq. (65) becomes

$$\partial_t \left(\varphi_e - \varphi_i - d_e^2\omega_e\right) + v_e \cdot \nabla \left(\varphi_e - \varphi_i - d_e^2\omega_e\right) + B \cdot \nabla j$$
$$= -\mu_e\nabla^2\omega_e \quad . \tag{69}$$

Equations (64), (67), (69) form a system of equations for ions and electrons generalizing the EMHD equation (64), (65). The total energy, conserved in the limit $\mu_e, \mu_i \to 0$, is

$$W = \tfrac{1}{2} \int \left[(\delta B_z)^2 + B_\perp^2 + d_e^2 \left(v_e^2 + j^2\right) + d_i^2 v_i^2\right] dV \quad . \tag{70}$$

Numerical simulations show that whistler-dominated reconnection dynamics persists in the region $< d_i$, while at larger distances ions and electrons move essentially together. A typical simulation run is shown in Fig. 16, with $d_i = 0.05, d_e = 0.0075$ and sufficiently weak dissipation, such that the conditions $L \gg d_i \gg d_e \gg \delta$ are satisfied. Scaling studies show that the reconnection rate depends only on d_i, i.e. m_i, being independent of d_e and the dissipation coefficients. For not too small d_i, $1 > d_i \gtrsim 0.1$ we find that reconnection proceeds essentially on the Alfvén time scale, while for still smaller d_i it appears to be somewhat slower than Alfvénic. In any case reconnection is much faster than in the absence of the Hall term. To demonstrate the importance of the latter a

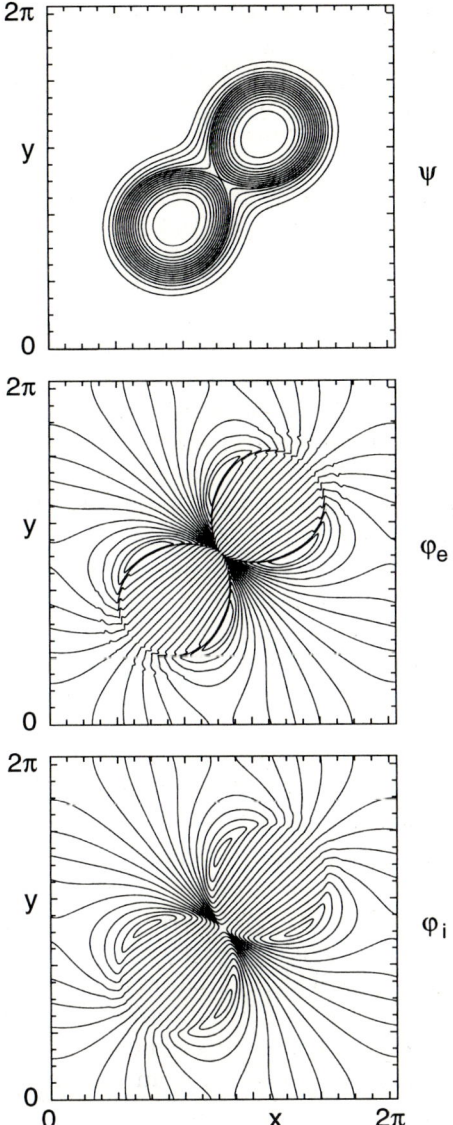

Fig. 16. Coalescence of magnetic flux bundles including the ion dynamics. Contours of ψ, φ_e, φ_i

number of simulations such as in Fig. 16 have been repeated with the Hall term switched off by setting $\varphi_e = \varphi_i$. Since the ion flow is now forced into a narrow layer of width d_e, the reconnection rate now is strongly reduced. Scaling studies reveal a strong dependence on d_e, $E \sim d_e$.

5 Conclusions

I have presented an overview of the present status of the theory of magnetic reconnection in plasmas. It has by now become rather common knowledge that reconnection in the framework of resistive, or more generally, dissipative MHD is inherently slow. For intermediate magnetic Reynolds numbers R_m reconnection takes place via current sheets of macroscopic length, Sweet-Parker sheets, with rates $E = O(\eta^{1/2})$, which if extrapolated to the large values of R_m in real, mostly astrophysical systems, is far too slow to explain the observed time scales. This puzzle is partly solved by the fact that for sufficiently large R_m current sheets become unstable and reconnection proceeds in an instationary way involving MHD turbulence, which in principle allows finite reconnection rates in the limit $\eta \to 0$. In fact, much of the modeling of real processes, e.g. magnetic substorms corresponding to plasmoid generation and ejection in the magnetotail of the earth, is being performed by introducing some anomalous, i.e. locally strongly enhanced resistivity to mimic in an ad hoc way the effect of small-scale turbulence.

The other and probably even more important aspect are quasi-collisionless reconnection processes, which have conventionally been neglected by the somewhat vague argument that reconnection process always requires finite dissipation and hence the nondissipative terms in Ohm's law are unimportant, though in almost collisionless plasmas such as in the solar corona or the magnetosphere these are formally much larger than the very weak collisional effects. The first part of the argument seems to be essentially correct, but the conclusions drawn from it are not. We have seen that the Hall term, which simply restates the conservation of magnetic flux following the electron flow instead of the mass flow, has a dramatic effect on the reconnection dynamics. It appears that the dispersive properties of the plasma – the Hall term introduces the highly dispersive whistler mode – are more important than the dissipative ones.

I have not considered kinetic effects, in particular collisionless dissipation due to wave-particle interaction such as Landau or cyclotron damping. Inverse Landau damping can give rise to micro-instabilities. A well-known example is the ion-sound instability driven by a sufficiently strong current in a plasma with temperature ratio $T_e/T_i > 1$, which generates small-scale turbulence $k\lambda_D \sim 1$ (λ_D = Debye length), acting as an anomalous resistivity. A further source of collisionless dissipation are chaotic particle motions, which arise preferentially in the vicinity of magnetic X-points and which can also be associated with collisionless reconnection.

References

Aydemir, A.Y. (1992): Phys. Fluids **B4** 3469-3472

Biskamp, D. (1986): Phys. Fluids **29** 1520-1531

Biskamp, D. (1993): Nonlinear Magnetohydrodynamics (Cambridge University Press, 1993), pp. 156-159

Biskamp, D. (1994): Phys. Rep. **237** 179-247

Biskamp, D., Welter, H. (1989): Phys. Fluids **B1** 1964-1979

Biskamp, D., Drake, J.F.: submitted to Phys. Plasmas

Bulanov, S.V. et al. (1986): Sov. J. Plasma Phys. **12** 180-189

Bulanov, S.V., Pegoro, F., Sakharov, A.S. (1992): Phys. Fluids **B4** 2499-2508

Cowley, S.W.H. (1975): J. Plasma Phys. **14** 475-490

Drake, J.F., Kleva, R.G., Mandt, M.E. (1994): Phys. Rev. Lett. **73** 1251-1254

Finn, J.M., Antonsen, T.M. (1985): Comments Plasma Phys. and Controlled Fusion **26** 111-136

Kaw, P.K., Valeo, E.J., Rutherford, P.H. (1979): Phys. Rev. Lett. **43** 1398-1401

Kingsep, A.S., Chukbar, K.V., Yan'kov, V.V.: in Reviews of Plasma Physics (Consultants Bureau, New York, 1990), Vol. 16

Kleva, R.G., Drake, J.F., Waelbroeck, F.L. (1995): Phys. Plasmas **2** 23-34

Lau, Y.T., Finn, J.M. (1990): Astrophys. J. **350** 672-691

Mikhailowskii, A.B.: Theory of Plasma Instabilities (Consultants Bureau, New York, 1974), Vol. 2

Moffat, H.K., Ricca, R.L. (1992): Proc. R. Soc. Lond. **A439** 411-429

Petschek, H.E. (1964): AAS/NASA Symposium on the Physics of Solar Flares, ed. W.N. Hess (NASA, Washington D.C) pp. 425-437

Pfister, H., Gekelman, W. (1990): Am J. Phys. **59** 497-502

Politano, H., Pouquet, A., Sulem, P.L. (1995): to be published in Phys. Plasmas

Schindler, K., Hesse, M., Birn, J. (1988): J. Geophys. Res. **93** 5547-5557

Syrovatskii, S.J. (1971): Sov. Phys. JETP **33** 933-940

Taylor, J.B. (1986): Rev. Mod. Phys. **53** 741-763

ROTATING MHD WINDS

Jean Heyvaerts

Observatoire de Strasbourg
Université Louis Pasteur
11 Rue de l'Université, 67000 Strasbourg, France

1. THE ISOTROPIC HYDRODYNAMIC WIND

1.1 ASTROPHYSICAL CONTEXT

1.1.1 The Solar Corona

It has been discovered only late that the sun possesses an extended atmosphere, the solar corona, visible at eclipses. It has been recognized in the 1940's that the corona consists of a very hot plasma at 10^6K. Its visible light is photospheric light scattered by coronal free electrons. Why the corona is so hot is a problem that is coming near to a solution only now. It appears that, at least in closed magnetic field regions, the solar corona is heated by the dissipation of electric currents (see for example Heyvaerts (1989)).

1.1.2 The Solar Wind

The question then arouse as to how the corona extends faraway from the sun and eventually merges with the local interstellar medium. The first idea was that the corona should be just an hydrostatic extension of the solar atmosphere. But it met with the difficulty that hydrostatic atmospheres have a pressure at infinity that is much larger than the expected pressure of the interstellar medium. This discrepancy lead Parker (1958) to the idea that the corona is in fact suffering hydrodynamic expansion. The idea of a permanent outflow from the sun was supported by the fact that comet tails are oriented radially with respect to it (Biermann (1957)). Finally, in the early 1960's, this expanding interplanetary fluid, the solar wind, has been observed in situ by spacecrafts (Gringauz et al. (1960)).

1.2 THE ISOTROPIC ISOTHERMAL HYDRODYNAMICAL WIND MODEL

1.2.1 The Simplest Model

The simplest possible model is one in which the wind flow is radial, isotropic, stationary, and moves only under the effect of pressure forces and gravity. Since there is negligible mass in the corona and in the wind, the gravity field is that one which the main body of the sun creates.

The energy balance in the wind may be quite complicated. For a simple first approach, we assume that the heat conductibility is so large that the temperature $\overline{T}$ is constant throughout the wind. This assumption replaces a more realistic energy equation. The wind is described as a fluid, assumed to be a perfect gas of fully ionized hydrogen.

1.2.2 Dimensional Wind Equations

Let $\overline{\rho}$, $\overline{T}$, $\overline{P}$, v, be the dimensional mass density, temperature, pressure and velocity. These quantities, under the isothermal assumption, obey the hydrodynamical equations:

$$\frac{\partial \overline{\rho}}{\partial t} + \operatorname{div}(\rho v) = 0 \tag{1}$$

$$\rho \left(\frac{\partial v}{\partial t} + (v.\nabla) v \right) = -\nabla \overline{P} - \overline{\rho} \left(\frac{GM}{r^2} \right) e_r \tag{2}$$

$$\overline{T}(r) = \overline{T} \tag{3}$$

$$\overline{P} = \overline{\rho} \left(\frac{2k_B T}{m_p + m_e} \right) = \overline{\rho} c^2 \tag{4}$$

Equation (1) expresses mass conservation, eq (2) expresses momentum conservation, eq (3) is the isothermality assumption, and eq (4) is the equation of state for a completely ionized hydrogen plasma. M is the mass of the sun, and c is the sound speed, in this case a constant by eq (3). These relations are then particularized to the case of stationary, spherically symmetric and radial flow. Writing $v = v(r) e_r$, they become:

$$\frac{1}{r^2} \frac{d}{dr} \left(r^2 \overline{\rho} v \right) = 0 \tag{5}$$

$$\overline{\rho} v \frac{dv}{dr} = -\frac{d\overline{P}}{dr} - \overline{\rho} \left(\frac{GM}{r^2} \right) \tag{6}$$

$$\overline{P} = \overline{\rho} c^2 \tag{7}$$

1.2.3 Remark on the Isothermal Assumption

The energy equation would in general be written as

$$\frac{\partial}{\partial t} \left(\overline{U} + \frac{1}{2} \overline{\rho} v^2 + \overline{\rho} \Phi(r) \right) + \operatorname{div} \left(\frac{1}{2} \rho v^2 v + \overline{\rho} \Phi(r) v + (\overline{U} + \overline{P}) v + q \right) = 0 \tag{8}$$

$\overline{U}$ is the internal energy density, $\Phi(r)$ the gravitational potential, and q the heat flux, given by the Fourier law $q = -\kappa \nabla T$. The isothermal assumption means that κ is so large that ∇T must be very nearly zero, otherwise a very large energy flux would result which would quickly restaure almost perfect isothermality.

We may however regard isothermality anotherway, cheating a little bit with actual microphysics. We know that gases made of molecules that have a very large number of internal degrees of freedom remain approximately isothermal under adiabatic transformations. For such supposedly perfect gases the adiabatic transformation law $P = k\rho^\gamma$ applies with $\gamma = 1$. We could pretend our solar wind to be made of such a gas, since the end result for the thermodynamics would just be the same. The corresponding energy equation would have $q = 0$ and $\gamma = 1$, and, in stationary state, it would appear in the form:

$$\text{div} \left(\frac{1}{2}\rho v^2 v + \overline{\rho}\Phi\, v + \left(\overline{U} + \overline{P}\right) v \right) = 0 \tag{9}$$

where now the enthalpy density $\left(\overline{U} + \overline{P}\right)$ is that of the isothermal gas (see below subsection 1.3.4).

1.2.4 Impossibility of an Hydrostatic Solution

A static corona satisfying the same assumptions and boundary conditions would have a density distributed according to the hydrostatic law :

$$c^2 \frac{d\overline{\rho}}{dr} = -\overline{\rho} \left(\frac{GM}{r^2} \right) \tag{9}$$

The solution which has $\overline{\rho} = \rho_0$, the density at the base of the solar corona, at $r = R$, the radius of the sun, is

$$\overline{\rho}_{stat} = \rho_0 \, \exp \left(\frac{GM}{Rc^2} \left(\frac{R}{r} - 1 \right) \right) \tag{10}$$

For $r \to \infty$ this density approaches the finite value $\rho_0 \, \exp\left(-GM/Rc^2\right)$, which can be checked to be much marger than the expected pressure of the interstellar medium.

1.2.5 Dimensionless Wind Equations

It is convenient to use suitable dimensionless variables. Since the wind extends from the solar surface to infinity, we normalize distances to the solar radius, R, and define

$$x = \left(\frac{r}{R} \right) \tag{11}$$

$\overline{\rho}$ is similarly normalized to the density at the base of the solar corona, defining a dimensionless ρ such that:

$$\overline{\rho} = \overline{\rho}_0 \, \rho \tag{12}$$

and the pressure is $\overline{\rho}c^2$ so that, given the isothermality relation:

$$\overline{P} = \overline{\rho}_0 c^2 \rho \tag{13}$$

The velocity is measured in units of the sound speed, that is, by a Mach number, M :

$$\bar{v} = cM \tag{14}$$

These normalizations introduce the following dimensionless number :

$$\lambda = \frac{GM}{Rc^2} \tag{15}$$

The ratio of the square of the escape speed from the solar surface to the square of the sound speed is 2λ. The escape speed equals the sound speed at the distance $2\lambda R$ from the star center. The larger λ, the better the confinement of the coronal gas in the gravitational potential well. Finally, the dimensionless equations of the problem are

$$\frac{d}{dx}\left(x^2 \rho M\right) = 0 \tag{16}$$

$$\rho M \frac{dM}{dx} + \frac{d\rho}{dx} + \frac{\lambda \rho}{x^2} = 0 \tag{17}$$

One boundary condition is $\rho = 1$ when $x = 1$. This is not enough for this system of two equations of first order. Another boundary condition will be discussed below.

1.3 SOLUTIONS TO ISOTHERMAL WIND EQUATIONS

1.3.1 Differential Equation for the Wind Velocity

Transforming the mass conservation equation (16) into a log derivative, and dividing the momentum one by ρ, we can eliminate it and obtain the following equation for M(x) alone:

$$\frac{1}{M^2}\frac{dM^2}{dx} = 2\,\frac{2/x - \lambda/x^2}{M^2 - 1} \tag{18}$$

This equation is singular when $M = 1$. When the solution reaches $M = 1$, the derivative becomes infinite, unless at the same time the numerator also vanishes. This happens when $x = x_c = \lambda/2$. Conversely, $dM^2/dx = 0$ at $x = x_c$, unless at the same time $M = 1$.

1.3.2 Critical Point

The point C of coordinates $(\lambda/2, 1)$ in the (x-M) plane, is called a critical point. This is because at this point the differential equation does not determine dM/dx, since the latter is given by an expression of the form 0/0. Although only one solution passes at normal points, two solutions pass at the critical point. This can be seen using l'Hopital's rule. In the vicinity of the critical point, the numerator N(x) and denominator D(x) of the rhs of eq(18) can be Taylor-expanded as

$$N(x) = 4\,(x - x_c)\,/\,x_c^2 \tag{19.a}$$

$$D(x) = (x - x_c)\,p \tag{19.b}$$

where p is the slope dM^2/dx calculated at the critical point. Using these expansions, eq(18) evaluated at x_c, gives for p the equation

$$p = \frac{16}{\lambda^2 p} \tag{20}$$

This is a second order equation for p, which can thus take at the critical point one of the two values $p = \pm 4/\lambda$

1.3.3 First Integrals of the Wind Equations

Equations (16) and (17) can be integrated to express the fact that the mass flux through a sphere is independant of radius, as is the total energy flux, a relation that can also be expressed as constancy, following the motion, of the specific energy (energy per gram of moving fluid):

$$x^2 \rho M = \mu \tag{21}$$

$$\frac{M^2}{2} + \ln \rho - \frac{\lambda}{x} = \epsilon \tag{22}$$

ϵ and μ are integration constants. Substituting for ρ from eq (21) into eq (22), we get a relation between Mach number and radius in the implicit form:

$$\frac{M^2}{2} - \frac{\lambda}{x} - \ln M - \ln x^2 = \epsilon - \ln \mu \tag{23}$$

1.3.4 Bernoulli equation and Energy Conservation

Note that we have obtained the Bernoulli relation, eq (22), from the momentum equation, but we could have obtained it from the energy equation for an isothermal fluid, eq (9). Actually for radial isothermal flow, eq (9) reduces to

$$\frac{d}{dx} \left(\frac{v^2}{2} - \frac{GM}{R} + \frac{\overline{U} + \overline{P}}{\rho} \right) = 0 \tag{24}$$

Note that $\left(\overline{U} + \overline{P} \right)/\rho$ is the specific enthalpy. Let us calculate it for an isothermal gas in terms of pressure and density. For any piece of material, the change of enthalpy, dH, is given in terms of its volume V, temperature T and entropy S by

$$dH = d(U + PV) = dU + PdV + VdP = TdS - PdV + PdV + VdP = TdS + VdP \tag{25}$$

If the piece of material happens to be a gram of matter, its volume is $1/\rho$. Noting its enthalpy as h, we find:

$$dh = TdS + \frac{dP}{\rho} \tag{26}$$

Integrating along a line of constant entropy gives

$$h = h_0 + \int_{(S)} \frac{dP}{\rho} \tag{27}$$

For an isothermal gas, $P = \rho c^2$, with c^2 a constant at constant entropy. The above general relation becomes

$$h = h_O + c^2 \ln \left(\frac{\overline{\rho}}{\overline{\rho}_0} \right) = k_O + c^2 \ln (\overline{\rho}) \tag{28}$$

For a perfect gas of adiabatic index γ larger than unity, pressure varies with density at constant entropy as $P = K(S) \rho^\gamma$, and we obtain instead

$$h = h_O(S) + K(S) \frac{\gamma}{\gamma - 1} \left(\overline{\rho}^{\gamma-1} - \overline{\rho}_O^{\gamma-1} \right) \tag{29}$$

Since we are interested in variations of enthalpy only, we can just take

$$h = \frac{\gamma}{\gamma - 1} \frac{P}{\rho} \tag{30}$$

Coming back to the isothermal case, we got from the energy equation the relation

$$\frac{v^2}{2} - \frac{GM}{r} + c^2 \ln \bar{\rho} = 0 \tag{31}$$

When expressed in dimensionless variables, this just gives the Bernoulli equation (22), as found earlier by other reasoning.

1.3.5 The Network of Bernoulli Curves

To sum up, the general solution for $M(x)$ is given by the implicit relation eq (23), a more convenient form of which is obtained by exponentiating, which gives

$$M \, e^{-M^2/2} = \frac{K}{x^2} \, e^{-\lambda/x} \tag{32}$$

The constant K is given by

$$\ln K = -\epsilon + \ln \mu \tag{33}$$

Equation (32) provides a convenient way to construct the solution graphically or numerically. For a given x, calculate its left hand side, and search for M's that make the right hand side equal to it (see figure (1)).

It can be seen that several cases can occur, according to the value of K. If the maximum of the function of x at the right hand side of eq (32), which is $4K/(e^2\lambda^2)$, is smaller than the maximum of the function of M on the left hand side, which is $1/\sqrt{e}$, there are two solutions for M for any given x, one supersonic and the other subsonic. Otherwise solutions do not exist for any x: there are no solutions in a certain vicinity of $x_c = \lambda/2$. When K equals the critical value K_c

$$K_c = \frac{1}{4} \lambda^2 e^{3/2} \tag{34}$$

the interval with no solution shrinks to nil, and a double root appears at M = 1.

The sign of dM/dx, knowing that dM/dx vanishes at $x = x_c$ and is infinite at $M = 1$ (except at the critical point) is easy to determine, and from it, a qualitative picture of the set of possible solutions can be drawn. This is summarized in fig (2).

1.3.6 The Critical Point as a Saddle Point of the Bernoulli Function

This graph shows that the critical point C is a saddle point of the Bernoulli function

$$B(x, M) = \frac{M^2}{2} - \ln M - \frac{\lambda}{x} - \ln x^2 \tag{35}$$

Actually the function B reaches a maximum at C following the line $x = x_c$, but reaches a minimum following the line $M = 1$. It can be checked that at the critical point the differential dB vanishes. Indeed:

$$dB = \frac{1}{2} \frac{M^2 - 1}{M^2} \, dM^2 - \left(\frac{2}{x} - \frac{\lambda}{x^2} \right) \, dx \qquad (36)$$

The vanishing of the differential is a property that does not depend on the variables used to represent the function. A critical point can be recognized by this property in any pair of variables, for example ρ and x instead of M and x. Near the critical point the Bernoulli function should be Taylor expanded to second order, which gives, with $x = \lambda/2 + \xi$ and $M = 1 + \eta$:

$$B \left(\frac{\lambda}{2} + \xi, M + \eta \right) = \ln \frac{4}{\lambda^2 e^{3/2}} + \eta^2 - \frac{4\xi^2}{\lambda^2} \qquad (37)$$

We find that near the critical point, isocontours of the function $B(x, M)$ are similar to hyperbolae. Solutions that pass at C locally degenerate into the pair of straight lines of slope $\pm 2/\lambda$.

1.4 THE WIND SOLUTION

1.4.1 Boundary Conditions and the Value of the First Integrals

In eqs (21) and (22), we obtained two first integrals of the motion, which involve two integration constants ϵ and μ, the value of which should be obtained from boundary conditions. One such condition is that ρ should equal 1 at $x = 1$. This indicates that μ, the dimensionless mass flux, is also $M(1)$, the Mach number at the base of the corona. The Bernoulli equation then relates it to the dimensionless specific energy ϵ, by

$$\mu^2 = 2(\epsilon + \lambda) \qquad (38)$$

It remains to find ϵ, or equivalently the related quantity K of eq (33) from another boundary condition. One could expect it to result from the necessity for the wind to smoothly merge into the interstellar medium at infinity by having $M \to 0$ as $x \to \infty$. In this case, the pressure of the wind would match that of the interstellar medium. Whereas many solutions meet the first requirement, none meets the second. As shown below, this is because the interstellar pressure is too small, or, said another way, because the pressure in the low corona is too large. However, there exists a unique smooth transsonic solution. Such a solution, which is supersonic at infinity, can meet the interstellar medium at a shock front, a possibility that is forbidden to asymptotically subsonic flows.

1.4.2 Elimination of Physically Inadequate Solutions

By examining their properties, most of the solutions of the Bernoulli equations can be eliminated as physically irrelevant. Such is the case of solutions with $K > K_c$, which are inadequate because they do not provide a solution for any x between 1 and ∞.

Solutions with $K < K_c$ are defined for any x, but the ones which are supersonic everywhere cannot be accepted, because the low corona is known from observation to be in a state of very subsonic motion. This argument also eliminates the decreasing critical solution. The solutions which are everywhere subsonic have too large a pressure at infinity, as we now show. This is expectable, since their velocity approaches zero at large distances, so that the situation does not differ very much from that in a static

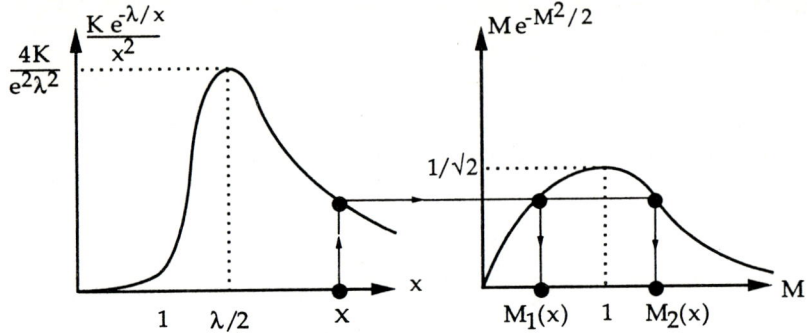

Fig. 1. A graphical construction of the solutions $M(x)$ of eq (32).

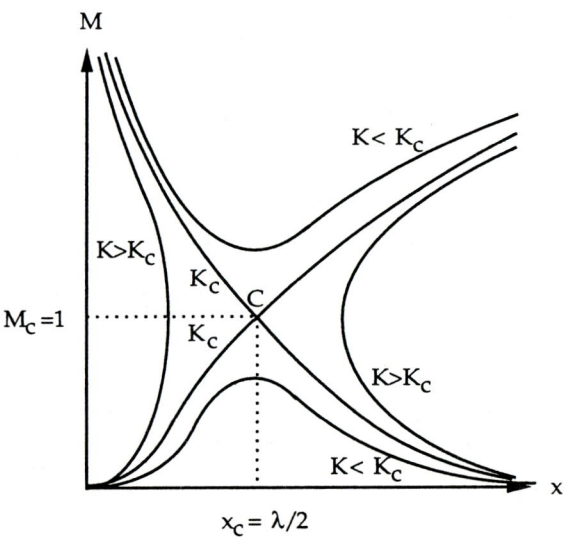

Fig. 2. A few Bernoulli curves in the (x,M) plane

medium. To be more specific, note that for small M and large x the Bernoulli relation, eq (32), reduces to

$$M \approx \frac{K}{x^2} \tag{39}$$

This shows that the product Mx^2 approaches K as $x \to \infty$. Since $Mx^2\rho = \mu$, we see that, at infinity, the dimensionless density approaches the finite value, ρ_∞, given by

$$\rho_\infty = \frac{\mu}{K} \tag{40}$$

Since μ is just M(1), it is related to K by eq (32), written for $x = 1$:

$$\mu e^{-\mu^2/2} = Ke^{-\lambda} \tag{41}$$

Then, from eq (40),

$$\rho_\infty = e^{-\lambda} e^{\mu^2/2} > e^{-\lambda} \tag{42}$$

This shows that the pressure at infinity in these asymptotically static solutions, called "breezes", is even larger than in completely static ones. We conclude that any asymptotically subsonic solar wind would be in overpressure to the interstellar medium too.

1.4.3 The Transsonic Wind Solution

The only solution left is the special one for which with $K = K_c$, which grows from subsonic to supersonic with distance. Since $M \to \infty$ for this solution, it can meet the interstellar medium through a shock (or a series of shocks), a possibility that is denied to breezes, because shocks from subsonic to subsonic flows do not exist. Of course the position of the shock is not stationary, so that, on the large scale, the wind is not stationary as assumed, but, confining one's attention to a region of the flow much smaller than the size of this large shock surface, the stationarity assumption is acceptable, and will become exact at very late times, when the shock is rejected to infinity.

It may look strange that the correct solution does not emerge from imposing boundary conditions at infinity, but from a selection procedure to which only the critical transsonic solution survives. Later on we shall expand more on the deep reasons why a criticality condition substitutes itself to boundary conditions, which have to do with hydrodynamic causality.

We have then shown that the solution is the critical one, i.e., the accelerating transsonic wind, which is associated with the value K_c of the integration constant K. The relation between Mach number and distance now takes the very definite form:

$$Me^{-M^2/2} = \frac{\lambda^2 e^{3/2}}{4} \frac{e^{-\lambda/x}}{x^2} \tag{43}$$

This implicit relation can be solved numerically. For example, at the terrestrial orbit, the dimensionless distance is $x_t = 214.28$, and M_t, the Mach number at the earth orbit, is given approximately, since x_t is large by

$$M_t \, e^{-M_t^2/2} = \frac{\lambda^2 e^{3/2}}{4x_t^2} \tag{44}$$

This gives M = 3.79, which corresponds, for a coronal temperature of 10^6K, to a velocity Mc_S of 498 km/s. The rate of mass loss results from the calculation of the integration constant μ, related to K by equation (41). For μ small and the appropriate value of K, it simply becomes

$$\mu \approx \frac{\lambda^2 e^{3/2}}{4} \, e^{-\lambda} \tag{45}$$

The dimensional mass loss rate is expressed in terms of this parameter μ as

$$\dot{M} = 4\pi \, R^2 \, \rho_0 \, c \, \mu \tag{46}$$

With μ given by eq (45), the mass loss $\dot{M}$ is found to be of order $10^{-14} M_O$ per year.

1.5 POLYTROPIC THERMODYNAMICS

1.5.1 Polytropic Wind Equations

Of course, the isothermal approximation is not excellent. Alternatives must be considered. For example (Parker, 1960) introduced a wind that is isothermal up to a certain distance, and adiabatic further up. Let us generalize the previous results to cases polytropic thermodynamics. We now assume that pressure and density are related by:

$$\overline{P} = K \, \overline{\rho}^{\,\gamma} \tag{47}$$

This ansatz replaces an energy balance equation, or equivalently it can be regarded as an adiabatic energy equation for a gas with this ratio of specific heats. The dimensional equations of the problem are

$$\frac{d}{dr}(r^2 \overline{\rho} v) = 0 \tag{48}$$

$$v \frac{dv}{dr} = -\frac{GM}{r^2} - \frac{1}{\overline{\rho}} \frac{d\overline{P}}{dr} = -\frac{GM}{r^2} - \frac{d}{dr} \left(\frac{\gamma}{\gamma - 1} K \, \overline{\rho}^{\,\gamma-1} \right) \tag{49}$$

We normalize these dimensional quantities as before, except velocities which we now normalize to the sound speed at the base of the solar corona, c_0, defined by $c_O^2 = \gamma \rho_0^{\gamma-1}$. Note that the sound speed is not constant in space anymore, as in the isothermal model. Its value at distance x, $c(x)$, is given by

$$c^2(x) = c_0^2 \, \rho^{\gamma-1} \tag{50}$$

It would now be inadequate to name the normalized velocity a Mach number. So, we switch to a new notation for it, u, defined by :

$$u = v/c_0 \tag{51}$$

The square of the Mach number at distance x is

$$M^2 = \frac{v^2}{c^2} = \frac{u^2}{\rho^{\gamma-1}} \tag{52}$$

The gravitational parameter is still defined by

$$\lambda = \frac{GM}{R\,c_0^2} \tag{53}$$

The dimensionless equations are then:

$$\frac{d}{dx}\left(x^2\rho u\right) = 0 \tag{54}$$

$$u\frac{du}{dx} = -\frac{\lambda}{x^2} - \frac{d}{dx}\left(\frac{\rho^{\gamma-1}}{\gamma-1}\right) \tag{55}$$

1.5.2 Static Solutions

The static solutions of these equations are non-self gravitating polytropic atmospheres, with dimensionless density given by

$$\rho(x) = \left(1 - \lambda\,(\gamma-1)\,(1-\frac{1}{x})\right)^{\frac{1}{\gamma-1}} \tag{56}$$

If $\lambda(\gamma-1)$ is larger than 1, the static solutions have a top, i.e. an altitude at which the density vanishes. If $\lambda(\gamma-1)$ is smaller than 1, the static polytropic atmosphere extends to infinity, where it asymptotes to a finite density, ρ_{stat} given by

$$\rho_{stat} = \left(1 - \lambda\,(\gamma-1)\right)^{\frac{1}{\gamma-1}} \tag{57}$$

1.5.3 Bernoulli Equation

When there is flow, the momentum equation (55) admits again a first integral:

$$\frac{v^2}{2} = \frac{GM}{r} - \frac{\gamma}{\gamma-1}\frac{\overline{P}}{\overline{\rho}} + E \tag{58}$$

The second term on the right hand side is again the specific enthalpy, and E, the total specific energy, is a constant. The dimensionless form of this equation is:

$$\frac{u^2}{2} - \frac{\lambda}{x} + \frac{\rho^{\gamma-1}}{\gamma-1} = \epsilon \tag{59}$$

Of course there is also constancy of the mass flux

$$x^2\,\rho\,u = \mu \tag{60}$$

and μ the dimensionless mass flux is still $u(1)$. Eliminating ρ in eq (59) by using eq (60), we get u as a function of x in the following implicit form:

$$\frac{u^2}{2} - \frac{\lambda}{x} + \frac{1}{\gamma-1}\left(\frac{\mu}{ux^2}\right)^{\gamma-1} = \epsilon \tag{61}$$

where again μ is related to ϵ by this very relation, written for $x = 1$:

$$\frac{\mu^2}{2} - \lambda + \frac{1}{\gamma-1} = \epsilon \tag{62}$$

Differentiating the Bernoulli equation (61), we express it in the form of a differential equation for $u(x)$:

$$\frac{du}{u}\left(u^2 - (\frac{\mu}{ux^2})^{\gamma-1}\right) - dx\left(\frac{2}{x}(\frac{\mu}{ux^2})^{\gamma-1} - \frac{\lambda}{x^2}\right) = 0 \tag{63}$$

1.5.4 Critical Point

The critical point is where the differential (63) vanishes. This happens at a point of coordinates (x_c, u_c) which satisfy equations

$$u_c^2 = \left(\frac{\mu}{u_c x_c^2}\right)^{\gamma-1} \tag{64.a}$$

$$\frac{\lambda}{2x_c} = \left(\frac{\mu}{u_c x_c^2}\right)^{\gamma-1} \tag{64.b}$$

Note that $(\mu/ux^2)^{\gamma-1}$ is the normalized value of the square of the local sound speed, $c^2(x)/c_0^2$. Therefore the critical point is again where the flow is transsonic as implied by eq (64.a). Note the relation, obtained by eliminating μ:

$$x_c u_c^2 = \lambda/2 \tag{65}$$

u_c and x_c are given eplicitly by

$$u_c = \left(\frac{4\mu}{\lambda^2}\right)^{\frac{\gamma-1}{5-3\gamma}} \tag{66.a}$$

$$x_c = \left(\frac{\lambda^{\gamma-1}}{2^{\gamma+1}\mu^{2(\gamma-1)}}\right)^{\frac{1}{5-3\gamma}} \tag{66.b}$$

1.5.5 Asymptotic Behaviour

The solutions as $x \to \infty$ can be discussed from eqs (61) and (62) combined to give:

$$\frac{u^2}{2} - \frac{\lambda}{x} + \frac{1}{\gamma-1}\left(\frac{\mu}{ux^2}\right)^{\gamma-1} = \frac{\mu^2}{2} - \lambda + \frac{1}{\gamma-1} \equiv \epsilon \tag{67}$$

A solution with $ux^2 \to \infty$ is approximately given by

$$u^2 \approx 2\epsilon \tag{68}$$

If on the other hand ux^2 does not approach infinity, equation (67) can be rewritten for large x as

$$\frac{1}{\gamma-1}\left(\frac{\mu}{ux^2}\right)^{\gamma-1} = \epsilon + \frac{\lambda}{x} - \frac{u^2 x^4}{2x^4} \approx (\epsilon + \frac{\lambda}{x}) \tag{69}$$

This gives the solution which vanishes at infinity approximately as

$$u \approx \frac{\mu}{x^2} \left(\frac{1}{(\gamma - 1)(\epsilon + \lambda/x)} \right)^{\frac{1}{\gamma - 1}} \tag{70}$$

Thus there is a solution such that $u \to 0$, and another one which approaches a finite value (instead of increasing indefinitely as in the isothermal case). Since, for the small solution (the so called "breeze"), ux^2 approaches a finite value, so do the density and the pressure. The finite pressure reached at infinity is usually too large to match the small pressure of the interstellar medium. By contrast, the finite velocity solutions have zero pressure at infinity.

1.5.6 Behaviour at Small Distances from the Origin

Let us now examine the behaviour of the solutions of the Bernoulli equation (61) as $x \to 0$. For this, we write this equation as

$$\frac{u^2}{2} = \epsilon + \frac{\lambda}{x} - \frac{1}{\gamma - 1} \left(\frac{\mu}{ux^2} \right)^{\gamma - 1} \tag{71}$$

If ux^2 were not to approach zero, the solution would be $u \approx (2\lambda/x)^{1/2}$, but such a solution is in fact inconsistent with the assumption that ux^2 does not approach zero. When ux^2 approaches zero, there are two subcases to consider, according to whether $\lambda/x >> (ux^2)^{\gamma-1}$, case-a, or $\lambda/x \approx (ux^2)^{\gamma-1}$, case-b.

In case-a the solution of eq(69) is approximately

$$u \approx (2\lambda/x)^{1/2} \tag{72}$$

which is consistent with the strong inequality characterizing this case, whatever γ (between 1 and 5/3). Near $x = 1$, this branch of solution is decelerating, and the square of its Mach number behaves as

$$M^2 = \frac{u^2}{\rho^{\gamma - 1}} = \frac{u^{\gamma + 1} x^{2(\gamma - 1)}}{\mu^{\gamma - 1}} \approx x^{-\frac{5 - 3\gamma}{2}} \tag{73}$$

This shows that this is a supersonic branch in the vicinity of $x = 0$ and presumably near $x = 1$ too.

In case-b, the solution balances the second and third terms of the right hand side of eq (69), giving to lowest order

$$u \approx \frac{\mu}{((\gamma - 1)\lambda)^{\frac{1}{\gamma - 1}}} x^{\frac{3 - 2\gamma}{\gamma - 1}} \tag{74}$$

It can be checked that in this case μ^2 is indeed much les than λ/x.

If γ is less than 3/2 this velocity is small near the origin, and we have an accelerated solution. But if γ is larger than 3/2, $u(x)$ goes to infinity when x goes to zero. This is then, near the origin, a decelerating solution, but still a subsonic one, since its Mach number is approximately given by

$$M^2(x) \approx u^{\gamma + 1} x^{2(\gamma - 1)} \approx x^{\frac{5 - 3\gamma}{\gamma - 1}} \tag{75}$$

In any case, whether the solution be accelerated or not, the case-b solution is subsonic near the origin, and its Mach number approaches zero as $x \to 0$. In his 1960 paper

Parker argued that the solution should necessarily accelerate near $x = 1$, which lead him to reject models having indices γ larger than $3/2$.

1.5.7 The Solution is the Critical Solution

To sum up this discussion, we may conclude that, as in the isothermal situation, the only acceptable solution is the critical transsonic accelerating solution. Actually, we have seen that entirely subsonic solutions have non zero and too large pressure at infinity, while non-critical solutions develop a singularity when they become transsonic. If, like Parker, we demand that $u(x)$ be an increasing function at small x, then only models with $\gamma < 3/2$ are acceptable. At infinity the velocity of the critical accelerating solution reaches an asymptotic value equal to $\sqrt{2\epsilon_{crit}}$.

2. HYDRODYNAMIC CAUSALITY AND CRITICAL POINTS

2.1 THE ISENTROPIC ONE-DIMENSIONAL MODEL

2.1.1 The Isentropic Gas Model

Consider an infinite homogeneous medium, unperturbed at times $t < 0$, with density ρ_0, pressure P_O, and specific entropy σ_0, all uniform in space, with no motion and sitting in zero gravity. At initial time, introduce a density perturbation ρ_1, a pressure perturbation P_1, and a velocity field v. Assume no entropy perturbation and that the further evolution is adiabatic. Assume v to be parallel to the direction x, and these perturbations to depend on space variable x and on time only. The isentropy of the motion implies that the pressure P is a function of density ρ only. Generally, pressure is a function of density and temperature, or, equivalently, of density and specific entropy. But here entropy is uniform in space and constant in time. It then simply parametrizes the dependance of P on ρ. So, here, pressure is only a function of density: $P = P(\rho)$. The square of the adiabatic sound velocity, c^2 is:

$$c^2 = \frac{dP}{d\rho} \tag{76}$$

Hence the sound speed, c, is a function of ρ too. For a gas of constant adiabatic index γ, $P = K(\sigma)\rho^\gamma$, so that $c^2 = \gamma P/\rho$. For an isothermal gas, $P = \rho c^2$, with a constant c^2.

2.1.2 Linear Perturbations

For simplicity we begin by considering small perturbations. It is well known that these propagate forward and backward as sound waves. The quantities associated with the unperturbed state are indicated by subscript 0. The linearized equations of motion can be written as

$$\frac{\partial \rho_1}{\partial t} + \rho_0 \frac{\partial v}{\partial x} = O \tag{77.a}$$

$$\rho_0 \frac{\partial v}{\partial t} + c_0^2 \frac{\partial \rho_1}{\partial x} = 0 \tag{77.b}$$

Elimination of ρ_1 gives for v the sound wave equation

$$\frac{\partial^2 v}{\partial t^2} - c_0^2 \frac{\partial^2 v}{\partial x^2} = 0 \tag{78}$$

the general solution of which is the sum of a forward propagating disturbance, v_+, and a backwards propagating one, v_-, so that

$$v(x,t) = v_+(x - c_0 t) + v_-(x + c_0 t) \tag{79}$$

and similarly for ρ_1. Let us stress the fact that v_+ is a function of only one variable, $(x - c_0 t)$, not of x and t independantly. Simlilarly for v_-. The functions $v_\pm$ can be found from the initial perturbation. The information that v takes some initial value $v(x, 0)$ at point x propagates away from x forward and backwards no faster than the sound speed c_0. At a time t later, only those points which are in the interval $(x \pm c_0 t)$ could have felt influence of the fact that the velocity was $v(x, 0)$ at x initially.

2.1.3 Linear Riemann Invariants

This can be seen another way, which will later be generalized to non linear situations. For homogeneity, let us define in place of ρ_1 a variable w having dimension of velocity:

$$w = c_0 \frac{\rho_1}{\rho_0} \tag{80}$$

Equations (77) transform into:

$$\frac{\partial w}{\partial t} + c_0 \frac{\partial v}{\partial x} = O \tag{81.a}$$

$$\frac{\partial v}{\partial t} + c_0 \frac{\partial w}{\partial x} = 0 \tag{81.b}$$

Adding and substracting, they become:

$$\frac{\partial}{\partial t}(v + w) + c_0 \frac{\partial}{\partial x}(v + w) = O \tag{82.a}$$

$$\frac{\partial}{\partial t}(v - w) - c_0 \frac{\partial}{\partial x}(v - w) = 0 \tag{82.b}$$

These equations express that the quantities J_{0+} and J_{0-}, called linear Riemann invariants, defined by

$$J_{0\pm} = v \pm w = v \pm c_0 \frac{\rho_1}{\rho_0} \tag{83}$$

do not vary following points that travel at velocity $\pm c_0$ resp. As a result, each linear Riemann invariant does not depend separately on x and t, but on only one variable, which can be taken as the initial abcissa of a point travelling at velocity $\pm c_0$. So,

$$J_{0+}(x, t) = J_{0+}(x - c_0 t) = J_{0+}(\xi_+) \tag{84.a}$$

$$J_{0-}(x, t) = J_{0-}(x + c_0 t) = J_{0-}(\xi_-) \tag{84.a}$$

2.1.4 Transfer of Information in Linear Sound Waves

We can now discuss more precisely the transfer of information in linear sound waves. The linear Riemann invariant functions are easily found from initial conditions, since at initial time the variables $\xi_\pm$ reduce to the variable x. Hence, the invariant functions are obtained explicitly from:

$$J_{0+}(\xi) = v(\xi, 0) + c_0 \frac{\rho_1(\xi, 0)}{\rho_0} \tag{85.a}$$

$$J_{0-}(\xi) = v(\xi, 0) - c_0 \frac{\rho_1(\xi, 0)}{\rho_0} \tag{85.b}$$

The velocity and density at later times are obtained by inverting eqs (83):

$$v(x, t) = \frac{1}{2}[J_{0+}(x, t) + J_{0-}(x, t)] \tag{86.a}$$

$$\rho_1(x, t) = \frac{\rho_0}{2c_0}[J_{0+}(x, t) - J_{0-}(x, t)] \tag{86.b}$$

This solution is written explicitly in terms of initial conditions using eqs (85):

$$v(x, t) = \frac{1}{2}[v(x - c_0 t, 0) + v(x + c_0 t, 0) + \frac{c_0}{\rho_0}\rho_1(x - c_0 t, 0) - \frac{c_0}{\rho_0}\rho_1(x + c_0 t, 0)] \tag{87.a}$$

$$\rho_1(x, t) = \frac{\rho_0}{2c_0}[v(x - c_0 t, 0) - v(x + c_0 t, 0) + \frac{c_0}{\rho_0}\rho_1(x - c_0 t, 0) + \frac{c_0}{\rho_0}\rho_1(x + c_0 t, 0)] \tag{87.b}$$

This shows that the linear perturbation at (x,t) depends only on the initial perturbations at the sound-connected points $(x \pm c_0 t)$. When the medium is not infinite, similar results apply, but they involve also boundary conditions, not just initial ones.

2.2 NON-LINEAR RIEMANN INVARIANTS, CHARACTERISTICS AND HYDRODYNAMIC CAUSALITY

2.2.1 Non-Linear Riemann Invariants

Let us now consider non-linear one-dimensional isentropic flows, which obey equations:

$$\frac{\partial \rho}{\partial t} + \rho \frac{\partial v}{\partial x} + v \frac{\partial \rho}{\partial x} = 0 \tag{88.a}$$

$$\rho \frac{\partial v}{\partial t} + \rho v \frac{\partial v}{\partial x} + \frac{\partial P}{\partial x} = 0 \tag{88.b}$$

We want to show that perturbations at (x, t) can be constructed from generalized Riemann invariants which remain constant following convected sound perturbations. Multiply the mass conservation equation by c/ρ, and, to homogeneize dimensions, divide the momentum one by ρ. We get:

$$\frac{c}{\rho} \frac{\partial \rho}{\partial t} + c \frac{\partial v}{\partial x} + \frac{cv}{\rho} \frac{\partial \rho}{\partial x} = 0 \tag{89.a}$$

$$\frac{\partial v}{\partial t} + v \frac{\partial v}{\partial x} + \frac{1}{\rho} \frac{\partial P}{\partial x} = 0 \tag{89.b}$$

Since P is a function of ρ, we can write

$$dP = \frac{dP}{d\rho}d\rho = c^2 d\rho \tag{90.a}$$

$$\frac{\partial P}{\partial t} = c^2 \frac{\partial \rho}{\partial t} \tag{90.b}$$

$$\frac{\partial P}{\partial x} = c^2 \frac{\partial \rho}{\partial x} \tag{90.c}$$

Note that c^2 is now a function of ρ, which depends on x, and not a mere constant, as was c_0^2 in the linear case. By eqs (90) we can replace all derivatives of ρ by derivatives of P, which results in the system

$$\frac{1}{\rho c}\frac{\partial P}{\partial t} + c\frac{\partial v}{\partial x} + \frac{v}{\rho c}\frac{\partial P}{\partial x} = O \tag{91.a}$$

$$\frac{\partial v}{\partial t} + v\frac{\partial v}{\partial x} + \frac{1}{\rho}\frac{\partial P}{\partial x} = 0 \tag{91.b}$$

This suggests to define a function of ρ, call it $W(\rho)$, by the differential relation

$$dW = \frac{dP}{\rho c} = \frac{cd\rho}{\rho} \tag{92}$$

which integrates into

$$W(\rho) = \int_{\rho i}^{\rho} \frac{c(\rho')d\rho'}{\rho'} \tag{93}$$

Since the lower bound of the integral, ρ_i, is arbitrary, the function $W(\rho)$ is defined up to an additive constant. For small motions, we could take ρ_i equal to ρ_0, the unperturbed density, and $W(\rho)$ would then reduce to the function w defined by eq (80). In terms of W and ρ, the flow equations take the form:

$$\frac{\partial W}{\partial t} + c\frac{\partial v}{\partial x} + v\frac{\partial W}{\partial x} = O \tag{94.a}$$

$$\frac{\partial v}{\partial t} + v\frac{\partial v}{\partial x} + c\frac{\partial W}{\partial x} = 0 \tag{94.b}$$

Adding and substracting as before gives

$$\frac{\partial}{\partial t}(v + W) + (v + c)\frac{\partial}{\partial x}(v + W) = O \tag{95.a}$$

$$\frac{\partial}{\partial t}(v - W) + (v - c)\frac{\partial}{\partial x}(v - W) = 0 \tag{95.b}$$

Hence, the so-called Riemann invariants J_+ and J_-, defined by

$$J_{\pm} = v \pm W \tag{96}$$

are constant following motions of instantaneous velocity $(v \pm c)$ resp.

2.2.2 Characteristics

This is a result similar to linear theory, but with important differences. Here too the Riemann invariants are transported with constant value along lines of the (x, t) plane followed by infinitesimally small sound disturbances, but account is taken of the fact that the fluid which supports these waves is moving. Sound is convected by the fluid. The paths in the (x, t) plane of convected sound disturbances are called characteristics. There exists "forward" characteristics, noted C_+, defined by the differential equation

$$\frac{dx}{dt} = v + c \qquad (97.a)$$

and "backwards" characteristics, C_-, which follow paths described by equation:

$$\frac{dx}{dt} = v - c \qquad (97.b)$$

Note that both v and c are not a priory known. The problem, in fact, is just to determine them. The linear situation was simpler, since v could be neglected compared to c and the sound speed could be taken as the unperturbed value. Characteristics in this case were simply straight lines of slope $dx/dt = c_0$. In the non linear situation, v is not negligible to c, which itself depends on the state of the fluid at (x, t). As a result, characteristics usually are complicated lines, described by the above differential equations, c and v being given in terms of the Riemann invariants carried unchanged along them. Characteristics of a non linear motion are not known a-priori, but should be solved for, at the same time as for the flow variables v and ρ.

Characteristics have some general properties. For example

1- At any point (x, t), pass two characteristics, one C_+ and one C_-.

2-Characteristics of similar type usually do not intersect, for that would mean that one Riemann invariant would have become doubled valued there. When this happens, a shock wave forms.

So, normally, the characteristics C_+ and C_- form a mutually intersecting, but non self-intersecting, network. It is in principle straightforward to compute the characteristics. Start with known initial conditions on line $t = 0$ of the (x, t) plane. Calculate, everywhere along this line the slope of the characteristics passing at each point. Then consider an infinitesimally later time, δt, say. At each point $(x, \delta t)$ pass a C_+ and a C_- characteristic which originates on the $t = 0$ axis at points that can be numerically found, since we know the tangents to characteristics all over the $t = 0$ axis. Since each characteristic carries its own Riemann invariant with it, and since the latter are known on the axis $t = 0$, we obtain the value of the Riemann invariants at any point on the axis $t = \delta t$, from which we calculate the flow variables, and the tangents to characteristic curves everywhere on this axis. The procedure is then reiterated, the axis $t = \delta t$ replacing the axis $t = 0$.

2.2.3 Hydrodynamic Causality

This construction shows how hydrodynamic causal relations establish in the flow. Let M be an event, i.e a point of the (x, t) plane. The flow variables there are obtained from the value of the Riemann invariants, which have been brought by the $C_+(M)$ and $C_-(M)$ characteristics which pass at event M. So, the flow variables at M result from the initial conditions existing at events P_0 and Q_0 on the $t = 0$ axis from which these two characteristics emanate (see figure (3)).

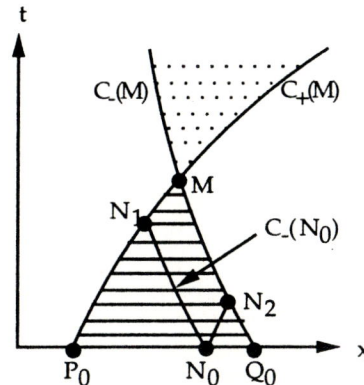

Fig. 3. The shape of the characteristics at M is under the influence of flow variables at events like N_1 and N_2 and is indirectly influenced only by events in the shaded "cone of the past". Initial events (at $t = 0$) which have an influence on M are between P_0 and Q_0.

But the characteristic curves are not a-priori known. So their very shape, and the positions of P_0 and Q_0, are influenced by the state of the flow met at any event between P_0 and M along $C_+(M)$ and between Q_0 and M along $C_-(M)$. Let N be one such event on characteristic P_0- M, say. The flow conditions at N result from the initial conditions at P_0 and at the event N_0 from which $C_-(N)$ emanates on axis $t = 0$. Initial conditions at N_0 then also influence indirectly the flow variables at M. We can extend this reasoning to any event like N. This shows that the situation at M is under the influence of the initial situation at all events like N_0, which form the segment $[P_0, Q_0]$ of the axis $t = 0$. More generally, only those events which are situated below M in the angle formed by the two characteristics $C_\pm(M)$ can have any hydrodynamical influence at M. This region constitutes an hydrodynamical cone of the past of M, shown hatched in figure (3).

Conversely, the situation at M can only have direct or indirect influence on those events which are above M in the angle subtended by the two characteristics $C_\pm(M)$. This region is the cone of hydrodynamical future of M.

An important point is to be made here. If the flow is somewhere supersonic, then both characteristics head to the same x-direction, since $v + c$ and $v - c$ have in that case the same sign (see figure (4)).

For definiteness, suppose this to be the direction of larger x's. Then the cone of the hydrodynamical future is in this case only in the region of larger x's. Events on the other side are not causally influenced by M. This is because the sound signals that could make the situation at M known elsewhere are blown downstream with the wind. Nothing is heard upstreams.

Suppose moreover the supersonic flow to be stationary. In that case the characteristics of a given type can be deduced from eachother by a translation parallel to the time axis.

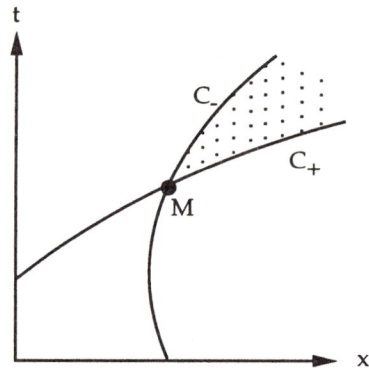

Fig. 4. Characteristics at an event M where the fluid flow is supersonic: the "cone of the future" of M is entirely towards larger x's. The state of the flow at M cannot influence events of smaller x's.

Then, at all events of the same position as M but of different times, the hydrodynamical cone of the future is tilted, say, to larger x's, while the cone of the past is towards smaller x's. This means that the flow at M can only be influenced by events of smaller x, and can only influence events of larger x. This is basically the reason why boundary conditions exerted at a place situated downstream of a zone of supersonic flow cannot have any effect on the flow upstream. Therefore boundary conditions in the downstream region cannot affect the flow in the upstream one.

2.3 CRITICAL POINTS AND CAUSALITY

From this discussion we realize that if, in a stationary one dimensional flow, the motion is subsonic at a given point x, the C_+ characteristic propagate forward there, and the C_- characteristic backwards. But if the velocity at x is supersonic, both characteristics propagate in the same direction. A critical point of the hydrodynamical flow sits at a place where $|v| = c$. At such a place, the characteristic which propagates backwards to the direction of flow has a vanishing net velocity. Its slope dt/dx becomes infinite, and since the flow is assumed to be stationary, this entire characteristic is a vertical line of the (x, t) plane.

Figure (5) represents the characteristics network of a stationary transsonic flow, accelerated in the positive x direction. Although the eye does not see it very well, the networks of C_+ and C_- characteristics are translationnally invariants in the t-direction. The feeling that the C_- network might not be is produced by the vertical asymptote at the transonic transition.

The important point is that any event downstream from the transonic point has a region of causal influence that never reaches to the subsonic region, because characteristics are dragged with the flow to larger x's. Another way to understand the suppression of

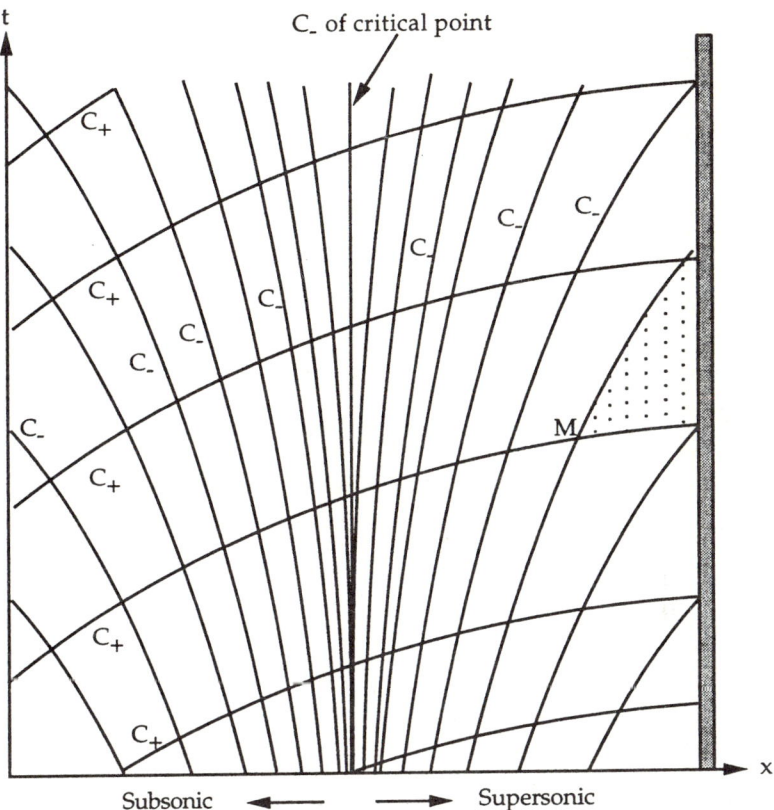

Fig. 5. Characteristics in a stationary transsonic flow. The critical point is at K. The region of the flow which is in the supersonic region has no influence on the subsonic region. As a result, boundary conditions in the supersonic region are irrelevant to the subsonic one.

communication with the subsonic region, is to imagine that a boundary is placed somewhere in the supersonic region. If the flow were indeed to be stationary, the boundary conditions at this place would have to match the "incident" J_+ and J_-, because both the C_+ and C_- characteristics "enter" the boundary. The boundary does not "emit" reflected characteristics, and so has no means to make its presence known to the upstream gas, then also no means to cause any sort of adaptation of the flow in the upstream region. If the boundary conditions imposed at this boundary were incompatible with the incoming values of the Riemann invariants, a shock would form and emanate from the boundary, non-linearly propagating in the counter stream direction. The situation then would cease to be stationary.

In the subsonic region of a stationary flow, events can only be consequences of other events also pertaining this region. This zone is bounded (see figure (5)) by the critical (transsonic) point. The subsonic region receives no influence from the supersonic one. Conversely, however, signals can pass from the subsonic zone to the supersonic one, so that the flow in the latter zone still results from the situation in the subsonic zone.

This sheds some light on an important aspect of transsonic flows: such flows cannot be determined in a subsonic region by any boundary condition imposed past the supersonic zone. This is why "boundary conditions at infinity" are completely irrelevant to them. The ultimate limit to which the subsonic region is causally connected sits at the transsonic point (the critical point). It is therefore no surprise that for such flows a criticality condition, which expresses a specific property of this place in the flow (i.e that $v = c$ there), must be substituted to irrelevant boundary conditions at infinity.

3. MHD ROTATING WINDS : FIRST INTEGRALS

3.1 ASTROPHYSICAL CONTEXT

3.1.1 Winds and Jets

The solar wind is only but one example of an astrophysical outflow. It is in fact much more complex than the model discussed above, being both anisotropic to some degree and time dependent, with a complex microphysics. A number of stars suffer mass loss to winds that are not always driven, as is the solar wind, by over-pressure in their lower atmosphere, but sometimes by the coupling of their own radiation to the plasma of their outer regions. Isotropy is not always realized. Young stars show so-called bipolar outflows, initially observed in molecular lines, and, now, also in some optical lines. These outflows are definitly anisotropic. Some compact objects have impressively collimated outflows, which reach relativistic speeds, like the jet of object SS 433.

Many active galaxies emit extremely collimated jets, made visible by the synchrotron radiation of high energy electrons present in them. These jets may also reach relativistic velocities, and extend to distances much larger than the size of the galaxy in which they originate. Galactic jets, jets from galactic compact sources and outflows from young stellar objects may emanate partly, or perhaps entirely, from an accretion disk orbiting the central object, be it a star or a galactic nucleus.

3.1.2 Possible Origins of Wind Anisotropies

Why should winds emitted from astrophysical sources be anisotropic, and even sometimes very collimated? An obvious reason might be that the conditions at the source of the wind would not themselves be isotropic. This is certainly so in the case of the sun, since it is known that the solar wind originates in those regions of the solar corona that are connected to open magnetic field lines, and not from active regions where the magnetic structure closes back to the solar surface. The medium around the outflow may also be intrinsically anisotropic, due to the presence of an accretion disk, for example, or because the molecular environment may have large density gradients, as in the case of young stellar objects forming near the edge of a molecular cloud. Very large collimations, though, should have a more intrinsic origin. The wind-emitting object may be rotating, in the case of disks as fast as the keplerian speed in the ambiant gravitational field of the accreting object. Magnetic fields anchored in rotating wind sources not only create further geometrical anisotropies, but they also exert on the flow Lorentz forces which may deeply influence its dynamics and even focus it.

This is one of the main ideas to be discussed in these lectures, having in mind the interesting conjecture that the large differences in morphology observed between these different winds may reflect different regimes of the coupling of MHD forces to rotation. We do not regard other possibilities as disproved or even unlikely. The rotating magnetized wind concept, though, is well defined and involves interesting MHD physics. It certainly deserves the attention paid to it in these lectures.

It is very difficult to treat analytically three-dimensional structures. Since rotation is involved, we shall limit our discussion to axisymmetrical systems, as do many other papers on this subject.

3.2 THE ROLE OF MAGNETIC FIELDS IN WINDS: A KINEMATICAL FIRST APPROACH

3.2.1 The Kinematic Approximation

The interaction of wind flow with the magnetic field generates electric currents of density

$$j = \frac{\text{rot}\,B}{\mu_0} \qquad (98)$$

which develop Lorentz forces $j \times B$, which act on the flow, in addition to pressure and gravity forces. As a simple first approximation, though too crude, let us nevertheless assume provisionnally that the plasma flow forms independently of these Lorentz forces, and discuss the associated magnetic configuration and currents. This kinematical description is not self-consistent. Later, we come back to complete self consistency. The kinematic approximation is valid when the Lorentz (MHD) forces are indeed negligible as compared to pressure forces and to inertia forces. This can be expressed by the following strong inequalities:

$$\frac{B^2}{2\mu_0} \ll P_{gas} \qquad \text{and} \qquad \frac{B^2}{2\mu_0} \ll \rho v^2 \qquad (99)$$

If the flow consists of a predominantly neutral gas, some attention should be given to the tightness of the collisional coupling between neutral atoms or molecules and the ionized component of the gas, which is never totally absent. Here we assume that this coupling is strong enough for the medium to behave as a single fluid of large electrical conductibility. For a known velocity field, the evolution of the magnetic field in perfect MHD is described by the equation

$$\frac{\partial B}{\partial t} = \text{rot}\,(v \times B) \qquad (100)$$

which for stationary flows reduces to

$$\text{rot}\,(v \times B) = O \qquad (101)$$

3.2.2 Magnetic Structure of a Radial Wind Emitted by a Rotating Object

From this, it is possible to calculate the magnetic structure in a radially flowing wind emitted by a rotating source. Since the field is frozen in the plasma, it is actually not possible to assume the fluid velocity to be radial from the surface of the source to infinity. We then refine the description by assuming that the fluid velocity evolves from almost complete corotation at the surface of the source to purely radial flow at large distances. For definiteness we shall assume this source to be a spherical star. It would be just as easy to account for any other type of object. Let R, ψ, φ be spherical coordinates (distance, colatitude and longitude resp.) centered on the star with the polar axis parallel to the axis of rotation. Assume the wind velocity to be representable in the form:

$$\boldsymbol{v} = v(R)\,\boldsymbol{e}_R + v_\varphi(R,\psi)\,\boldsymbol{e}_\varphi \tag{102}$$

The azimuthal component of the velocity $v_\varphi(R,\psi)$ matches the star's rotation at the star's surface $R = R_\star$:

$$v_\varphi(R_\star,\psi) = \Omega(\psi)\,R_\star\,sin\psi \tag{103}$$

It then declines rapidly to zero for larger R's, so that the flow smoothly converges to an almost radial and isotropic pattern. According to eq (102) any fluid element which leaves the star at colatitude ψ flows on a cone of half opening angle ψ. Because of flux freezing, this fluid element remains on one and a given field line, along which it moves from surface to infinity. Then this field line must be drawn on this cone, and the ψ-component of $\boldsymbol{B}$ must be zero, so that $\boldsymbol{B}$ can be represented as

$$\boldsymbol{B} = B_R(R,\psi)\,\boldsymbol{e}_R + B_\varphi(R,\psi)\,\boldsymbol{e}_\varphi \tag{104}$$

It is then a simple matter to write the components of eq (101), the only non trivial one being:

$$\frac{\partial}{\partial R}\,(\,R\,(v_\varphi\,B_R - v_R\,B_\varphi)\,) = 0 \tag{105}$$

The quantity in the parenthses is constant with R. We can calculate it at the surface of the star, where the radial flow velocity can be neglected, and $B_R(R_\star,\psi) = B_\star(\psi)$. At large distances, on the contrary, the azimuthal velocity vanishes. In these regions, then:

$$-R\,v(R)\,B_\varphi = +R_\star^2\,\Omega(\psi)\,sin\psi\,B_\star(\psi) \tag{106}$$

From flux conservation between two conical surfaces of half opening angles ψ and $\psi + d\psi$ we find that B_R varies as $1/R^2$. By eq (106) the azimuthal component varies in $1/R$:

$$B_R(R,\psi) = B_\star(\psi)\,(R_\star^2/R^2) \tag{107.a}$$

$$B_\psi(R,\psi) = 0 \tag{107.b}$$

$$B_\varphi(R,\psi) = -R_\star^2\,\Omega(\psi)\,sin\psi\,B_\star(\psi)\,/\,(R\,v(R)) \tag{107.c}$$

This shows that on the cone on which they are drawn, the field lines assume, because of rotation, a spiral shape wound backwards with respect to the sense of rotation.

3.2.3 MHD Forces on a Radial Wind in the Kinematical Approximation

Ampere's law gives the current density, as

$$\mathrm{rot}\,\boldsymbol{B} = \mu_0 \boldsymbol{j} \tag{108}$$

from which the components of the Lorentz force density $\boldsymbol{f} = \boldsymbol{j} \times \boldsymbol{B}$ can be calculated. This gives after little algebra:

$$f_R = \frac{B_R^2}{\mu_0 R}\,\frac{R^2\Omega^2\,sin^2\,\psi}{v^2}\,\frac{R}{v}\,\frac{\partial v}{\partial R} \tag{109.a}$$

$$f_\psi = -\frac{1}{R\,sin^2\,\psi}\,\frac{\partial}{\partial\psi}\left(\frac{B_R^2}{2\mu_0}\,\frac{\Omega^2 R^2\,sin^4\,\psi}{v^2}\right) - \frac{1}{R}\,\frac{\partial}{\partial\psi}\left(\frac{B_R^2}{2\mu_0}\right) \tag{109.b}$$

$$f_\varphi = \frac{B_R^2}{\mu_0 R}\,\frac{\Omega R\,sin\,\psi}{v}\,\frac{R}{v}\,\frac{\partial v}{\partial R} \tag{109.c}$$

3.2.4 Role of MHD Forces in Wind Acceleration

Since there is a radial component of the Lorentz force, we see that the magnetic field contributes to the dynamics of the rotating wind.

3.2.5 The Collimating Hoop Stress

There is also a latitudinal component of the force, with two contributions (eq (109.b). The second is due to the gradient of the poloidal magnetic pressure, while the first one is due to the tension of the azimuthal field. At large distances, this tension force dominates, as can be seen from eqs (107) and (109.b). This is because B_φ decreases more slowly with R than B_R. It can be checked that this force, a component of the vector product $\mathbf{B}_\varphi \times \mathbf{j}_P$, is oriented such as to pinch the fluid towards the rotation axis. It is called the hoop stress, or the pinch force. The hoop stress would collimate the flow to the axis if it were to asymptotically dominate over the defocusing forces, like outward pressure gradients or centrifugal forces.

3.2.6 Angular Momentum Transport to Infinity by Magnetized Winds

There is a component of the force in the azimuthal direction. The flow being axisymmetric, the resultant of these forces exerts a torque on the outflowing plasma. For an accelerated wind, the sign of azimuthal forces is in the sense of rotation: the fluid is azimuthally accelerated by MHD forces. In a stationary state, the magnetic field does not store the angular momentum exchanged with the matter, so this quantity must be taken from the star, to which the magnetic field is connected. By the action-reaction theorem, the star eventually provides for the angular momentum given to the escaping wind. This rotational braking makes its effects felt on a long time scale only, because the star is a large reservoir of angular momentum. Of course the star would loose angular momentum even in the absence of field, because matter would then leave with its initial angular momentum, $\Omega^2 R_*^2 \sin^2 \psi$, since in that case no torques are exerted on the wind fluid, and each bit of escaping matter conserves its angular momentum when travelling from source to infinity. The magnetic torques add substancially to this loss.

It is possible to see that the force which brakes the rotation of the star is in fact exerted next to its surface, in the region where the wind flow changes from corotation to radial outflow. In this region field lines are progressively bent backwards to rotation, so that a large tension force is exerted. For simplicity, let us regard this transition region as a thin layer near R_*, accross which B_φ jumps from zero inside the star to approximately (eq (107.c)):

$$B_{\varphi\,ext}(\psi) = -\frac{R_* \, \Omega(\psi) \, \sin\psi \, B_*(\psi)}{v_*} \tag{110}$$

where v_* is the radial flow velocity at the top of the transition layer. The jump in B_φ from zero inside the star to $B_{\varphi\,ext}$ at the top of the layer is accounted for by a current flowing in the ψ-direction, which we treat as a surface current $\mathbf{i}$:

$$\mathbf{i} = -\frac{B_{\varphi\,ext}}{\mu_0} \, \mathbf{e}_\psi = \frac{R_* \, \Omega(\psi) \, \sin\psi \, B_*(\psi)}{\mu_0 \, v_*} \, \mathbf{e}_\psi \tag{111}$$

The surfacic force density associated with it has a φ-component given by

$$i \times B_\star(\psi)\, e_R = -\frac{R_\star\, \Omega(\psi)\, \sin\psi\, B_\star^2(\psi)}{\mu_0\, v_\star}\, e_\varphi \tag{112}$$

It can be seen that this force is exerted opposite to the rotation. This rotational braking may be very important, both for stars, the rotation of which decreases secularly due to the continuous application of this torque, and for accretion disks, because this angular momentum loss favours accretion, and may even drive it.

How large is this magnetic torque as compared to the torque suffered by stars blowing non-magnetic winds? In non magnetized systems, each piece of wind escapes with its initial angular momentum. So each gram escaping at colatitude ψ carries away an angular momentum $\Omega(\psi)\, R_\star^2\, \sin^2\psi$. The torque per unit surface at colatitude ψ is the amount of angular momentum that leaves per unit area and unit time at this colatitude. It is the product of mass flux per unit area by the above mentioned specific angular momentum. For non magnetized systems this amounts to:

$$T_{\text{matter}} = -\rho_\star\, v_\star\, \Omega(\psi)\, R_\star^2\, \sin^2\psi \tag{113}$$

In magnetized systems, the force per unit area is given by eq (112), and the associated torque per unit surface is then

$$T_{\text{field}} = -\frac{R_\star^2\, \Omega(\psi)\, \sin^2\psi\, B_\star^2(\psi)}{\mu_0\, v_\star} \tag{114}$$

The ratio of (114) to (113) is approximately

$$\frac{T_{\text{field}}}{T_{\text{matter}}} = \frac{B_\star^2}{\mu_0\, \rho_\star\, v_\star^2} \tag{115}$$

If the field energy density dominates over the wind kinetic energy density at the surface of the star, the magnetic rotational braking dominates. Of course such a situation escapes the kinematical approximation! Nevertheless the above estimation shows that magnetic rotational braking may be very efficient.

3.3 FIRST INTEGRALS

We now want to be more ambitious and discuss rotating MHD winds in full self consistency. This means that we do not want only to analyse the effect of a given flow on the magnetic structure, but we also want to take into account the MHD forces exerted on it, which both accelerate and shape it.

The assumptions of axial symmetry and stationarity give rise to the conservation, following the fluid motion, of a number of quantities. Since, as we shall see, the fluid flows on magnetic surfaces, such quantities are constants on them. This is why they are sometimes named surface functions, or first integrals, or constants of the motion. This subsection is devoted to finding and describing these surface functions.

3.3.1 Representation of the Magnetic Field

We first establish a convenient representation of the magnetic field. It is usual for ax-isymmetrical systems to separate vectors into a poloidal part, in the meridian plane, and a toroidal part, in the azimuthal direction. These parts are indicated by subscripts P and T resp. Let us adopt cylindrical coordinates, r, θ, z based on the axis of rotation and centered at the star. So,

$$\boldsymbol{B}_P = B_r \, \boldsymbol{e}_r + B_z \, \boldsymbol{e}_z \tag{116.a}$$

$$\boldsymbol{B}_T = B_\theta \, \boldsymbol{e}_\theta \tag{116.b}$$

None of these components depends on the angle θ. This implies that the poloidal field is itself solenoidal, $\mathrm{div}\,\boldsymbol{B}_P = 0$, and derives from a separate vector potential,

$$\boldsymbol{B}_P = \mathrm{rot}\,\boldsymbol{A} \tag{117}$$

that can be taken in the azimuthal direction, and be written as:

$$\boldsymbol{A} = \frac{a(r, z)}{r} \, \boldsymbol{e}_\theta \tag{118}$$

Then

$$B_r = -\frac{1}{r}\frac{\partial a}{\partial z} \qquad B_z = +\frac{1}{r}\frac{\partial a}{\partial r} \qquad \boldsymbol{B}_P = \frac{1}{r}\nabla a \times \boldsymbol{e}_\theta \tag{119}$$

The function $a(r, z)$ is constant following a poloidal field line. This is seen by writing the equation of such a line as

$$\frac{dr}{B_r} = \frac{dz}{B_z} \quad \Longrightarrow \quad \frac{\partial a}{\partial r}\,dr + \frac{\partial a}{\partial z}\,dz = 0 \tag{120}$$

This shows that $a(r, z)$ is constant along a poloidal field line. The total field $\boldsymbol{B} = \boldsymbol{B}_P + \boldsymbol{B}_T$ is in the plane defined by the tangent to the line of constant $a(r, z)$ in the meridional plane and to the azimuthal direction. This plane is tangent to the surface generated by the rotation about the axis of the poloidal line of constant $a(r, z)$. Such a surface is called a magnetic surface. It is a flux tube. Field lines of the total magnetic field are drawn on magnetic surfaces, on which they wrap helically, as we shall see. The flux through a magnetic surface which is simply the flux through the circle defined by the intersection of the magnetic surface $a(r, z) = A$ with a plane perpendicular to the rotation axis, is $2\pi A$. We assume here that the function $a(r, z)$ is conventionnally taken to be zero on the polar axis.

3.3.2 Constancy of the Mass Flux to Magnetic Flux Ratio

Our first surface function is the mass flux to magnetic flux ratio. To show this, consider the stationary form of the magnetic field evolution equation, eq(101), which integrates into

$$\boldsymbol{v} \times \boldsymbol{B} = \nabla \Phi \tag{121}$$

Since there is axial symmetry, $\partial/\partial\theta = 0$, and $\nabla \Phi$ has no azimuthal component. This shows that $\boldsymbol{v}_P$ and $\boldsymbol{B}_P$ are parallel, because the θ-component of eq (121) is

$$\boldsymbol{v}_P \times \boldsymbol{B}_P = 0 \tag{122}$$

The poloidal velocity follows the poloidal field lines, and, reasoning as before for B, we conclude that the total velocity $v = v_P + v_T$ is in the tangent plane to magnetic surfaces. In brief, flow lines are drawn on magnetic surfaces. From the alignement of the poloidal parts of the field and of the velocity, we can infer that there exists a function $\alpha(r, z)$ such that

$$\rho v_P = \alpha B_P \tag{123}$$

By the equation of mass conservation in stationary state div $(\rho v) = 0$, eq (123) gives rise to the following relation

$$B_P.\nabla\alpha = 0 \tag{124}$$

which shows that the function α is constant on a magnetic surface. So $\alpha(r, z)$ depends in fact on r and z indirectly through the value of $a(r, z)$:

$$\alpha(r, z) \equiv \alpha(a(r, z)) \equiv \alpha(a) \tag{125}$$

To sum up, α, the mass flux to magnetic flux ratio, is a surface function.

3.3.3 Ferraro's Isorotation Law

Our second surface function is the electric potential, or an associated function, the rate of rotation of the magnetic surface. Hereafter, we shall refer briefly to a surface on which $a(r, z) = a$ as "the surface a". To see why the magnetic surfaces are equipotential, consider the poloidal component of equation (121), which, using eq (123), can be written and arranged as follows:

$$\nabla\Phi = v_T \times B_P + v_P \times B_T = v_T \times B_P - B_T \times \frac{\alpha B_P}{\rho} = (v_T - \frac{\alpha B_T}{\rho}) \times B_P \tag{126}$$

Using eq (119), eq (126) becomes

$$\nabla\Phi = (v_\theta - \frac{\alpha B_\theta}{\rho})\frac{\nabla a}{r} \tag{127}$$

This shows that $\nabla\Phi$ and ∇a are everywhere parallel to eachother, or else that the lines of iso-value of functions a and Φ have everywhere the same normal. This implies that one is a function of the other. Then the electric potential Φ is a surface function:

$$\Phi(r, z) \equiv \Phi(a(r, z)) \equiv \Phi(a) \tag{128}$$

Using this information in eq (127), we first obtain

$$\frac{\partial\Phi}{da}\nabla a = (v_\theta - \frac{\alpha B_\theta}{\rho})\frac{\nabla a}{r} \tag{129}$$

Since ∇a does not vanish, except perhaps at isolated points, this relation can be written as:

$$v_\theta = r\frac{\partial\Phi}{da} + \frac{\alpha B_\theta}{\rho} \tag{130}$$

A straightforward dimensional analysis shows that the derivative of electric potential with respect to magnetic flux is an angular velocity, which we call $\Omega(a)$:

$$\frac{d\Phi}{da} = \Omega(a) \tag{131}$$

Then

$$v_\theta = r\Omega(a) + \frac{\alpha B_\theta}{\rho} \tag{132}$$

Grouping this with eq (123), we find that the fluid velocity can be written as

$$v = \frac{\alpha B}{\rho} + r\Omega(a) \; e_\theta \tag{133}$$

This equation says that the fluid motion on a magnetic surface is made of a flow aligned to the field lines supplemented by the rotation $r\Omega(a)e_\theta$. Since Ω depends only on a, this indicates that this rotation is like a solid body rotation of the magnetic surface. This results is often refered to as Ferraro's isorotation law. In brief, the flow is as if the matter were flowing along rigid magnetic pipes, themselves entrained in a rotation which is like solid body for each magnetic surface. However different magnetic surfaces may have different rotations. Some authors refer to $\Omega(a)$ as being the rotation of the field, although this expression is, in full rigour, meaningless. Note that $\Omega(a)$ is not the angular velocity of the fluid, since the latter should be defined as v_θ/r, which equals

$$v_\theta/r = \Omega(a) + \frac{\alpha B_\theta}{r\rho} \neq \Omega(a) \tag{134}$$

Very near the wind source, though, the density may be very large, and the field-aligned velocity very small. In that case v_θ/r and Ω coincide approximately.

3.3.4 Total Specific Angular Momentum Conservation Law

Our third surface function is the total specific angular momentum. Consider the equation of motion in stationary state :

$$\rho \, (v.\nabla)v = -\nabla P + \rho g + j \times B \tag{135}$$

This equation has three components. Consider the azimuthal one, which, using the vector identity

$$(v.\nabla)v = \nabla\frac{v^2}{2} - v \times \mathrm{rot}v \tag{136}$$

can be cast into the form

$$\rho v_P \times \mathrm{rot} \; v_T - \frac{1}{\mu_0}B_P \times \mathrm{rot}B_T = 0 \tag{137}$$

It is easily calculated that

$$\mathrm{rot} \; v_T = \nabla(rv_\theta) \times \frac{e_\theta}{r} \tag{138}$$

$$\mathrm{rot} \; B_T = \nabla(rB_\theta) \times \frac{e_\theta}{r} \tag{139}$$

Using finally the relation $\rho v_P = \alpha B_P$ further algebra reduces eq (137) to

$$B_P.\nabla. \left(rv_\theta - \frac{rB_\theta}{\mu_0\alpha} \right) = 0 \tag{140}$$

This indicates that the quantity $(rv_\theta - rB_\theta/\mu_0\alpha)$ is constant on a magnetic surface. It is a surface quantity, the total specific angular momentum, $L(a)$:

$$\left(rv_\theta - \frac{rB_\theta}{\mu_0\alpha}\right) = L(a) \tag{141}$$

Indeed we see that rv_θ is the specific angular momentum (i.e. angular momentum per unit mass) of the matter. Eq (141) indicates that the matter angular momentum is not conserved following the motion. This is no surprise since we have seen that magnetic fields exert a torque on the plasma, so rv_θ cannot be a constant of the motion. Angular momentum is exchanged between matter and field by these torques. This is why we find that some sort of total angular momentum per unit mass of matter, $L(a)$, combining the matter specific angular momentum with a quantity that we could name the field's equivalent specific angular momentum, $-rB_\theta/\mu_0\alpha$, is conserved.

Why is it that such a concept as specific angular momentum of the magnetic field could make sense? First let us explain how it is possible to define a flux of angular momentum of the magnetic field. It is well known that the density q of any quantity Q, whatever its nature, obeys an equation of the conservation form, which involves a flux of the quantity, F, and a rate of volumic creation, s, so that

$$\frac{\partial q}{\partial t} + \operatorname{div} F = s \tag{142}$$

This simply expresses that the total quantity of Q included in a certain volume varies in time because of in situ creation or destruction of Q (s term) or because of incomes and outcomes through the border (flux term). If instead of being a scalar quantity Q is a vector quantity, its density and rate of creation are vectors, q and s, while the flux becomes a second order tensor $\overline{\overline{F}}$. The amount of Q that goes through an oriented surface element dS per second is, as for scalar quantities, given by the dot product $\overline{\overline{F}} \cdot dS$, the result being in this case a vector. To define a linear momentum density of the electromagnetic field, and an associated tensor flux, it is possible to proceed as follows. Obviously the Lorentz force density $(\hat{q}E + j \times B)$ is the volumic rate of creation of momentum of matter due to the interaction with the electromagnetic field. Conversely, its opposite $-(\hat{q}E + j \times B)$ should be the rate of creation of momentum of the electromagnetic field, s_{em}, due to the interaction with plasma. By cleverly manipulating the Maxwells equations, it is possible to show that $-(\hat{q}E + j \times B)$ can be written as:

$$-(\hat{q}E + j \times B) = \frac{\partial}{\partial t}\left(\frac{E \times B}{\mu_0 c^2}\right) + \nabla \cdot \left(\left(\frac{\epsilon_0 E^2}{2} + \frac{B^2}{2\mu_0}\right)\overline{\overline{\delta}} - \epsilon_0 \overline{\overline{EE}} - \frac{\overline{\overline{BB}}}{\mu_0}\right) \tag{143}$$

Hence we see that if the left hand side is to be understood as being the rate of creation of momentum of the electromagnetic field, then $E \times B/\mu_0 c^2$ is to be regarded as its momentum density and the so called Maxwell stress tensor

$$\overline{\overline{M}} = (\frac{\epsilon_0 E^2}{2} + \frac{B^2}{2\mu_0})\overline{\overline{\delta}} - \epsilon_0 \overline{\overline{EE}} - \frac{\overline{\overline{BB}}}{\mu_0} \tag{144}$$

is its flux of momentum tensor. In non relativistic MHD, the momentum density of the field is negligible, as are also the electric force density and the electric terms in the Maxwell stress tensor. Then the latter reduces to

$$\overline{\overline{M}} = \left(\frac{B^2}{2\mu_0} \overline{\overline{\delta}} - \frac{\overline{\overline{BB}}}{\mu_0} \right) \tag{145}$$

The quantity of linear momentum carried per second through a surface element dS by the electromagnetic field is then

$$\frac{dp}{dt} = dS \cdot \overline{\overline{M}} \tag{146}$$

Particularizing to a surface element perpendicular to the poloidal field, B_P, $dS = dS\,(B_P/B_P)$ we obtain

$$\frac{dp}{dt} = \left(\frac{dS}{B_P} \right) B_P \cdot \left(\frac{B^2}{2\mu_0} \overline{\overline{\delta}} - \frac{\overline{\overline{BB}}}{\mu_0} \right) \tag{147}$$

$$\frac{dp}{dt} = \left(\frac{dS}{B_P} \right) \left(\frac{B^2}{2\mu_0} B_P - \frac{B_P^2}{\mu_0} B \right) \tag{148}$$

The θ-component of this is

$$\frac{dp_\theta}{dt} = -\frac{B_P B_\theta}{\mu_0} dS \tag{149}$$

Then, the angular momentum carried by the magnetic field through dS per second, dl/dt, is obtained by multiplying this by r:

$$\frac{dl}{dt} = -\frac{r B_\theta}{\mu_0} B_P\, dS \tag{150}$$

We can define an equivalent specific angular momentum carried by the magnetic field by dividing dl/dt by the mass $dm/dt = \rho\, v_P\, dS$ that goes per second through the same surface element. Finally the equivalent specific angular momentum of the magnetic field appears to be

$$\frac{dl/dt}{dm/dt} = -\frac{r B_\theta}{\mu_0} \left(\frac{B_P\, dS}{\rho\, v_P\, dS} \right) = -\frac{r B_\theta}{\mu_0 \alpha} \tag{151}$$

We have used the relation $\rho v_P = \alpha B_P$. This completes our understanding of the second term on the left hand side of eq (141).

3.3.5 Polytropic Thermodynamics and Specific Entropy Conservation Law

A fourth surface function results from the assumption that the gas thermodynamics is ruled by a polytropic law. The pressure of an element of fluid is assumed to be related to its density by

$$P = k\rho^\gamma \tag{152}$$

So, k remains constant following this fluid parcel. It is related to its specific entropy. Since a given piece of gas follows a given magnetic surface, k should be the same all over such a surface. However, the specific entropy may be different on different surfaces. Therefore, the constant k is in fact a function of a, and we can write it as $k = Q(a)$, so that :

$$P = Q(a)\rho^\gamma \tag{153}$$

3.3.6 The Poloidal Equation of Motion

A fifth, and most important, first integral comes from the equation of motion projected on the direction of the poloidal magnetic field. Separating the toroidal part from the poloidal part in the equation of motion (eq (135)), and introducing the gravitational potential G (de fined by $g = -\nabla G$), we get:

$$\nabla\left(\frac{v^2}{2}\right) + \frac{\nabla P}{\rho} + \nabla G = v_P \times \mathrm{rot}\, v_P + v_T \times \mathrm{rot}\, v_T + \frac{\mathrm{rot}\, B_P}{\mu_0 \rho} \times B_P + \frac{\mathrm{rot}\, B_T}{\mu_0 \rho} \times B_T \quad (154)$$

It can be calculated that

$$v_T \times \mathrm{rot}\, v_T = \frac{v_\theta}{r}\, \nabla(r v_\theta) \quad (155.a)$$

$$B_T \times \mathrm{rot}\, B_T = \frac{B_\theta}{r}\, \nabla(r B_\theta) \quad (155.b)$$

Since $P = Q(a)\,\rho^\gamma$, the term $\nabla P/\rho$ can be manipulated into

$$\frac{\nabla P}{\rho} = \frac{\gamma}{\gamma - 1}\, \nabla(Q\rho^{\gamma-1}) - Q'\, \frac{\rho^{\gamma-1}}{\gamma - 1}\, \nabla a \quad (156)$$

At this point the poloidal part of the equation of motion is written as

$$\frac{\mathrm{rot}\, B_P}{\mu_0 \rho} \times B_P - \mathrm{rot}\, v_P \times v_P = \nabla\left(\frac{v^2}{2} + \frac{\gamma}{\gamma - 1}Q\rho^{\gamma-1} + G\right)$$
$$- Q'\, \frac{\rho^{\gamma-1}}{\gamma - 1}\, \nabla a - \frac{v_\theta}{r}\, \nabla\,(r v_\theta) + \frac{B_\theta}{\mu_0 \rho r}\, \nabla\,(r B_\theta) \quad (157)$$

We now arrange the v_θ and B_θ terms substituting for $r v_\theta$ the expression obtained from the first integral of specific angular momentum

$$r v_\theta = L(a) + \frac{r B_\theta}{\mu_0 \alpha} \quad (158)$$

and for (v_θ/r) the one derived from Ferraro's isorotation law

$$\frac{v_\theta}{r} = \Omega(a) + \frac{\alpha B_\theta}{\rho r} \quad (159)$$

Some further algebra brings the poloidal part of the equation of motion to the form

$$\frac{\mathrm{rot}\, B_P}{\mu_0 \rho} \times B_P - \mathrm{rot}\, v_P \times v_P = \nabla\left(\frac{v^2}{2} + \frac{\gamma}{\gamma - 1}Q\rho^{\gamma-1} + G - \frac{r\Omega B_\theta}{\mu_0 \alpha}\right)$$
$$- \nabla a\left(Q'\, \frac{\rho^{\gamma-1}}{\gamma - 1} + \Omega L' - \Omega'\, \frac{r B_\theta}{\mu_0 \alpha} + L'\, \frac{\alpha B_\theta}{\rho r} - \frac{\alpha'}{\alpha}\, \frac{B_\theta^2}{\mu_0 \rho}\right) \quad (160)$$

3.3.7 Conservation of Total Specific Energy : Bernoulli Law

From eq (119) it can be seen that ∇a is perpendicular to $\boldsymbol{B}_P$, and we remember that $\boldsymbol{v}_P$ is aligned to $\boldsymbol{B}_P$. Then many terms disappear in projecting eq (160) on $\boldsymbol{B}_P$. We are left with

$$\boldsymbol{B}_P \cdot \nabla \left(\frac{v^2}{2} + \frac{\gamma}{\gamma - 1} Q \rho^{\gamma - 1} + G - \frac{r \Omega B_\theta}{\mu_0 \alpha} \right) = 0 \tag{161}$$

When $\gamma = 1$, $\rho^{\gamma - 1}/(\gamma - 1)$ should be replaced by $\ln \rho$. Equation (161) identifies yet another surface function, the total specific energy, $E(a)$, defined by:

$$\frac{v_P^2 + v_\theta^2}{2} + \frac{\gamma}{\gamma - 1} Q \rho^{\gamma - 1} + G - \frac{r \Omega B_\theta}{\mu_0 \alpha} = E(a) \tag{162}$$

This equation is often refered to as the Bernoulli equation. Here again the specific energy does not reduce to the energy of matter alone, but contains a magnetic contribution. The interpretation is similar to the one given above for specific angular momentum. There exists a flux of electromagnetic energy, the well known Poynting flux, $\boldsymbol{S}$, defined by:

$$\boldsymbol{S} = \frac{\boldsymbol{E} \times \boldsymbol{B}}{\mu_0} \tag{163}$$

The electric field is electrostatic because of stationarity, and, thanks to eq (128) and (131), can be expressed as

$$\boldsymbol{E} = -\nabla \Phi = -\frac{d\Phi}{da} \nabla a = -\Omega(a) \nabla a \tag{164}$$

Then we calculate

$$\boldsymbol{S} = -r \, \Omega \, \frac{B_\theta \, \boldsymbol{B}_P}{\mu_0} + r \, \Omega \, \frac{B_P^2}{\mu_0} \, \boldsymbol{e}_\theta \tag{165}$$

The electromagnetic energy flux through a surface element perpendicular to the poloidal field, $d\boldsymbol{S} = dS \, (\boldsymbol{B}_P/B_P)$, is

$$dw = \boldsymbol{S} \cdot d\boldsymbol{S} \left(\frac{\boldsymbol{B}_P}{B_P} \right) = -dS \, \frac{\Omega \, r \, B_\theta}{\mu_0} \, B_P \tag{166}$$

Dividing this by the mass flux dm through the same surface element, we obtain an expression for the equivalent specific enrgy of the electromagnetic energy flux

$$\frac{dw}{dm} = -\frac{\Omega r B_\theta B_P}{\mu_0 \, \rho \, v_P} = -\frac{\Omega r B_\theta}{\mu_0 \, \alpha} \tag{167}$$

This shows that the last term on the left hand side of eq (162) represents the Poynting flux.

3.4 THE TRANSFIELD EQUATION

3.4.1 Equation of Motion Projected Perpendicular to Magnetic Surfaces

Using the result of eq (162) in eq (160), the equation of motion reduces to only one component perpendicular to the poloidal magnetic field, since eq (162) expresses the vanishing of the field-aligned component. Using equation (119) which defines $\boldsymbol{B}_P$ in terms of the flux function a, and the relation, eq (123), $\rho \boldsymbol{v}_P = \alpha(a)\boldsymbol{B}_P$, the left hand side of eq (160) can be expressed in terms of the function a as:

$$\frac{\mathrm{rot}\,\boldsymbol{B}_P}{\mu_0 \rho} \times \boldsymbol{B}_P - \mathrm{rot}\,\boldsymbol{v}_P \times \boldsymbol{v}_P =$$
$$\frac{\alpha}{\rho r} \nabla a \left(\frac{\partial}{\partial r} \frac{\alpha}{\rho r} \frac{\partial a}{\partial r} + \frac{\partial}{\partial z} \frac{\alpha}{\rho r} \frac{\partial a}{\partial z} \right) - \frac{1}{\mu_0 \rho r} \nabla a \left(\frac{\partial}{\partial r} \frac{1}{r} \frac{\partial a}{\partial r} + \frac{\partial}{\partial z} \frac{1}{r} \frac{\partial a}{\partial z} \right) \qquad (168)$$

Then eq (160) becomes a partial differential equation for the flux function a, which involves the first integrals identified above, the density ρ, and, at this stage, B_θ:

$$\frac{\alpha}{\rho r} \left(\frac{\partial}{\partial r} \frac{\alpha}{\rho r} \frac{\partial a}{\partial r} + \frac{\partial}{\partial z} \frac{\alpha}{\rho r} \frac{\partial a}{\partial z} \right) - \frac{1}{\mu_0 \rho r} \left(\frac{\partial}{\partial r} \frac{1}{r} \frac{\partial a}{\partial r} + \frac{\partial}{\partial z} \frac{1}{r} \frac{\partial a}{\partial z} \right) =$$
$$E' - Q' \frac{\rho^{\gamma-1}}{\gamma-1} - L'\,\Omega + \Omega' \frac{r B_\theta}{\mu_0 \alpha} - L' \frac{\alpha B_\theta}{\rho r} + \frac{\alpha'}{\alpha} \frac{B_\theta^2}{\mu_0 \rho} \qquad (169)$$

3.4.2 Alfvèn Radius and Alfvèn Density

By the angular momentum and isorotation laws, eq (132) and (141), the azimuthal components of $\boldsymbol{v}$ and $\boldsymbol{B}$ can be expressed in terms of r, of the density, ρ and of the first integrals $\alpha(a)$, $\Omega(a)$ and $L(a)$:

$$B_\theta = \frac{\rho \mu_0 \alpha}{r} \frac{L - r^2 \Omega}{\mu_0 \alpha^2 - \rho} \qquad (170)$$

$$v_\theta = \frac{L}{r} + \frac{\rho}{r} \frac{L - r^2 \Omega}{\mu_0 \alpha^2 - \rho} = \Omega r + \frac{\mu_0 \alpha^2}{r} \frac{L - r^2 \Omega}{\mu_0 \alpha^2 - \rho} \qquad (171)$$

B_θ and v_θ would diverge as ρ assumes the value $\mu_0 \alpha^2$, unless at the same time r^2 is equal to L/Ω. The Alfvèn radius on magnetic surface a, $r_A(a)$ is defined by this property:

$$r_A^2(a) = \frac{L(a)}{\Omega(a)} \qquad (172)$$

The Alfvèn density on magnetic surface a, $\rho_A(a)$ is defined by

$$\rho_A(a) = \mu_0 \alpha^2(a) \qquad (173)$$

Then, for regularity, when the density equals the Alfvèn density, the radius must equal the Alfvèn radius

$$\rho = \rho_A \quad \Longrightarrow \quad r = r_A \qquad (174)$$

but the converse is not true (the density does not need to equal the Alfèn density when $r = r_A$).

The reason for this terminology is that when $\rho = \rho_A$ the poloidal velocity of the fluid equals the Alfvèn velocity calculated with the poloidal field. Note that this latter velocity is also the shear Alfvèn wave phase velocity calculated for the total magnetic field and for a propagation direction which is the same as the direction of the fluid poloidal velocity. Let us show this. Since $\rho v_P = \alpha B_P$, the fluid poloidal velocity is given successively , when $\rho = \rho_A$, by the following expressions:

$$v_P^2 = \frac{\alpha^2 B_P^2}{\rho_A^2} = \frac{\alpha^2 B_P^2}{\mu_0^2 \alpha^4} = \frac{B_P^2}{\mu_0 \rho_A} = v_{AP}^2 \tag{175}$$

3.4.3 Final Form of the Transfield and Bernoulli Equations

Using the results of paragraph (3.4.2), the azimuthal components of B and v can be eliminated from eqs (162) and (169). The poloidal fluid velocity can be expressed in terms of ∇a by the relation $\rho v_P = \alpha v_P$. Finally the system of eqs (162) and (169) is reduced to a pair of equations for the functions $\rho(r, z)$ and $a(r, z)$. Eq (162), the Bernoulli equation, assumes, after elimination of the azimuthal variables, the form :

$$\frac{1}{2} \frac{\alpha^2 \nabla a^2}{\rho^2 r^2} = E(a) - G(r, z) - \frac{\gamma}{\gamma - 1} Q \rho^{\gamma - 1}$$
$$+ \rho \Omega \frac{L - r^2 \Omega}{\mu_0 \alpha^2 - \rho} \frac{1}{2} \left(\frac{L}{r} + \frac{\rho}{r} \frac{L - r^2 \Omega}{\mu_0 \alpha^2 - \rho} \right)^2 \tag{176}$$

while eq (169), the so called transfield equation (because it expresses force balance in the direction transverse to the field), assumes the final form:

$$\frac{\alpha}{\rho r} \left(\frac{\partial}{\partial r} \frac{\alpha}{\rho r} \frac{\partial a}{\partial r} + \frac{\partial}{\partial z} \frac{\alpha}{\rho r} \frac{\partial a}{\partial z} \right) - \frac{1}{\mu_0 \rho r} \left(\frac{\partial}{\partial r} \frac{1}{r} \frac{\partial a}{\partial r} + \frac{\partial}{\partial z} \frac{1}{r} \frac{\partial a}{\partial z} \right) =$$
$$E' - Q' \frac{\rho^{\gamma - 1}}{\gamma - 1} + \frac{\alpha'}{\alpha} \frac{\rho \mu_0 \alpha^2}{r^2} \left(\frac{L - r^2 \Omega}{\mu_0 \alpha^2 - \rho} \right)^2 - \frac{\rho}{r^2} \frac{(L' - r^2 \Omega') (L - r^2 \Omega)}{\mu_0 \alpha^2 - \rho} - \frac{LL'}{r^2} \tag{177}$$

3.5 SUMMARY

The dynamics of a rotating, magnetized, polytropic, stationary MHD wind is ruled by the existence of a numbrer of first integrals of the motion, which are
1- The mass flux to magnetic flux ratio $\alpha(a)$, defined by:

$$\rho v_P = \alpha(a) B_P \tag{178}$$

2- The rotation rate of magnetic surfaces, which is the derivative with respect to flux a of the electric potential, $\Phi(a)$, and is related to azimuthal velocity and field by:

$$r \Omega(a) = v_\theta - \frac{\alpha(a) B_\theta}{\rho} \tag{179}$$

3- The total specific angular momentum, $L(a)$ defined by:

$$L(a) = r v_\theta - \frac{r B_\theta}{\mu_0 \alpha(a)}$$ (180)

4- The specific entropy Q(a) of the polytropic gas of adiabatic index γ , defined by:

$$P = Q(a) \, \rho^\gamma$$ (181)

5- The total specific energy, $E(a)$, defined by :

$$E(a) = \frac{v_P^2 + v_\theta^2}{2} + \frac{\gamma}{\gamma - 1} Q(a) \, \rho^{\gamma - 1} + G - \frac{\Omega(a) \, r \, B_\theta}{\mu_0 \alpha(a)}$$ (182)

and by the transfield equation, eq (177), which gives the shape of magnetic surfaces.

4. THE WEBER DAVIS MAGNETIZED WIND

4.1 THE FIXED POLOIDAL FIELD MODEL

The magnetized wind equations provide a mean of calculating both the velocity of the fluid along the magnetic surfaces, and the shape they assume as a result of trans-field force balance. Since however the transfield equation (177) looks formidable, a useful first approach, though non self-consistent, is to assume the shape of these surfaces to be known a-priori. This reduces the problem to solving the Bernoulli equation, and to determine the energy integration constant that enters it. Weber and Davis (1967) assumed conical magnetic surfaces near the eccliptic to model solar wind flow.

4.2 THE BERNOULLI EQUATION FOR A MAGNETIZED WIND

4.2.1 Elimination of Azimuthal Field and Velocity

The azimuthal field , B_θ, and velocity, v_θ can be eliminated by solving for them from eq (179) and (180), with the result

$$v_\theta = \frac{L}{r} + \frac{\rho}{r} \frac{L - r^2 \Omega}{\mu_0 \alpha^2 - \rho}$$ (183)

$$B_\theta = \frac{\mu_0 \, \alpha \, \rho}{r} \frac{L - r^2 \Omega}{\mu_0 \alpha^2 - \rho}$$ (184)

Assuming provisionally all the first integrals to be known, this can be substituted into te Bernoulli equation (162) to give eq (176), an algebraic equation to determine the density ρ as a function of r on that particular magnetic surface.

4.2.2 Other Variables in the Bernoulli Equation

Other variables than ρ and r could be used. For example, Weber and Davis prefered the alfvènic Mach number, M, defined by

$$M^2 = v_P^2 / \left(\frac{B_P^2}{\mu_0 \rho}\right) \tag{185}$$

It is easily shown that

$$M^2 = (\rho_A / \rho) \tag{185}$$

In alfvènic Mach number and radius variables, the Bernoulli equation can be written as

$$E = \frac{1}{2} \frac{B_P^2(r) M^4}{\mu_0 \rho_A} + G + \frac{\gamma}{\gamma - 1} \frac{Q \rho_A^{\gamma-1}}{M^{2(\gamma-1)}} + \frac{1}{2} \frac{\Omega^2 r_A^4}{r^2} \left(1 + \frac{(2M^2 - 1)(r_A^2 - r^2)^2}{r_A^4 (M^2 - 1)^2}\right) \tag{186}$$

One could prefer dimensionless variables. Define a dimensionless radius, x, by :

$$x = \frac{r}{r_A} \tag{187}$$

and refer velocities to the poloidal Alfvèn velocity at the Alfvèn point, given by

$$v_{PA}^2 = \frac{B_P^2(r_A)}{\mu_0 \rho_A} \tag{188}$$

This brings the following dimensionless quantities, all positive:

$$\epsilon = \frac{2E}{v_{PA}^2} \qquad g = -2\frac{G(r_A, z_A)}{v_{PA}^2} \qquad \beta = \frac{2\gamma}{\gamma - 1} \frac{Q \rho_A^{\gamma-1}}{v_{PA}^2} \qquad \omega^2 = \frac{\Omega^2 r_A^2}{v_{PA}^2} \tag{189}$$

Defining finally the functions $\Phi(x)$ and $s(x)$ by

$$\frac{2G(r, z)}{v_{PA}^2} = -g \, \Phi(x) \qquad \frac{r^2 \, B_P(r)}{r_A^2 \, B_P(r_A)} = s(x) \tag{190}$$

For conical magnetic surfaces $s(x) = 1$ and for a point source of gravitational field, $\Phi(x) = 1/x$. Using these variables, the completely dimensionless Bernoulli equation reduces to the following relation between x and M :

$$\epsilon - \frac{s^2(x) \, M^4}{x^4} + g \, \Phi(x) - \frac{\beta}{M^{2(\gamma-1)}} = \frac{\omega^2}{x^2} \left(1 + \frac{(2M^2 - 1)(x^2 - 1)^2}{(M^2 - 1)^2}\right) \tag{191}$$

Other variables could be used, like $y = \rho/\rho_A$ or $z = x^2 y$, which give yet other forms of the Bernoulli equation.

4.2.3 Differential Form and Critical Points

All these forms of the equation can be converted into differential equations. Whichever the variables, this equation can be written in the general form

$$\epsilon = B(x, y) \tag{192}$$

This gives by differentiation:

$$\frac{\partial B}{\partial x}dx + \frac{\partial B}{\partial y}dy = 0 \qquad \longrightarrow \qquad \frac{dy}{dx} = -\frac{\partial B/\partial x}{\partial B/\partial y} \tag{193}$$

The derivative, dy/dx is infinite whenever $\partial B/\partial y = 0$, unless at the same time the numerator $\partial B/\partial x$ also vanishes. At places such that $\partial B/\partial y = 0$ the condition for the solution to be regular is then that the numerator of eq (193) also vanishes. The system of equations

$$\frac{\partial B}{\partial x} = 0 \qquad\qquad \frac{\partial B}{\partial y} = 0 \tag{194}$$

defines points of the (x, y) plane called critical points. At such points the differential of the Bernoulli function $B(x, y)$ vanishes. This property identifies them, whichever variables are used. Since the first order part of the Taylor expansion of $B(x, y)$ is zero at critical points, a local representation calls for a second order expansion. As a result, we find that the slope p $= dy/dx$ is given by the second order equation

$$p^2 \frac{\partial^2 B}{\partial y^2} + 2p \frac{\partial^2 B}{\partial x\,\partial y} + \frac{\partial^2 B}{\partial x^2} = 0 \tag{195}$$

If the solutions for p are not real, the critical point is of the O-type, and $B(x, y)$ is at a maximum or at a minimum, while if the solutions are real, the critical point is of X-type, or saddle point type. In this case there are two, in general different, possible slopes for a solution which passes at the critical point, which reflects the existence of two different branches of the curve $B(x, y) = \epsilon$ which cross eachother at the critical point.

4.3 THE BERNOULLI CURVES

4.3.1 General Aspect of Solution Curves

At a given x, the Bernoulli equation (191) can be regarded as an equation for the Alfvèn Mach number, M, which can be conveniently written as:

$$\epsilon + g\,\Phi(x) - \frac{s^2(x)\,M^4}{x^4} - \frac{\beta}{M^{2\,(\gamma-1)}} - \frac{\omega^2}{x^2} = \frac{\omega^2}{x^2}\,\frac{(2M^2-1)\,(1-x^2)^2}{(M^2-1)^2} \tag{196}$$

Let us call the left hand side of eq (196) $P(x, M)$ and the right hand side $R(x, M)$.

$$P(x, M) = \epsilon + g\,\Phi(x) - \frac{s^2(x)\,M^4}{x^4} - \frac{\beta}{M^{2\,(\gamma-1)}} - \frac{\omega^2}{x^2} \tag{197.a}$$

$$R(x, M) = \frac{\omega^2}{x^2}\,\frac{(2M^2-1)\,(1-x^2)^2}{(M^2-1)^2} \tag{197.b}$$

In the absence of rotation, R vanishes and the equation reduces to $P(x, M) = 0$, which is the Bernoulli equation for the polytropic Parker wind. The function $P(x, M)$ regarded as a function of M for fixed x is concave, and approaches $-\infty$ as $M \to 0$ or $M \to \infty$. P has a maximum at

$$M_{max} = \frac{(\gamma-1)\,\beta\,x^4}{2\,s^2(x)} \tag{198}$$

where it reaches the value

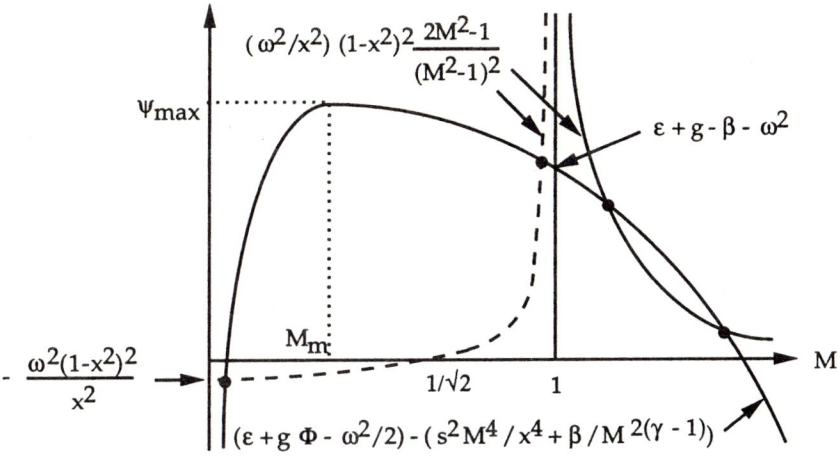

Fig. 6. The functions defined by equations (197.a) and (197.b) and the roots of eq (196)

$$P_{max} = \epsilon + g\,\Phi(x) - \frac{\omega^2}{x^2} - \frac{\gamma+1}{\gamma-1}\left(\frac{s^2(x)}{x^2}\right)^{\frac{\gamma-1}{\gamma+1}}\left(\frac{\beta(\gamma-1)}{2}\right)^{\frac{2}{\gamma+1}} \tag{199}$$

The value of P at $x-1$ is

$$P_1 = \epsilon + g - 1 - \beta - \omega^2 \tag{200}$$

The function $R(x, M)$, regarded as a function of M for fixed x, is convex, negative for $M^2 < 1/2$ but positive otherwise, and has a positive singularity at $M = 1$. Figure (6) shows these two functions of M and their intersections.

By concavity arguments, it can be shown that there are at most four solutions for M to the equation (196). Two of them are just alterations due to rotation of the roots of equation $P = 0$. This is obvious from the slow rotator limit, $\omega \to 0$, since in this case the curve R sticks to the horizontal axis and on its vertical asymptote. The two other solutions are related to the $M = 1$ singularity.

All these four solutions may not be present together at any x. This can be seen by considering, at fixed x, curves of different values of ϵ. Varying ϵ causes the curve $P(x, M)$ to shift down when ϵ is decreased, without changing its shape. For small enough ϵ, the two superalfvènic solutions disappear after having merged into a double root. For still smaller ϵ, the two subalfvènic solutions may disappear in the same way. For ϵ becoming large, on the other hand, we find always four solutions. It is difficult to study how these solutions vary at fixed ϵ when x varies, except in the limits of large or small x, or x near 1, which we consider below.

4.3.2 Near the Alfvèn Point

When $x \to 1$, the function $R(x, M) \to 0$, but not uniformly, because the singularity at $M = 1$ remains whatsoever. As $x \to 1$ the curve representing $R(x, M)$ as a function of M at fixed x sticks closer and closer to the axis $M = 0$ and to the vertical asymptote at $M = 1$. Therefore two roots of equation (196) converge to $M = 1$ when $x \to 1$, provided the value of function $P(x, M)$ at $x = 1$ be positive, which requires that

$$\epsilon > 1 + \beta + \omega^2 - g \tag{201}$$

When $x \to 1$, there are also two other solutions, very near the non-rotating Parker solutions of equation $P = 0$, one of them being superalfvènic and the other subalfvènic when inequality (201) is satisfied, and both being super- or subalfvènic in the opposite case.

From this discussion it appears that all Bernoulli curves pass at the Alfvèn point, whichever ϵ, subject only to the condition (201). This means that an infinity of Bernoulli curves pass in the (x, M) plane at the Alfvèn point $(1, 1)$. The Alfvèn point is a star point. To discuss how Bernoulli curves of different ϵ behave near this point, let us expand around $x = 1$ and $M = 1$ by writing:

$$x = 1 + \xi \qquad\qquad M = 1 + \mu \tag{202}$$

Then, eq (196) becomes approximately:

$$\epsilon - 1 + g - \beta = \omega^2 \left(1 + \frac{\xi^2}{\mu^2}\right) \tag{203}$$

which shows that the slope $p = dM/dx$ of the solutions with an energy parameter ϵ which pass at the Alfvèn point can take one of the two values :

$$p = \pm \frac{\omega}{\sqrt{\epsilon + g - 1 - \beta - \omega^2}} \tag{204}$$

The existence of two solutions indicates, as we have seen above, that two different branches of the curve defined by eq (196) cross at the Alfvèn point. There are of course no such branches when (201) is not satisfied.

4.3.3 Near the Wind Source

At small distances $x \ll 1$ the function $R(x, M)$ diverges as $x \to 0$ for any finite M. However, for very small M, this divergence is matched by a divergence of $P(x, M)$ on the left hand side of the equation, due to the behaviour of $-\omega^2/x^2$ (eq (197.a)). $P(x, M)$ diverges negatively as $x \to 0$ for any finite M, because of the $-s^2 M^4/x^4$ term, and its maximum (eq (198)) shifts to smaller M as $x \to 0$. For definiteness, let us consider the case when $\varPhi = 1/x$ and $s(x) = 1$. Rearranging terms, the Bernoulli equation (196) takes the form

$$\epsilon + \frac{g}{x} = \frac{\beta}{M^{2(\gamma-1)}} + \frac{M^4}{x^4} + \frac{\omega^2 M^4}{(M^2 - 1)^2 x^2} - 2\omega^2 \frac{2M^2 - 1}{(M^2 - 1)^2} + \omega^2 x^2 \frac{2M^2 - 1}{(M^2 - 1)^2} \tag{205}$$

We see again that, if the solution for M is not to approach zero as $x \to 0$, the most diverging term on the right hand side, M^4/x^4, cannot be matched by the g/x term on the left hand side. If on the other hand $M \to 0$ as $x \to 0$ the potentially most diverging terms on the right hand side are the β term and the M^4/x^4 term. The equation then simplifies to:

$$\frac{g}{x} = \frac{\beta}{M^{2(\gamma-1)}} + \frac{M^4}{x^4} \tag{206}$$

This is the same, in this limit $x \to 0$, as the Parker polytropic non rotating wind equation. There are then two branches of solutions, according to which term dominates on the right hand side of eq (206) :

$$M = (g \, x^3)^{1/4} \tag{207.a}$$

$$M = \left(\frac{\beta \, x}{g}\right)^{\frac{1}{2(\gamma-1)}} \tag{207.b}$$

It can be checked that both solutions are consistent with the ordering assumed to derive them, provided $\gamma < 5/3$. The corresponding wind velocity is recovered by making use of the relations $\rho v_P = \alpha B_P$ and $M^2 = \rho_A/\rho$. This gives, in conical geometry:

$$v_P = \frac{\alpha r^2 B_P}{\rho_A \, r_A^2} \frac{M^2}{x^2} = v_{PA} \frac{M^2}{x^2} \tag{208}$$

This shows that solutions (207.a) have $v_P \to \infty$ as $x \to 0$, while solutions (207.b) behave like $v_P \approx x^{(3-2\gamma)/(\gamma-1)}$. For them $v_P \to 0$ as $x \to 0$ if $\gamma < 3/2$, but $v_P \to \infty$ if $\gamma > 3/2$.

4.3.4 At Large Distances

For large x, the Bernoulli equation (196) assumes the form:

$$\epsilon - \frac{M^4 s^2}{x^4} - \frac{\beta}{M^{2(\gamma-1)}} = \omega^2 x^2 \frac{2M^2 - 1}{(M^2 - 1)^2} - 2\omega^2 \frac{2M^2 - 1}{(M^2 - 1)^2} \tag{209}$$

We are still assuming conical geometry, and we have neglected terms that become small as $x \to \infty$. It can be seen that, unless $2M^2 - 1 = 0$, there can be no solution with M remaining finite as $x \to \infty$, because $\omega^2 x^2$ would diverge. Expanding around the special value $M = 1/\sqrt{2}$, we find a first class of solutions

$$M^2 = \frac{1}{2} + \frac{\epsilon - 2^{(\gamma-1)}\beta}{8\omega^2 x^2} \tag{210}$$

This solution is subalfvènic, and its velocity (given by eq(208)) approaches zero as $x \to \infty$, because M^2/x^2 approaches zero. The density approaches a finite value $\rho = \rho_A/M^2 = 2\rho_A$.

There are solutions with small M, which are given approximately by

$$\frac{\beta}{M^{2(\gamma-1)}} = \epsilon - 2\,\omega^2 + \omega^2 \, x^2 \tag{211}$$

which solves into

$$M = \left(\frac{\beta}{\omega^2 \, x^2 + \epsilon - 2\omega^2}\right)^{\frac{1}{2(\gamma-1)}} \tag{212}$$

This is a very subalfvènic solution, for which the velocity goes to zero and the density to infinity when x goes to infinity.

Let us finally explore solutions for which $M \gg 1$. Keeping all potentially non vanishing terms in this limit, equation (209) simplifies to:

$$\epsilon = \frac{M^4}{x^4} + \frac{2\,\omega^2\,x^2}{M^2} \tag{213}$$

which suggests to pass to the velocity variable

$$u \equiv \frac{M^2}{x^2} \equiv \frac{v_\infty}{v_{PA}} \tag{214}$$

in terms of which eq (213) can be written as :

$$\epsilon = u^2 + \frac{2\omega^2}{u} \tag{215}$$

The equivalent dimensional form of eq (215) is :

$$E = \frac{v_\infty^2}{2} + \frac{\Omega^2 r_A^2 v_{PA}}{v_\infty} \tag{216}$$

The quantity $\Omega^2 r_A^2 v_{PA}$ is the cube of a velocity, v_M, called the Michel's velocity. In this conical model, $r^2 B_P$ is constant, so the Michel's velocity can be equivalently defined by

$$v_M^3 = \Omega^2\, r_A^2\, v_{PA} = \Omega^2\, r_A^2\, \frac{B_P(r_A)}{\sqrt{\mu_0\,\rho_A}} = \frac{\Omega^2\, r^2\, B_P(r)}{\mu_0\,\alpha} = \frac{\Omega^2\, r^4\, B_P^2}{\mu_0\,\rho\, r^2\, v_P} \tag{217}$$

Note that $\rho r^2\, v_P$ is also independant of r in this geometry. The dimensionless value of the Michel's velocity is

$$u_M = \omega^{2/3} \tag{218}$$

A minimum of $\epsilon = u^2 + 2\omega^2/u$ is reached at $u^3 = \omega^2$, whose value is $\epsilon_{min} = 3\omega^{4/3}$, or, dimensionally, E given by:

$$E_{min} = \frac{3}{2}\, v_M^2 \tag{219}$$

If $E < E_{min}$, the Bernoulli curves have no superalfvènic branch at infinity. If $E > E_{min}$, there are two solutions for v_∞ to the cubic equation (216) (or else two solutions for u to eq (215)), both superalfvènic, since u approaches a finite value, while M, which is related to it by $M^2 = u\,x^2$ goes to infinity. As a result, the density, which is proportional to ρ_A/M^2, goes to zero, as does the pressure, which equals $Q\rho^\gamma$. Note that one solution for u is larger than $\omega^{2/3}$, while the other is smaller.

4.3.5 Superfast Solution at Infinity

Let us compare the asymptotic velocity of the wind, v_∞, to the fast mode speed, v_f, for the superalfvènic solutions (for the other solutions, we know that v_∞ is smaller than v_f). For the superalfvènic solutions, the sound speed approaches zero at infinity because the density does, and so, the dispersion relation of the magnetosonic modes reduces to the fast mode one, the phase velocity of which is then given, for zero sound speed, by:

$$v_f^2 = \frac{(B_\theta^2 + B_P^2)}{\mu_0 \rho} \tag{220}$$

However, as r goes to infinity, B_P varies as $1/r^2$, while B_θ varies as $1/r$, and $\rho r^2 v_P$ approaches, for conical magnetic surfaces, a constant value. So, the fast mode velocity approaches

$$v_f^2 \approx \frac{B_\theta^2}{\mu_0 \rho} \tag{221}$$

Using eq (184), this becomes

$$v_f^2 = \Omega^2 r^2 \frac{\rho}{\rho_A} = \frac{\Omega^2 r^2}{M^2} \tag{222}$$

and finally the fast mode value of the variable u defined in eq (214) approaches

$$u_f^2 = \frac{v_f^2}{v_{PA}^2} = \frac{x^2}{M^2} \frac{\Omega^2 r_A^2}{v_{PA}^2} = \frac{\omega^2 x^2}{M^2} = \frac{\omega^2}{u} \tag{223}$$

The ratio of wind velocity to the fast mode speed (calculated for a direction of propagation that is the same as the direction of the poloidal flow) is the so called fast mode Mach number, M_f. It is given by:

$$M_f^2 \equiv \frac{u^2}{u_f^2} = \frac{u^3}{\omega^2} \tag{224}$$

This shows that the solution for which u is larger than $\omega^{2/3}$ is faster than the fast mode speed at infinity while the other one is slower. Reexpressed in dimensional terms, this means that only the solution faster than the Michel's velocity is super-fast at infinity.

4.3.6 Summary of Bernoulli Curves in the (x-M) Plane

A typical Bernoulli curve starts at $x \ll 1$ with two branches for which $M \to 0$ as $x \to 0$. As x increases, a new branch with superalfvènic solutions may appear, if ϵ is large enough. Otherwise, the two original solutions may either persist or disappear.

For large enough ϵ's, one superalfvènic and one subalfvènic branch of solution meet at the Alfvèn point, becoming transalfvènic there.

For still larger x's, the four roots may persist or else the two superalfvènic solutions may disappear by merging. At infinity, there are four solutions if $\epsilon > 3\omega^{4/3}$, one of them only being faster than the fast mode speed, another one being still suparalvènic but not super-fast, and the two others being subalfvènic. If $\epsilon < 3\omega^{3/4}$, only the subalfvènic solutions exist at infinity, where they have non-zero pressure. They should be rejected, since they cannot match smoothly to a very rarefied medium.

The figures 7.a and 7.b illustrate a few Bernoulli curves.

4.4 CRITICAL POINTS AND CRITICALITY CONDITIONS

4.4.1 Equations for the Critical Points

For later convenience, let us use in this chapter physical variables, $x = r$ and $y = \rho$. In these variables, the Bernoulli function is

$$B(r, \rho) = \frac{\alpha^2 B_P^2(r)}{2 \rho^2} + \frac{\gamma}{\gamma - 1} Q \rho^{\gamma - 1} + G - \frac{\Omega^2 r^2}{2} + \frac{1}{2} \left(\frac{\rho_A}{r} \frac{L - r^2 \Omega}{\rho_A - \rho} \right)^2 + \Omega L \tag{225}$$

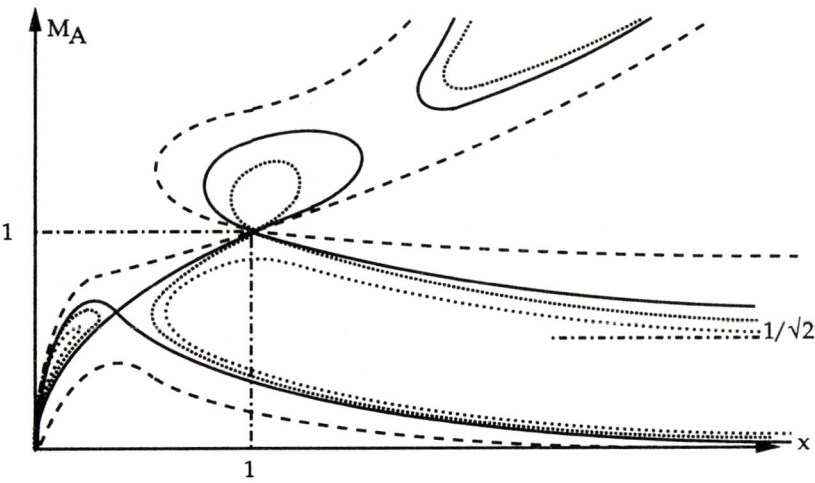

Fig. 7.a. Qualitative representation of a few Bernoulli curves for magnetized wind flow in the reduced position and alfvènic Mach number plane. A solution passing at a slow mode critical point is represented (full line)

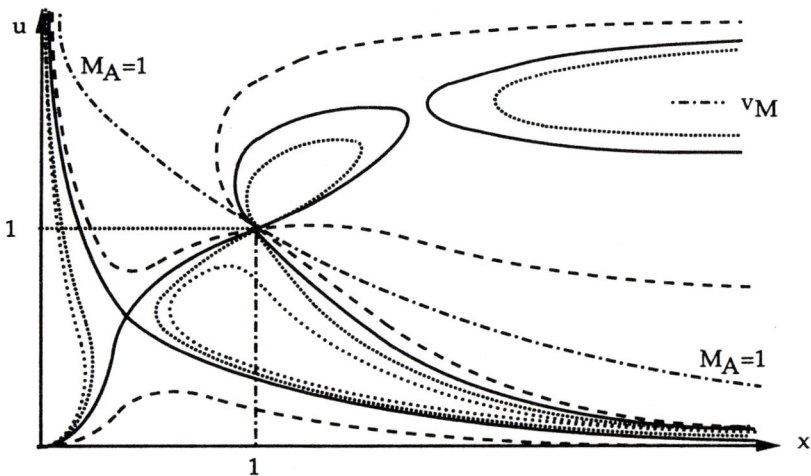

Fig. 7.b. The same Bernoulli curves as in figure (7.a), represented in the reduced position - reduced velocity plane.

The vanishing of the differential of $B(r, \rho)$ gives the equations for the critical points

$$r \frac{\partial B}{\partial r} = -\frac{1}{2} \frac{\alpha^2}{\rho^2} r \frac{dB_P^2}{dr} - r \frac{dG}{dr} + \Omega^2 r^2 + \frac{\rho_A^2}{r^2} \frac{L^2 - \Omega^2 r^4}{(\rho_A - \rho)^2} = 0 \qquad (226)$$

$$\rho \frac{\partial B}{\partial \rho} = -\frac{\alpha^2 B_P^2}{\rho^2} + \gamma Q \rho^{\gamma-1} + \frac{\rho \rho_A^2}{r^2} \frac{(L - \Omega r^2)^2}{(\rho_A - \rho)^3} = 0 \tag{227}$$

4.4.2 Critical Points and MHD Modes Velocities

The critical points are situated where the poloidal velocity equals one of the two magnetosonic mode speeds. Since the fast and slow modes have velocities that depend on the direction of propagation, it should be precised that we refer to a direction which is parallel to the poloidal fluid velocity. Since the field has a θ-component, there is an angle between B and v_P. The propagation angle, call it ψ, is the angle between the total magnetic field B and its poloidal component B_P. That the poloidal velocity equals some mode speed at critical points can be proved from eq (227), in which a few characteristic velocities appear. The square of the sound speed is $c_S^2 = \gamma Q \rho^{\gamma-1}$, while the square of poloidal Alfvèn speed is $B_P^2/(\mu_0 \rho)$. The Alfvèn speed associated with the azimuthal field, B_θ is given by:

$$v_{A\theta}^2 = \frac{B_\theta^2}{\mu_0 \rho} = \frac{1}{\mu_0 \rho} \frac{\mu_0^2 \alpha^2 \rho^2}{r^2} \left(\frac{L - r^2 \Omega}{\rho_A - \rho} \right)^2 = \frac{\rho \, \rho_A}{r^2} \left(\frac{L - r^2 \Omega}{\rho_A - \rho} \right)^2 \tag{228}$$

Then eq (227) can be written as

$$\frac{\alpha^2 B_P^2}{\rho^2} = c_S^2 + \frac{\rho_A}{\rho_A - \rho} v_{A\theta}^2 \tag{229}$$

Since $\alpha B_F/\rho = v_P$, and $\rho_A/\rho = M^2 = v_P^2/v_{AP}^2$, eq (229) can be written as

$$v_P^2 \left(1 - \frac{v_{AP}^2}{v_P^2} \right) = c_S^2 \left(1 - \frac{v_{AP}^2}{v_P^2} \right) + v_{A\theta}^2 \tag{230}$$

which can be rearranged into

$$v_P^4 - v_P^2 (v_{AP}^2 + v_{A\theta}^2 + c_S^2) + v_{AP}^2 c_S^2 = 0 \tag{231}$$

The square of the total Alfvèn velocity is $v_A^2 = v_{AP}^2 + v_{A\theta}^2$ and the poloidal Alfven speed, v_{AP}, equals $v_A \cos \psi$. So eq (231) can be reduced to

$$v_P^4 - v_P^2 (v_A^2 + c_S^2) + v_A^2 c_S^2 \cos^2 \psi = 0 \tag{231}$$

This is just the same as the magnetosonic mode dispersion equation, v_P replacing the phase velocity of the mode. This proves that at critical points the poloidal velocity equals one of the two magnetosonic mode speeds.

4.4.3 Necessity for the Wind to be Transalfvènic and Transfast

If we seek a solution such that the pressure goes to zero at infinity, the results summarized in (4.3.6) show that we should pick up one which is superalfvènic at infinity. Since the wind is assumed to start at very low alfvènic and sound Mach numbers, it must then also go through a slow mode critical point. The wind must be super-fast at infinity. Otherwise, boundary conditions concerning the magnetic structure at infinity should be imposed, since fast mode signals could be propagated upwind from infinity, bringing information about the magnetic structure there back to the source. Then, we would have to demand that the magnetic structure of the wind merges smoothly into that of the non-wind medium at infinity. But in the wind $B_\theta/B_r \to \infty$ while a non rotating medium at infinity would have $B_\theta/B_r = 0$. Therefore, the matching is impossible, and a trans-fast solution is needed to allow connection between these two regions through a fast shock.

4.4.4 Boundary Conditions and Criticality Conditions

Up to now, we have regarded $L(a)$, $\alpha(a)$, $Q(a)$, $\Omega(a)$ as known, and $E(a)$ as a parameter. In fact these quantities are not known, except for $Q(a)$ which is fixed by the entropy of the gas at the source, and $\Omega(a)$, which is fixed by the velocity of rotation at the base of the field lines. Strictly speaking $\Omega(a)$ is not the angular velocity of the fluid at the source, but, as eq (183) shows, it coincides with it in the limit of very large density, which is often the physical situation prevailing at the origin of the wind. The other first integrals $E(a)$, $L(a)$, $\alpha(a)$, which have to do with the total value and distribution with flux of the energy flux, torque and mass flux are not known from the boundary conditions, but should result from the calculation. We have seen before that, when the wind speed passes one of the magnetosonic velocities, a part of the communication with boundary conditions at infinity is broken, and these boundary conditions are replaced by a criticality condition, which expresses the fact that such a critical transition can only occur where equations (194) are satisfied. Indeed the coordinates (x_S, M_S) and (x_f, M_f) of these slow and fast critical points in, say, the (x, M) plane, are functions of L, α, Q and Ω, as can be seen from eqs (226) and (227). Writing the Bernoulli equation at these points gives rise to two new relations between E, L, α, Q and Ω, namely:

$$E = B\left(x_S(L, \alpha, Q, \Omega), M_S(L, \alpha, Q, \Omega)\right) \tag{232.a}$$

$$E = B\left(x_f(L, \alpha, Q, \Omega), M_f(L, \alpha, Q, \Omega)\right) \tag{232.b}$$

These relations are the so-called criticality conditions. They express the requirement that the solution passes through both the slow and the fast critical point, which necessitates some tuning of the values of L, α and E, for given Ω and Q. Since the solution goes continuously from slow to fast mode velocity, it needs to be somewhere transalfvènic. This however imposes no other condition than just inequality (201).

The figures (8.a) and (8.b) illustrate the general aspect of the desired solution. The physical branch is the unique one that ends up with super-fast velocity at infinity. These figures show qualitatively the double-critical solution, both in the (x, M) plane and in the (x, u) plane. Of course solving for the critical points is not an easy task, and uniqueness is not guaranteed. We do not expand on this here.

4.4.5 Boundary Conditions in Complete Self Consistency

Note that, as we have imposed two criticality conditions, one of the three first integrals that were not defined by boundary conditions at the source still remains undetermined in the fixed field model. Those authors who remain in this framework usually impose some extra condition, for example that the density of the fluid be known at some place in the wind. Since, after solving the Bernoulli equation, the wind velocity is in principle known on each magnetic surface in terms of the first integrals, the requirement that the

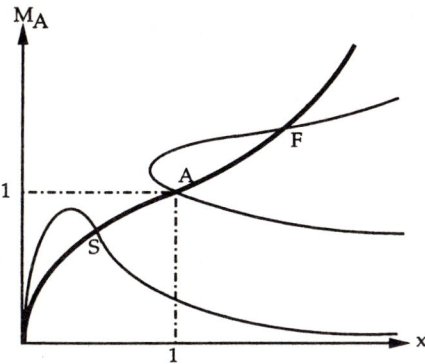

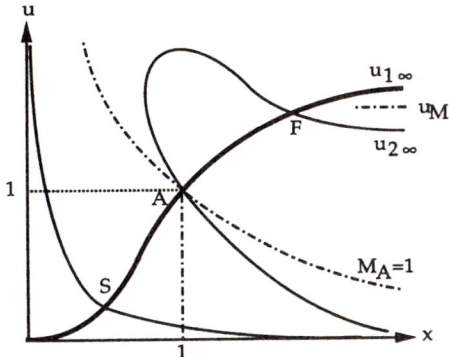

Fig. 8.a. A double-critical solution in the reduced position - alfvènic Mach number plane.

Fig. 8.b. A double-critical solution in the reduced position - reduced velocity plane.

density be known somewhere imposes, by the relation $\rho v_P = \alpha B_P$, yet another relation between the first integrals, which determines the one that was left undetermined, for example the value of the mass flux per flux tube, i.e $\alpha(a)$. However the fixed field model is not self-consistent. It regards the MHD wind as flowing in rigid magnetic pipes. But in reality these pipes are flexible and inflatable. Their shape is not given a-priori, but adapts, by the transfield force balance, to the flow they convey. In full consistency, the Bernoulli equation cannot be solved until the shape of magnetic surfaces is known from this force balance, and conversely this balance depends on solution for the flow along each magnetic surface. Bernoulli equation and transfield equation are coupled.

As a result, one cannot prescribe simultaneously the entropy and the density of the gas at the wind source, because the shapes of magnetic surfaces adapt to the flow and react on it. This feedback will control, for example, the density at the source of the wind for a given entropy $Q(a)$. This means that in a completely self-consistent calculation the mass flux to magnetic flux ratio should be a result of the calculation, not a function given by boundary conditions. How complete self-consistency may be achieved will be described in the next paragraph, where it will be shown that a new condition for regularity of the solution of the transfield equation at the Alfvèn surface (the locus of Alfvèn points) emerges. This regularity condition expresses as a differential equation. Therefore it will impose the variation with flux a of some first integral, but a global integration constant will still remain, which will allow to fix arbitrarily, for example, the total mass flux blown by the source object, but not its distribution with flux a.

Before closing this paragraph, let us make a small remark. We have regarded $Q(a)$ as known from boundary conditions, rather than any other thermodynamical variable describing the state of the plasma at the wind source. This may be too simple. We should rather have written and solved an energy balance equation for the quasistatic part of the wind near to the source object, which would have described in some detail the rates of heating and cooling of this gas, the wind itself being a particular term in this balance. The result of such a calculation could have been, for example, some complicated relation between $Q(a)$, $\alpha(a)$ and $E(a)$. Here we disregard such complications.

5. PROPERTIES OF THE TRANSFIELD EQUATION

5.1 TOWARDS SELF CONSISTENCY

As discussed in the preceding paragraph, the full problem consists in solving the transfield equation, coupled to the Bernoulli equation. Rigourous calculation of the angular momentum loss, for example, require such a solution. Some spectacular properties of MHD winds, such as the possibility of axial self-focalization of the flow by the azimuthal magnetic field, can only be described in the framework of the transfield equation (hereafter TF equation for brevity).

5.2 CURVATURE OF POLOIDAL FIELD LINES

In eq (160) the TF equation has been written as

$$\frac{\text{rot}\boldsymbol{B}_P \times \boldsymbol{B}_P}{\mu_0 \rho} - \text{rot}\boldsymbol{v}_P \times \boldsymbol{v}_P =$$
$$\nabla a \left(E'(a) - Q'\frac{\rho^{\gamma-1}}{\gamma-1} - \Omega L' + \frac{rB_\theta}{\mu_0 \alpha}\Omega' - \frac{\alpha B_\theta}{\rho r}L' + \frac{\alpha'}{\alpha}\frac{B_\theta^2}{\mu_0 \rho} \right) \quad (233)$$

The parenthesis on the right hand side, S, can be expressed entirely in terms of density, using eq (170) for B_θ. This gives

$$S \equiv E'(a) - Q'\frac{\rho^{\gamma-1}}{\gamma-1} - \Omega L' + \frac{rB_\theta}{\mu_0 \alpha}\Omega' - \frac{\alpha B_\theta}{\rho r}L' + \frac{\alpha'}{\alpha}\frac{B_\theta^2}{\mu_0 \rho} =$$
$$E'(a) - Q'\frac{\rho^{\gamma-1}}{\gamma-1} + \frac{\alpha'}{\alpha}\frac{\rho\rho_A}{r^2}\left(\frac{L - r^2\Omega}{\rho_A - \rho}\right)^2 - \frac{\rho}{r^2}\frac{(L - r^2\Omega')(L - r^2\Omega)}{\rho_A - \rho} - \frac{LL'}{r^2} \quad (234)$$

Since the Bernoulli equation gives, at least implicitly, $|\boldsymbol{v}_P|$, noted simply as v_P, the TF equation gives the information about the direction of $\boldsymbol{v}_P$. Some simple manipulations make this information explicit. Let us define a unit vector $\boldsymbol{t}$ tangent to magnetic surfaces in the meridian plane such that

$$\boldsymbol{v}_P = v_P\,\boldsymbol{t} \qquad \text{with} \qquad \boldsymbol{t} = \cos\psi\boldsymbol{e}_r + \sin\psi\boldsymbol{e}_z \quad (235)$$

Using the vector calculus identity $\text{rot}\boldsymbol{v}_P \times \boldsymbol{v}_P = -\nabla v_P^2/2 + (\boldsymbol{v}_P.\nabla)\,\boldsymbol{v}_P$ this can be transformed into

$$\text{rot}\boldsymbol{v}_P \times \boldsymbol{v}_P = -\nabla\frac{v_P^2}{2} + (\boldsymbol{t}.\nabla(\frac{v_P^2}{2}))\,\boldsymbol{t} + v_P^2\left((\boldsymbol{t}.\nabla)\,\boldsymbol{t}\right) \quad (236)$$

and the Fresnet formulae for curvature give $(\boldsymbol{t}.\nabla)\,\boldsymbol{t}$ in terms of the variations, following poloidal field lines, of the anle ψ:

$$(\boldsymbol{t}.\nabla)\,\boldsymbol{t} = \boldsymbol{n}\frac{d\psi}{ds} \quad (237)$$

where s is the curvilinear abcissa along poloidal field lines and n is the unit normal vector:

$$n = -\sin\psi e_r + \cos\psi e_z = -\frac{\nabla a}{|\nabla a|} \tag{238}$$

$d\psi/ds$ is the inverse of the curvature radius. From this we finally get

$$n.(\text{rot} v_P \times v_P) = -(n.\nabla)\frac{v_P^2}{2} + v_P^2\frac{d\psi}{ds} \tag{239.a}$$

$$n.(\text{rot} B_P \times B_P) = -(n.\nabla)\frac{B_P^2}{2} + B_P^2\frac{d\psi}{ds} \tag{239.b}$$

The dot product of the TF equation with the normal vector n then becomes an equation that gives the poloidal field line curvature :

$$(v_P^2 - \frac{B_P^2}{\mu_0\rho})\frac{d\psi}{ds} = (n.\nabla)(v_P^2/2) - (1/\mu_0\rho)(n.\nabla)(B_P^2/2) + |\nabla a|S \tag{240}$$

where S is the expression defined by eq (234). From $\rho v_P = \alpha B_P$, this can be further arranged into :

$$\frac{d\psi}{ds} = \frac{(\alpha^2/\rho^2)(1-\rho/\rho_A)(n.\nabla)(B_P^2/2) + (B_P^2/2)(n.\nabla)(\alpha^2/\rho^2) + |\nabla a|S}{v_P^2(1-\rho/\rho_A)} \tag{241}$$

5.3 THE ALFVEN REGULARITY CONDITION

5.3.1 A Regularity Condition for the Transfield Equation

It is readily seen that when the flow becomes transalfvènic, i.e., when $v_P^2 = B_P^2/\mu_0\rho$, or equivalently when $\rho = \rho_A$, the denominator for the expression of the curvature in eq (241) vanishes. Unless the numerator also vanishes, the curvature would be infinite, which means that the angle ψ would jump at the Alfvèn point, or else that the magnetic surfaces would be kinked there. To avoid such an irregularity in the solution, the vanishing of the numerator of the right hand side of eq (241) must be demanded. This introduces yet another regularity condition, specific to the TF equation, which can be written, expliciting S, as:

$$\lim_{r \to r_A}\left(\frac{B_P^2}{2|\nabla a|}(n.\nabla)\frac{\alpha^2}{\rho^2} + E'(a) - Q'\frac{\rho^{\gamma-1}}{\gamma-1}\right)$$

$$+ \lim_{r \to r_A}\left(\frac{\alpha'}{\alpha}\frac{\rho\rho_A}{r^2}\left(\frac{L-r^2\Omega}{\rho_A-\rho}\right)^2 - \frac{\rho}{r^2}\frac{(L'-r^2\Omega')(L-r^2\Omega)}{\rho_A-\rho} - \frac{LL'}{r^2}\right) = 0 \tag{242}$$

This Alfvèn regularity condition expresses the fact that there should be no standing rotational discontinuity in the wind. A rotational discontinuity is to Alfvèn waves what shocks are to other wave modes. From a more mathematical standpoint, it can be seen that the Alfvèn surface is a place where the transfield equation becomes singular in the sense that its higher order derivative terms all vanish. Eq (240) makes it apparent that the left hand side, which obviously consists of second order derivative terms, vanishes. However, the right hand side still contains potentially second order derivative terms, namely the $(\boldsymbol{n}.\nabla)\,(v_P^2/2)$ and $(\boldsymbol{n}.\nabla)\,(B_P^2/2)$ terms, since B_P, or v_P which is proportional to it, is proportional to $|\nabla a|$. It can be checked that

$$
(\boldsymbol{n}.\nabla)\,(v_P^2/2) - (1/\mu_0\rho)\,(\boldsymbol{n}.\nabla)\,(B_P^2/2) =
$$
$$
\frac{\alpha^2}{\rho^2}\,(1 - \frac{\rho}{\rho_A})\,(\boldsymbol{n}.\nabla)\,(\frac{B_P^2}{2}) - \alpha\alpha'\,\frac{B_P^2}{\rho^2}|\nabla a| - \frac{B_P^2\alpha^2}{\rho^3}\,(\boldsymbol{n}.\nabla\rho) \qquad (243)
$$

The second order derivative terms proportional to $\nabla B_P^2/2$ are factored by $(1 - \rho/\rho_A)$ which vanish at the Alfvèn point. The term $(\boldsymbol{n}.\nabla\rho)$ might still contain second order derivatives of function a. However these second order derivative terms also come to vanish at the Alfvèn point. This can be checked by just taking the grad of the Bernoulli relation, which we do in the next subsection.

5.3.2 Explicit Writing of the Alfvèn Regularity Condition

We need to express more explicitly the regularity condition (242),in which the quantity $(L - r^2\Omega)/(\rho - \rho_A)$ is to be evaluated in a limit sense, since numerator and denominator approach zero as the Alfvèn point is approached. This limit can be taken following a given magnetic surface, where the relation between ρ and r is given by the Bernoulli equation. Equation (242) involves the quantity $(\boldsymbol{n}.\nabla\rho)$ calculated at the Alfvèn point, which we can also calculate from the Bernoulli equation. Let us introduce a number characterizing the slope at the Alfvèn point of the function $\rho(r)$ as seen when following a magnetic surface. Define the "slope at the Alfvèn point", q, by

$$
q = \lim_{r \to r_A}\ \frac{r_A}{\rho_A}\left(\frac{\rho - \rho_A}{r - r_A}\right) = \left(\frac{\partial \ln \rho}{\partial \ln r}\right)_A \qquad (244)
$$

The slope q can be expressed in terms of the first integrals of the motion, or else in terms of the parameters introduced in equations (189). Actually, we calculated in section (4.3.2) that the slope $p = dM/dx$ at the Alfvèn point is given by eq (204). Since M is related to ρ by equation (185), it is easily checked that slopes p and q are related by $q = -2p$, an expression for p being given in eq (204). In terms of the slope q, we can calculate

$$
\lim_{r \to r_A}\left(\frac{L - r^2\Omega}{\rho_A - \rho}\right) = \frac{1}{q}\frac{2L}{\rho_A} \qquad (245)
$$

This makes for the limit of most terms in eq (242). We now calculate the limit of $(\boldsymbol{n}.\nabla\rho)$ at the Alfvèn point. Quite generally, ρ is given implicitly by the Bernoulli equation. Substituting in eq(162) the expressions (170) and (171) for B_θ and v_θ, we get it in the form :

$$
\frac{1}{2}\frac{\alpha^2|\nabla a|^2}{\rho^2 r^2} + G + \frac{\gamma}{\gamma - 1}Q\rho^{\gamma-1} + \frac{1}{2}\frac{L^2}{r^2} + \frac{\rho}{r^2}\frac{(L - r^2\Omega)^2}{\rho_A - \rho} + \frac{1}{2}\frac{\rho^2}{r^2}\frac{(L - r^2\Omega)^2}{(\rho_A - \rho)^2} - E = 0 \quad (246)
$$

This relates the density field $\rho(r, z)$ to the flux function $a(r, z)$, and of course also to r and z. Taking the gradient of this relation, we obtain $\nabla\rho$ in terms of $a(r, z)$, its gradient and its second derivatives. After some algebra we obtain:

$$\frac{a|\nabla a|}{\rho r}\nabla\left(\frac{a|\nabla a|}{\rho r}\right) + \nabla G + \frac{\gamma}{\gamma-1}Q\rho^{\gamma-1}\left(\frac{Q'}{Q}\nabla a + (\gamma-1)\frac{\nabla\rho}{\rho}\right) + \frac{L}{r}\nabla(\frac{L}{r})$$

$$+\frac{\rho}{r^2}\frac{(L-r^2\Omega)^2}{\rho_A-\rho}\left(\frac{\nabla\rho}{\rho} - 2\frac{\nabla r}{r} + 2\frac{\nabla(L-r^2\Omega)}{L-r^2\Omega} - \frac{\nabla(\rho_A-\rho)}{\rho_A-\rho}\right)$$

$$+\frac{\rho^2}{r^2}\frac{(L-r^2\Omega)^2}{(\rho_A-\rho)^2}\left(\frac{\nabla\rho}{\rho} - \frac{\nabla r}{r} + \frac{\nabla(L-r^2\Omega)}{L-r^2\Omega} - \frac{\nabla(\rho_A-\rho)}{\rho_A-\rho}\right) - E'\nabla a = 0 \qquad (247)$$

This equation could be solved in general to give $n.\nabla\rho$ in terms of function a and derivatives. Note in particular the second order derivative terms associated with the first term of eq (247). In the limit where the Alfvèn point is approached, a few terms on the second and third line of eq (247) diverge, but multiplying it by $(\rho_A - \rho)$ makes all terms regular in this limit, with most terms indeed vanishing, because $(L - r^2\Omega)/(\rho_A - \rho)$ becomes finite. Note in particular that the second order derivative terms vanish in this limit as said in subsection (5.3.1).The remaining non vanishing terms are:

$$\frac{\rho_A^2}{r_A^2}\left(\lim_{r\to r_A}\frac{(L-r^2\Omega)^2}{(\rho_A-\rho)^2}\right)\left((\lim_{r\to r_A}(\frac{\rho_A-\rho}{L-r^2\Omega}))\ \nabla(L-r^2\Omega) - \nabla\rho_A - \nabla\rho\right) = 0 \qquad (248)$$

Using eq (245) and $n = -\nabla a/|\nabla a|$, we obtain after little algebra the simple relation

$$(n.\nabla\rho)_A = \rho_A\,|\nabla a|_A\left(-2\frac{\alpha'}{\alpha} - \frac{q\sin\psi_A}{r_A\,|\nabla a|_A} + q\frac{r_A'}{r_A}\right) \qquad (249)$$

Where ψ_A is the value at the Alfvèn point of the angle ψ of the poloidal field line to the equator as defined by eqs (235) and (238).

Using eq (249) and (245) in eq (242), we finally get the following explicit form of the Alfvèn regularity condition:

$$\frac{\alpha'}{\alpha} - q\frac{r_A'}{r_A} + q\frac{\sin\psi_A}{r_A\,|\nabla a|_A} + \frac{E'}{E}\frac{E}{v_{PA}^2}$$

$$-\frac{1}{\gamma-1}\frac{Q\rho_A^{\gamma-1}}{v_{PA}^2}\frac{Q'}{Q} + \frac{\Omega^2 r_A^2}{v_{PA}^2}\left(\frac{4}{q^2}\frac{\alpha'}{\alpha} + \frac{2(2+q)}{q}\frac{r_A'}{r_A} - \frac{\Omega'}{\Omega}\right) = 0 \qquad (250)$$

Since the slope at the Alfvèn point, q, is a function of the first integrals, as is v_{PA}^2, the Alfvèn regularity condition is but an ordinary differential equation which relates the variations with flux, a, of the various first integrals. To within an arbitrary integration constant, it precises the variation with a of the last first integral left undetermined by the criticality relations (subsection (4.4.4)).

5.3 THE SHAPE OF THE FLOW AT INFINITY

5.3.1 The Bernoulli Equation at Infinity

For completely self-consistent MHD winds, the shape of the flow is not a priori given. This implies that the behaviour at infinity of v_P and ρ is not known a-priori. Nevertheless some general relations hold. One of them can be derived from the Bernoulli equation, which can be written as:

$$\frac{1}{2}\frac{\alpha^2|\nabla a|^2}{\rho^2 r^2} = E - G(r,z) - \frac{\gamma}{\gamma-1}\frac{Q\rho^{\gamma-1}}{\rho-\rho_A} + \rho\Omega^2\frac{r^2 - r_A^2}{\rho-\rho_A} - \frac{1}{2}\frac{\Omega^2 r_A^4}{r^2}\left(1 + \frac{\rho\left(r_A^2 - r^2\right)}{r_A^2\left(\rho_A - \rho\right)}\right)^2 \quad (251)$$

If we go to infinite distance away from the source on a given field line, r will in general approach infinity, unless the magnetic surface becomes asymptotically cylindrical. Let us suppose here that this is not so. Keeping only potentially dominant terms in this limit, the above equation simplifies to:

$$\frac{1}{2}\frac{\alpha^2|\nabla a|^2}{\rho^2 r^2} = E - \frac{\gamma Q\rho^{\gamma-1}}{\gamma-1} - \frac{\rho r^2\Omega^2}{\rho_A - \rho} - \frac{1}{2}\frac{\rho^2\Omega^2 r^2}{(\rho_A - \rho)^2} \quad (252)$$

Since the flow is necessarily superalfvènic at infinity, ρ remains bounded by ρ_A. Examining eq (252), we see that ρr^2 cannot diverge as $r \to \infty$, for otherwise the right hand side would not remain positive. Note however that this conclusion does not hold for an isothermal wind because in this case $\rho^{\gamma-1}$ is replaced in eq (252) by a term proportional to $\ln \rho$, which becomes infinitly negative for vanishing ρ. Except when $\gamma = 1$, we conclude from this remark that ρ necessarily approaches zero at least as fast as $1/r^2$, and that the asymptotic form of the Bernoulli equation is

$$\frac{1}{2}\frac{\alpha^2|\nabla a|^2}{\rho^2 r^2} = E - \frac{\rho r^2\Omega^2}{\rho_A - \rho} \quad (253)$$

which implies that

$$\lim_{\infty/a}\left(\frac{\rho r^2\Omega^2}{\rho_A}\right) < E \quad (254)$$

The notation $\lim_{\infty/a}$ means limit at constant a when $r \to \infty$. Physically this inequality expresses the fact that the Poynting flux per unit mass flux cannot exceed the total energy per escaping gram, E. We also read on the Bernoulli equation that the asymptotic poloidal flow speed, $v_\infty(a)$ is such that

$$v_\infty(a) = \lim_{\infty/a}(v_P) < \sqrt{2E} \quad (255)$$

We can further transform the inequality concerning the finiteness of ρr^2 into an inequality for $r\nabla a$. Namely we have:

$$\lim_{\infty/a}\left(\frac{\rho r^2\Omega^2 v_P}{\rho_A}\right) < \sqrt{2}E^{\frac{3}{2}} \quad (256)$$

Using $\rho v_P = \alpha B_P$ and $B_P = |\nabla a|/r$, this can be transformed into

$$\lim_{\infty/a} r|\nabla a| < \frac{\sqrt{2}\mu_0\alpha E^{\frac{3}{2}}}{\Omega^2} = \lambda(a) \quad (257)$$

This constrains, as we shall see, the shape of the flow at infinity.

We read on the form (253) of the Bernoulli equation that polytropic flows which have a finite limit of ρr^2 as $r \to \infty$ on a given field line, or equivalently have a finite limit of $r|\nabla a|$, are flows which carry a finite, non vanishing, Poynting flux at infinity.

5.3.2 Polytropic Flows Cannot Asymptotically Flatten to the Equator

The inequality $r|\nabla a| < \lambda(a)$ can be converted into a constraint on the shape of field lines. Actually, we note that

$$|\nabla a| \geq |\frac{\partial a}{\partial z}| \tag{258}$$

With our conventions, $\partial a/\partial z$ is negative, and inequality (257) can be weakened into

$$-r\frac{\partial a}{\partial z} < \lambda(a) \tag{259}$$

Consider a fixed large value of r, and let $a = A$ be the equatorial value of the flux function a. At this given r the altitude z is a function of a, call it $z(r, a)$. From inequality (259), this function is constrained by

$$-\frac{\partial a}{\partial z} > \frac{r}{\lambda(a)} \tag{260}$$

which integrates, at fixed r, into

$$z(r,a) > r \int_a^A \frac{da'}{\lambda(a')} \tag{261}$$

This shows that asymptotically the magnetic surface a is situated somewhere above some minimal cone, defined by (261). This implies that z/r cannot approach zero as $r \to \infty$. In brief, magnetic surfaces cannot flatten towards the equator.

5.3.3 Polytropic Winds with Vanishing Poynting Flux at Infinity Focus Parabolically

The constraints are more stringent for those flows for which $lim \ (\rho r^2) = 0$. Such flows carry no Poynting flux to infinity, and they have the correlative property (from eq(170) as $r \to \infty$) that also $lim(rB_\theta) = 0$ By Ampere's law, $2\pi r B_\theta$ is the current enclosed in the circle of radius r at height z. So, if ρr^2 approaches zero, the total poloidal current enclosed in the magnetic surface approaches zero as $z \to \infty$. For such flows, the inequalities (256) and (257) are very loose, since they don't use the information that ρr^2 approaches zero as $r \to \infty$. We can however strengthen them by saying that in this case a function $\lambda_1(r,a)$ exists such that, at given r

$$\rho r^2 < \lambda_1(r,a) \tag{262}$$

the function λ_1 approaching zero as $r \to \infty$. Converting, as above, the inequality (262) into one that limits $r|\nabla a|$, and making use of inequality (259), we can derive an inequality for $z(a,r)$:

$$z(r,a) > r \int_a^A \frac{da'}{\lambda_1(a',r)} \tag{263}$$

Since $\lambda_1(a',r) \to 0$ as $r \to \infty$, this shows that z/r can diverge as $r \to \infty$, or else that the flow in this case acquires asymptotically a parabolic shape. In brief, flows that have asymptotically vanishing poloidal current (and Poynting flux) converge to a parabolic form, in the sense that

$$\lim_{\infty/a} (\frac{z}{r}) = \infty \tag{264}$$

The inverse property, i.e that any asymptotically paraboloidal flow would carry no Poynting flux to infinity, is not true. Counterexamples (i.e. parabolidal law with a constant exponent) can be found.

5.3.4 Asymptotically Cylindrical Flows Carry Current and Poynting Flux

It has been shown in (5.3.3) that flows which carry no Poynting flux at infinity are more focused than some "minimal" paraboloid. Could they assume a cylindrical shape asymptotically ? The answer is negative, as we now show. Asymptotically cylindrical flows have a finite limit of $r(a, z)$, $r_\infty(a)$. For such flows the poloidal field at infinity becomes purely vertical and approaches

$$B_{P\infty} = \frac{1}{r_\infty} \frac{da}{dr_\infty} = \frac{1}{r_\infty^2} \frac{1}{(d \ln r_\infty/da)} \tag{265}$$

while the density ρ approaches

$$\rho_\infty(a) = \frac{\alpha}{v_\infty} \frac{1}{r_\infty} \frac{da}{dr_\infty} \tag{266}$$

Since $v_\infty^2 < 2E$, ρ_∞ cannot be smaller than $\rho_{\infty min}$, defined as

$$\rho_{\infty min} \equiv \frac{\alpha}{\sqrt{2E}\, r_\infty^2 (d \ln r_\infty/da)} \tag{267}$$

As a result, the specific energy associated with the Poynting flux has a lower limit, which, in the case $r_\infty \gg r_A$ is given by:

$$\frac{\rho r^2 \Omega^2}{\rho_A} > \frac{\Omega^2}{\mu_0 \alpha \sqrt{2E}(d \ln r_\infty/da)} \tag{268}$$

The right hand side of eq (268) is not zero, which indicates that asymptotically cylindrical flows carry a non vanishing poloidal current to infinity, as well as a non vanishing Poynting flux.

5.4 TRANSFIELD EQUATION AT INFINITY FOR FLARING GEOMETRY

5.4.1 Asymptotic Form of the Transfield Equation

Let us come back to the case of magnetic surfaces which fan out at infinity, i.e. for which $lim_{\infty/a}(r) = \infty$. We refer to such surfaces as "flaring" ones. Following them to infinity, ρr^2 cannot become larger than some finite value, and the poloidal velocity approaches a limit value, v_∞, given by the asymptotic form of the Bernoulli equation (253). Let us adopt the notation

$$\lim_{\infty/a} \left(\frac{\rho r^2}{\rho_A} \right) = R_\infty^2(a) \tag{269}$$

Collecting the potentially dominant terms in the transfield equation (240), we obtain after some algebra the asymptotic form of this equation as:

$$\frac{v_\infty^2}{|\nabla a|} \frac{d\psi}{ds} = 2\Omega\Omega' R_\infty^2 + 2\Omega^2 R_\infty R_\infty' + \frac{\alpha'}{\alpha} \Omega^2 R_\infty^2 \tag{270}$$

Since $r|\nabla a| = \rho r^2 v_\infty/\alpha \approx \mu_0 \alpha v_\infty R_\infty^2$, this can be finally written as:

$$r\frac{d\psi}{ds} = \frac{\mu_0 \alpha \Omega^2 R_\infty^4}{v_\infty}\left(\frac{\Omega'}{\Omega} + \frac{\alpha'}{\alpha} + 2\frac{R'_\infty}{R_\infty}\right) \tag{271}$$

The right hand side is a function of a that obviously vanishes when R_∞ does, i.e. when the flow carries no Poynting flux to infinity. Otherwise it approaches a constant of unknown value along each magnetic surface at infinity. Then equation (271) gives the shape of magnetic surfaces in this asymptotic regime, since the curvature of a given poloidal field line a, is given by an equation of the form

$$r\frac{d\psi}{ds} = k(a) \tag{272}$$

5.4.2 Asymptotic Vanishing of the Electric Current Density in Flaring Geometry

This differential equation can be integrated for r(s) and z(s), which are also related by $ds^2 = dr^2 + dz^2$. So doing, it has been found (Heyvaerts and Norman (1989)) that there is no physically meaningful solutions, unless $k(a) = 0$. Therefore the TF equation reduces, at infinity, just to the vanishing of $k(a)$:

$$\frac{d}{da}\left(\lim_{\infty/a}(\frac{\rho r^2 \Omega}{\alpha})\right) = 0 \tag{273}$$

which integrates into

$$\lim_{\infty/a}(\frac{\rho r^2 \Omega}{\alpha}) = \text{Constant} \tag{274}$$

The physical meaning of this result can be best understood by noting that as $r \to \infty$, the poloidal electric current $I(z, a)$ enclosed by the circle defined in the plane of altitude z by the magnetic surface a approaches for large z

$$I_\infty(a) = \lim_{\infty/a}\left(\frac{2\pi r B_\theta}{\mu_0}\right) = -\lim_{\infty/a} -\left(\frac{2\pi \rho r^2 \Omega}{\mu_0 \alpha}\right) \tag{275}$$

Then , the TF equation says that, asymptotically, all magnetic surfaces enclose the same total poloidal current, which means that they carry no current between them. This means that the poloidal current density approaches zero on any magnetic surfaces that flares out to infinite r's as $z \to \infty$.

Looking at equation (275) again, we see that if ρr^2 approaches a non-zero value as $r \to \infty$, then the flaring magnetic surfaces must enclose a non zero total poloidal current, although they carry no current between them. If so, the current must be carried by a set of non-flaring magnetic surfaces around the polar axis. To sum up, flows which carry a non vanishing Poynting flux to infinity must have at least an inner core of asymptotically cylindrical magnetic surfaces. Or they could assume an entirely cylindrical geometry at infinity. We have shown in (5.3.3) that flows which carry no Poynting flux to infinity must be entirely paraboloidal asymptotically.

5.4.3 The Polar and Equatorial Boundary Layers

In the polarmost part of the current carrying cylindrical jet (if any!), it could be that r/r_A would not become large as $z \to \infty$. On the polar axis itself, both r and r_A are zero. Therefore the validity of the asymptotic transfield equation derived above breaks down, and neglected terms, like pressure or poloidal magnetic pressure, must be considered. A similar situation is found near the equator. If the field's polarity is antisymmetric between the northern and southern hemisphere, there must be a reversal of both B_r and B_θ as the equator is crossed. If, as $a \to A$, the equatorial flux value, the electric current, $2\pi r B_\theta/\mu_0$ does not vanish, the equator carries back the return current of the wind from infinity, and must have the structure of a sheet pinch. This is not in contradiction with the previous asymptotic analysis, because the equator contains a layer where the magnetic field vanishes, and there is then a region in it where the ordering that gave rise to eq (271) does not apply. Pressure cannot be neglected, in this region, in the transfield equation. This shows that for winds which carry a net Poynting flux to infinity, there must be polar and equatorial boundary layers in which pressure forces are still non negligible, even at large distances. Asymptotically, electric current flows exclusively in these boundary layers. Of course it is at this point not yet proved that winds which carry Poynting flux to infinity really do exist. Any proposed solution should be checked a posteriori for selfconsistency. Indeed it remains an interesting goal for research to precise the nature of the asymptotic wind regime in terms of properties of boundary conditions. Outside of boundary layers, equation (274) holds at large distances from the source. If a certain general form of magnetic surfaces is assumed, then eq (274) makes it possible to precise it entirely. For example, if it is assumed that magnetic surfaces are conical at infinity, so that they can be approximately represented by an equation of the form :

$$z - \tan \psi_\infty(a)\, r = 0 \tag{276}$$

we can find from eq (274) the distribution with flux of $\psi_\infty(a)$. $|\nabla a|$ can be calculated by differentiating eq (276) and then eq (274) gives, calling i the integration constant:

$$\frac{1}{\cos \psi_\infty} \frac{d\psi_\infty}{da} = \frac{1}{\mu_0 i} \frac{\Omega}{\sqrt{E - i\Omega/\alpha}} \tag{277}$$

This solution, if indeed the flow is partly conical, which at this stage must be checked a-posteriori for selfconsistency, is valid away from the pole and the equator. It must be matched smoothly to boundary layer solutions in these regions. The derivation of such synthetic solutions is in progress.

6. SPECIAL EXACT SOLUTIONS

6.1 WHY TO EXPLORE SPECIAL SOLUTIONS?

In view of the difficulty to produce a general solution of the full MHD wind problem, certain researchers have tried to produce special solutions, in order, not to model real winds, but to check whether these solutions can exhibit the properties that we expect from them. A main objective has indeed been to justify the existence of focusing solutions. The interest of such special solutions is that they are exact, not just approximate ones.

6.2 HOW TO PRODUCE SPECIAL SIMPLE SOLUTIONS?

The most effective way to obtain exact solutions, though particular ones, is to reduce the dimensionality of the problem by assuming some special variation of the physical quantities with distance, or with spherical angle. The problem so reduced boils down to a set of ordinary differential equations, which can be solved more easily, most often numerically though. The price to pay for this gain in simplicity is that the model cannot admit general boundary conditions, and sometimes it cannot satisfy a-priori imposed thermodynamics. If the realistic character of such solutions can be questioned, it is good to remember that their purpose is not to be realistic, but to be interesting.

Different types of self-similarity have been considered in the literature. The model of Chang and Henriksen (1980), assumes the magnetic surfaces to be deducible from eachother by an homothety in the cylindrical radial direction, so that:

$$r(a, z) = r_0(a)f(z) \qquad (278)$$

while the models of Blandford and Payne (1982) and Contopoulos and Lovelace (1994) assume spherical self similarity. With ψ the colatitude and R the spherical radius, Blandford and Payne describe magnetic surfaces as

$$R(a, \psi) = r_0(a)g(\psi) \qquad (279)$$

Most self similar models aim at modeling winds from accretion disks. They frequently assume the wind source to be the infinite equatorial plane. Sometimes the consequence of this assumption is that the magnetic flux spanned by the wind source is infinite.

Sauty and Tsinganos ( ), however, use spherical coordinates and assume some sort of angular self similarity. In their models, the flux is not bound to be infinite, and the wind source can be regarded as punctual, if wished. More precisely, they assume physical quantities to be expressible as linear functions of flux a. This is an extension to the full flow of forms of dependances that would be in general only valid near the polar axis. They assume the Alfvèn surface to be spherical, and more generally the alfvènic Mach number, M, to be only a function of spherical radius

$$M = M(R) \qquad (280)$$

This, and the assumption of linear dependance in a imply that the density and pressure must be of the form

$$\rho(R, a) = \rho_0(R) \, (1 + k_1 \, a) \qquad (281.a)$$

$$P(R, a) = P_0(R) \, (1 + k_2 \, a) \qquad (281.b)$$

and the shape of magnetic surfaces is expressed in a formula like

$$a(R, \psi) = f(R) \sin^2 \psi \qquad (282)$$

Their assumptions are entirely consistent with the equation of motion, but cannot accomodate an a-priori given equation of energy. The heating or cooling rate of the plasma that makes the energy balance consistent has to be found a-posteriori. Their model reduces to ordinary differential equations for the functions $\rho_0(R)$ and $P_0(R)$.

6.3 THE BLANDFORD AND PAYNE MODEL

We shall describe here in some more detail the model of Blandford and Payne (1982), hereafter "BP model" for short, which, to some extent, examplifies the features of all similar models. Their emphasis is on the structure of centrifugally driven winds from accretion disks. Therefore, they ignored gas pressure.

6.3.1 Description of the Blandford and Payne Model

This is a spherically selfsimilar model. The shape of flux surfaces is assumed to be representable in the form

$$r(a, z) = r_0(a) \; x \left(\frac{z}{r_0(a)} \right) \tag{283}$$

where $r(a, z)$ is the cylindrical distance to the polar axis for surface a at altitude z, and $r_0(a)$ is this distance in the equatorial plane. Let us use the variable

$$t \equiv \frac{z}{r_0(a)} \tag{284}$$

By the above definitions, $x(0) = 1$. Variables (r_0, t) constitute a pair of spatial coordinates which can replace (r, z), to which they are related by :

$$r = r_0 \; x(t) \tag{285.a}$$

$$z = r_0 \; t \tag{285.b}$$

At this point the function $x(t)$, which describes the shape of field lines, is not yet known, but this should not obscure the notion that (r_0, t) are basically spatial coordinates. The coordinate line of constant t is a straight line through the origin, since, eliminating r_0 from equations (285) gives

$$z = \frac{t}{x(t)} \; r \tag{286}$$

So, t is some sort of angular variable similar to a tangent, ranging from 0 to ∞, while r_0 is a radial variable. Field lines of different r_0 are deduced from each other by a mere homothety centered at the origin. This is why the model is said to be spherically self similar. The function $x(t)$ is the main unknown of the problem: it describes the shape of magnetic surfaces.

6.3.2 Keplerian Scaling

The intent of the model is to describe winds from keplerian accretion disks orbiting an object of mass $M_\star$, say. Therefore, the azimuthal velocity in the equatorial plane, v_θ, is assumed to be given by the keplerian law:

$$v_\theta(r_0) = \sqrt{\frac{GM_\star}{r_0}} \tag{287}$$

The poloidal velocity at the disk plane is supposedly very small, and then Ferraro's isorotation law identifies Ω as almost exactly equal to the angular velocity of the disk's gas, so that:

$$\Omega(r_0) = \sqrt{\frac{GM_\star}{r_0^3}} \tag{288}$$

Later on subscripts "0" refer to quantities measured in the equatorial plane. The field-aligned velocity, $\boldsymbol{v}_P$, is along poloidal field lines and can then be written as

$$\boldsymbol{v}_P = \sqrt{\frac{GM_\star}{r_0}}\ u(t)\ (x'(t)\ \boldsymbol{e}_r + \boldsymbol{e}_z) \tag{289}$$

$u(t)$ describes the variation along field lines of this velocity, which has been assumed to be small near the accretion disk, so $u(0)$ should be small. Similarly the azimuthal velocity v_θ can be written as

$$\boldsymbol{v}_\theta = \sqrt{\frac{GM_\star}{r_0}}\ w(t)\ \boldsymbol{e}_\theta \tag{290}$$

Since the flow is keplerian in the equatorial plane, $w(0) = 1$. Finally the magnetic field is written as

$$\boldsymbol{B} = B_{z0}\ (\ b(t)\ x'(t)\ \boldsymbol{e}_r + b_\theta(t)\ \boldsymbol{e}_\theta + b(t)\ \boldsymbol{e}_z) \tag{291}$$

When writing eq (291) we assumed that B_θ scales with r_0 as B_P, which is a natural assumption here, because it extends to the full vector magnetic field the spherical self-similarity, as defined by eq (283) for the shape of magnetic surfaces. The density is also written in the same form, i.e. as:

$$\rho = \rho_0(r_0)\ \nu(t) \tag{292}$$

and $\nu(0) = 1$ if $\rho_0(r_0)$ is to be the density at the equatorial plane.

6.3.3 First Integrals and Keplerian Scaling

A number of quantities, the so called first integrals, are conseved following the fluid motion. These are the mass flux to magnetic flux ratio, $\alpha(a)$, the magnetic surface rotation, $\Omega(a)$, the specific angular momentum $L(a)$, energy $E(a)$ and entropy $Q(a)$, here taken to be zero. In the coordinate system (r_0, t) flux surfaces are represented by lines of constant r_0, so the functions α, Ω, L, E must depend on r_0 only. This imposes some relations between the quantities introduced in subsection (6.3.2). Consider for example the mass to magnetic flux ratio, $\alpha(a)$:

$$\alpha = \frac{\rho v_z}{B_z} = \frac{\rho_0(r_0)}{B_{z0}}\ \sqrt{\frac{GM_\star}{r_0}}\ \frac{\nu(t)u(t)}{b(t)} \tag{293}$$

Since α must depend on r_0 only, we must have

$$\nu(t)\ u(t)\ /\ b(t) = \text{constant} = u(0) \tag{294}$$

Then

$$\alpha = \frac{\rho_0\ u_0}{B_{z0}}\ \sqrt{\frac{GM_\star}{r_0}} \tag{295}$$

Consider now the specific energy, E, which, for cold winds, equals

$$E = \frac{v_P^2 + v_\theta^2}{2}\ -\ \frac{GM_\star}{\sqrt{r^2 + z2}}\ -\ \frac{r\Omega B_\theta}{\mu_0 \alpha} \tag{296}$$

Introducing the keplerain scaling for these quantities from eqs (289), (290), (291), (292), we get

$$E = \frac{GM_\star}{r_0} \left(\frac{u^2}{2}(1 + x'^2) + \frac{w^2}{2} \right) - \frac{GM_\star}{r_0} \frac{1}{\sqrt{r^2 + z^2}} - \frac{B_{z0}^2}{\mu_0 \, \rho_0 u_0} \, x \, b_\theta(t) \qquad (297)$$

For the right hand side to be a function of r_0 alone, $B_{z0}^2/\mu_0\rho_0 u_0$ must be proportional to $GM_\star/r_0$. Let us introduce a number k, such that:

$$k = \frac{GM_\star}{r_0} \frac{\mu_0 \rho_0 u_0}{B_{z0}^2} \qquad (298)$$

Remembering that u_0 is a mere number, this shows that the Alfvèn speed must scale on the equator as the keplerian speed. Then we can write the specific energy E in the form

$$E = \frac{GM_\star}{r_0} \left(\frac{u^2}{2}(1 + x'^2) + \frac{w^2}{2} - \frac{1}{\sqrt{r^2 + z^2}} - \frac{x \, b_\theta}{k} \right) \qquad (299)$$

Since E must be a function of r_0 only, the parenthesis in eq (299) must be a constant number, call it e:

$$e = \frac{u^2}{2}(1 + x'^2) + \frac{w^2}{2} - \frac{1}{\sqrt{r^2 + z^2}} - \frac{x \, b_\theta}{k} \qquad (300)$$

Similarly, the specific angular momentum is defined, and can be written, as

$$L = rv_\theta - \frac{r B_\theta}{\mu_0 \alpha} = \sqrt{GM_\star r_0} \left(x \, w - \frac{x \, b_\theta}{k} \right) \qquad (301)$$

Since L must be a function of r_0 only, the parenthesis must be a constant number, call it λ :

$$\lambda = x \, w - \frac{x \, b_\theta}{k} \qquad (302)$$

Scaling considerations are not sufficient to specify the dependence of ρ_0, and B_{z0} on r_0. All we know is that B_{z0}^2/ρ_0 must scale as $1/r_0$. Different dependences of B_{z0} on r_0 are conceivable (Contopoulos and Lovelace (1994)), although the range of possibilities is limited, by the compatibility with the self similarity requirement, to power laws (see section (6.5)).Blandford and Payne make the specific assumption that $r_0^2\rho_0 v_{z0}$ is independent on r_0. Since $v_{z0} \approx r_0^{-1/2}$, this leads to

$$\rho_0(r_0) \approx r_0^{-3/2} \qquad (303.a)$$

$$B_{z0}(r_0) \approx r_0^{-5/4} \qquad (303.b)$$

At this point, we note an unfortunate limitation of the model: $2\pi r_0 B_{\theta 0}/\mu_0$ is the current enclosed by a circle of radius r_0. From eqs (291) and (303.b) this current diverges as $r_0 \to 0$ as:

$$I(r_0) \approx r_0^{-\frac{1}{4}} \qquad (304)$$

There is then an infinite electric current returning to the disk plane along the polar axis! Similarly ρ_0, B_{z0} and $B_{\theta 0}$ diverge as $r_0 \to 0$. For $r_0 \to \infty$, the total poloidal current, including the infinite one on the polar axis, approaches zero. The disk plane emits to the wind an infinite poloidal current which returns back to it in a thin polar wire. It is an unfortunate general feature of any self-similar model that the sense of the electric current that leaves the surface of the wind source cannot reverse at any place on the wind-emitting boundary. As a result all the return current must flow back on some singularity, for example here on a polar singularity. By contrast, solutions which would have all their current returning to the source in the wind flow, with no net poloidal current reaching to infinity, should have normal components of the current on the wind-emitting surface of both signs, according to which part of this surface is considered.

6.3.4 Parameters of the Blandford Payne Model

The parameters of the BP model are the numbers e, λ and k. The derivative at the equatorial plane of the function $x(t)$, call it x'_0, is also a parameter, since the equation for $x(t)$ will turn out to be of second order. However these parameters are not independent, as we shall see shortly. This is a consequence of the regularity conditions which must be satisfied, both the criticality ones and the Alfvèn regularity condition. We show below that $e = \lambda - 3/2$, and that x'_0 must be chosen adequately to ensure that the solution be regular at the Alfvèn point. Altogheter, the solutions are parametrized only by two numbers, which can be chosen to be k and λ. The parameter k, is obviously a magnetization parameter, while the parameter λ measures how much Poynting flux escapes the system, since when specific angular momentum is concentrated in matter only $\lambda = 1$.

6.4 EQUATIONS OF THE MODEL AND FIELD ALIGNED DYNAMICS

We want to reduce the MHD equations of this model to a single differential equation for the shape of magnetic surfaces, described by the function $x(t)$. We then need to express other physical quantities in termes of it.

6.4.1 Poloidal Field in Terms of the Shape of Magnetic Surfaces

Variations of the poloidal field, B_P, are ruled by equation $\mathrm{div}\, B_P = 0$. This is equivalent to saying that the flux between two magnetic surfaces of parameter r_0 and $r_0 + dr_0$ is independant of position along poloidal field lines. So:

$$2\pi r_0 dr_0 B_{z0}(r_0) = 2\pi r dr B_z(r_0, t) \tag{305}$$

Take into account the self-similarity ansatz, $B_z(r_0, t) = B_{z0}(r_0)\, b(t)$, and express dr, the difference of radii of flux surfaces r_0 and $r_0 + dr_0$ at the same height z, in terms of $r_{0,,}$ dr_0, t, $x(t)$. These quantities being related by the condition that z be the same:

$$z = r_0\, t = (r_0 + dr_0)\, (t + dt) \tag{306.a}$$

we get

$$r_0\, dt + t\, dr_0 = 0 \tag{306.b}$$

Then eq (305) leads to

$$b(t) = \frac{1}{x(t)\, (x(t) - t\, x'(t))} \tag{307}$$

The quantity $(x(t) - t\, x'(t))$, the jacobian of the coordinate transformation from (r_0, t) to (r, z), deserves the special notation

$$J(t) = x(t) - t\, x'(t) \tag{308}$$

6.4.2 Poloidal Mass Flux in Terms of the Shape of Magnetic Surfaces

Similarly we find that ρv_z varies with t as:

$$\rho v_z = \rho_0 \sqrt{\frac{GM_\star}{r_0}} \frac{u_0}{x(t) \, J(t)} \tag{309}$$

so that the density is given by

$$\rho = \rho_0 \, \nu(t) = \rho_0 \, \frac{u_0}{x(t) \, J(t) \, u(t)} \tag{310}$$

6.4.3 Alfvènic Mach Number in Terms of the Shape of Magnetic Surfaces

The alfvènic Mach number plays an important role. Its square is expressed in the present model as

$$m \equiv M^2 = v_P^2/v_{PA}^2 = \mu_0 \alpha^2/\rho = k \, x(t) \, u(t) \, J(t) \tag{311}$$

This shows that in the BP model, surfaces of constant alfvènic Mach number are cones of constant t. In particular, the Alfvèn surface, $m = 1$, is one of them.

6.4.4 Azimuthal Variables in Terms of the Shape of Magnetic Surfaces

From the isorotation law and specific angular momentum conservation law, we have obtained v_θ and B_θ in terms of the other flow variables (eqs (170) and (171)). In the present context, these relations particularize to

$$w = \frac{\lambda}{x} + \frac{(\lambda - x^2)}{x \, (m - 1)} \tag{312.a}$$

$$b_\theta = \frac{k \, (\lambda - x^2)}{x \, (m - 1)} \tag{312.b}$$

6.4.5 Bernoulli Equation in the Blandford and Payne Model

We obtained in eq (300) the Bernoulli equation in a form involving $x(t)$ as well as other flow variables that have been now expressed in terms of it. Equation (300) can be transformed, using (312.b) into

$$e = \frac{u^2}{2} \, (1 + x'^2) + \frac{(w - x)^2}{2} - \frac{1}{\sqrt{x^2 + t^2}} + \lambda - \frac{x^2}{2} \tag{313}$$

This is the model's version of the Bernoulli equation written in the rotating frame :

$$E = \frac{v_P^2}{2} + \frac{(v_\theta - \Omega r)^2}{2} - \frac{GM_\star}{\sqrt{r^2 + z^2}} + \Omega L - \frac{\Omega^2 r^2}{2} \tag{314}$$

In this form the Bernoulli equation expresses the conservation of the sum of the kinetic energy in the frame rotating with the magnetic surface, $(1/2) \, (v_P^2 + (v_\theta - \Omega r)^2)$, and the gravitational plus centrifugal potential energy, $\Phi_{eff} = -GM_\star/\sqrt{r^2 + z^2} - \Omega^2 r^2/2$. This sum is $E - \Omega L$.

If the plasma were not cold, the requirement that the energy parameter, E, be such that the solution passes at the slow mode critical point would have to be imposed. For a cold plasma, the slow mode speed is zero, and the corresponding critical point is passed very near the wind-source, at $t = 0$. In place of the slow mode criticality condition, we can then use the requirement that the Bernoulli equation be satisfied at the base of the wind, at $t = 0$, where $x = 1$, $w = 1$, and u is very small. At this point, eq (313) reduces to

$$e = \lambda - \frac{3}{2} \tag{315}$$

This transforms eq (313) into

$$x^2 + \frac{2}{\sqrt{x^2 + t^2}} - 3 = u^2(1 + x'^2) + (w - x)^2 \tag{316}$$

Expressing u and w, by eqs (311) and (312.a), in terms of the alfvènic Mach number and of the shape function $x(t)$ and its derivatives, eq (316) is transformed into an equation for the field aligned dynamics, namely for the square, $m(t)$, of the alfvènic Mach number:

$$x^2 + \frac{2}{\sqrt{x^2 + t^2}} - 3 = \frac{m^2(1 + x'^2)}{k^2 x^2 J^2} + \frac{m^2(\lambda - x^2)^2}{x^2(m - 1)^2} \tag{317}$$

This is a fourth degree equation for m, when t and $x(t)$ are known. We find again the same result as for Weber-Davis winds, that there are at most four solutions for the alfvènic Mach number at any position along a given field line.

6.4.6 Minimal Inclination of Field Lines with Respect to the Vertical

In equations (316) and (317) the right hand side is positive. This implies that the left hand side should be positive too for any t, and imposes on the shape function $x(t)$ the constraint that $x^2 + 2/\sqrt{x^2 + t^2} - 3$ be always positive. Physically this is easily understood by considering eq (314): since the wind plasma is assumed to be cold, it suffers no pressure push, and can only leave the disk from its state of rest in the rotating frame if, in this frame, it follows a path of decreasing effective potential:

$$\Phi_{eff}(r_0, 0) - \Phi_{eff}(r, z) > 0 \tag{318}$$

This is just the condition that the left hand side of eq (316) be positive. At the equatorial plane, $t = 0$, the function $x^2 + 2/\sqrt{x^2 + t^2} - 3$ vanishes as does its first derivative. The condition (318) then reduces in this neigbourhood to the requirement that the second order derivative be positive, which leads to:

$$x_0'^2 - \frac{1}{3} > 0 \tag{319}$$

This means that the angle of the tangent to the poloidal field line at its foot point to the vertical to the disk must be larger than $30°$.

6.4.7 Bernoulli Equation near the Alfvèn Point

The Alfvèn point is at $m = 1$, and $x = \sqrt{\lambda}$ (eqs (312)). At this point the poloidal fluid velocity is given, from eq (311), by

$$u_A = \frac{1}{kx_A J_A} = \frac{1}{k\sqrt{\lambda} J_A} \tag{320}$$

where J_A is given by eq (308), i.e. $J_A = (\sqrt{\lambda} - t_A x_A')$. The Bernoulli equation written at this point gives the value of the slope $(dm/dx)_A$ at the Alfvèn point, since when this point is approached

$$\frac{m-1}{\lambda - x^2} \rightarrow -\frac{1}{2\sqrt{\lambda}}(dm/dx)_A = -\frac{1}{2\sqrt{\lambda}}\frac{(dm/dt)_A}{(dx/dt)_A} \tag{321}$$

Solving equation (317) in this limit gives

$$(dm/dt)_A = \frac{2\,x_A'}{\left(\lambda - 3 + (\lambda + t_A^2)^{-1/2} - \frac{1 + x_A'^2}{k^2\,\lambda\,J_A^2}\right)^{1/2}} \tag{322}$$

6.5 TRANFIELD EQUATION IN THE BLANDFORD AND PAYNE MODEL

The transfield equation (177) can be translated in terms of the variables used in this model as follows: the flux function $a(r, z)$ is, by definition, a function of r_0 only, which can be expressed in terms of the field normal to the surface of the disk as

$$a(r_0) = \int_0^{r_0} B_{z0}(\xi)\,d\xi \tag{323}$$

The derivatives of a with respect to r and z can be calculated by the usual rules for changing from variables (r_0, t) to variables (r, z):

$$\frac{\partial}{\partial r} = \frac{1}{J}\frac{\partial}{\partial r_0} - \frac{t}{r_0 J}\frac{\partial}{\partial t} \tag{324.a}$$

$$\frac{\partial}{\partial z} = -\frac{x'}{J}\frac{\partial}{\partial r_0} + \frac{x}{r_0 J}\frac{\partial}{\partial t} \tag{324.b}$$

which gives, from subsection (6.3.2),

$$\frac{\partial a}{\partial r} = \frac{1}{J}r_0\,B_{z0}(r_0) \qquad\qquad \frac{\partial a}{\partial z} = -\frac{x'}{J}r_0\,B_{z0}(r_0) \tag{325}$$

$$\frac{\alpha}{\rho r} = \sqrt{\frac{GM_\star}{r_0^3}\frac{J(t)u(t)}{B_{z0}(r_0)}} \qquad\qquad \frac{1}{\mu_0\rho r} = \frac{J(t)u(t)}{\mu_0\,\rho_0(r_0)\,r_0\,u_0} \tag{326}$$

The right hand side of the transfield equation is expressed in terms of derivatives with respect to flux, a, of first integrals such as, say, $E'(a)$. In this description, these are expressed as functions of r_0. For example, from eqs (299) and (300), $E(r_0) = e(GM_\star/r_0)$. The chain rule for derivatives gives the required quantities, with, from eq (323):

$$\frac{da}{dr_0} = r_0 B_{z0}(r_0) \tag{327}$$

After some algebra involving eqs (177), (295), (299), (300), (301), (302), (311), (324), (325), the transfield equation (177) can be converted into the following, quite different looking, equation for the function $x(t)$, $m(t)$ being implicitly related to it by eq (317).

$$u(t) \left(\frac{u}{2}(1+x'^2) + t\frac{du}{dt} + x\frac{d}{dt}(x'u) - x\frac{d}{dt}(\frac{ux'}{m}) - t\frac{d}{dt}(\frac{u}{m}) + \frac{u(1+x'^2)}{2m}\frac{d\ln(\rho_0/r_0)}{d\ln(r_0)} \right)$$

$$= e + \frac{3}{4}\frac{m(\lambda-x^2)^2}{x^2(1-m)^2} + \frac{\lambda^2}{2x^2} + \frac{1}{2}\frac{(\lambda+3x^2)(\lambda-x^2)}{x^2(m-1)} \tag{328}$$

For selfsimilarity, this equation should be an ordinary differential equation for functions of t only. Hence the function of r_0 which still appears in eq (328) should be a constant:

$$\frac{d\ln(\rho_0/r_0)}{d\ln r_0} = \text{constant} \tag{329}$$

This means that $\rho_0(r_0)$ must be a power law function. For the reasons discussed above, Blandford and Payne have choosen it as indicated by eq (303.a). Taking eq (315) into account, the final form of the transfield equation in this model is

$$\frac{u^2}{2}(1+x'^2) + ut\frac{du}{dt} - ut\frac{d}{dt}(\frac{u}{m}) + ux\frac{d}{dt}(x'u) - ux\frac{d}{dt}(\frac{ux'}{m}) - \frac{5}{4}\frac{u^2}{m}(1+x'^2)$$

$$= \lambda - \frac{3}{2} + \frac{3}{4}\frac{m(\lambda-x^2)^2}{x^2(1-m)^2} + \frac{\lambda^2}{2x^2} + \frac{1}{2}\frac{(\lambda+3x^2)(\lambda-x^2)}{x^2(m-1)} \tag{330}$$

This equation is to be coupled to eqs (308) and (311), which together give $u(t)$ in terms of $m(t)$ and $x(t)$, and to eq (317) which relates, although implicitly, $m(t)$ to $x(t)$.

In this form the transfield equation is a second order ordinary differential equation. This is seen by noting that $m(t)$ and $u(t)$ depend on $x(t)$ and $x'(t)$, as is evident from eq (317). The highest derivative terms are all terms involving derivatives with respect to t on the left hand side of eq (330).

6.6 ALFVEN REGULARITY CONDITION IN THE BLANDFORD AND PAYNE MODEL

The vanishing at the Alfvèn point of the highest order derivative terms of the transfield equation is a general property (see (5.3.1) and (5.3.2)), and so is it in this model too. Gathering the highest derivative terms on its left hand side, eq (330) can be written:

$$ut(1-\frac{1}{m})\frac{du}{dt} + ux(1-\frac{1}{m})\frac{d}{dt}(x'u) - u^2(t+xx')\frac{d}{dt}(\frac{1}{m}) =$$

$$\frac{u^2}{2}(1+x'^2)(\frac{5}{2m}-1) + \lambda - \frac{3}{2} + \frac{3m(\lambda-x^2)^2}{4x^2(1-m)^2} + \frac{\lambda^2}{2x^2} + \frac{(\lambda+3x^2)(\lambda-x^2)}{2x^2(m-1)} \tag{331}$$

As $m \to 1$, the first two terms of the left hand side of eq (331) vanish. It is not obvious by mere inspection that the second order derivative x'' implicit in the factor dm/dt vanishes at the Alfvèn point. However m'$_A$ ($\equiv$ (dm/dt)$_A$) has been calculated in eq (322), and it turned out not to involve the second derivative, x"(t). The Alfvèn regularity condition is but the expression of the transfield equation at the Alfvèn surface, where m = m$_A$ =1 and x = x$_A$ = $\sqrt{\lambda}$. From eq (331), using l'Hopital's rule to calculate terms which appear in the indefinite form 0/0, the Alfvèn regularity condition can be written as:

$$u_A^2 \, (t_A + \sqrt{\lambda} \, x'_A) \, m'_A = \frac{3 \, u_A^2}{4} \, (1 + x_A'^2) + \lambda - 1 + 3 \, \frac{x_A'^2}{m_A'^2} + 4\sqrt{\lambda} \, \frac{x_A'}{m_A'} \tag{332}$$

The slope at the Alfvèn point, m'_A, is given by eq (322) in terms of x'_A, t_A and J_A, itself expressed by eq (308) as:

$$J_A = \sqrt{\lambda} - t_A \, x'_A \tag{333}$$

Similarly, from eq (311),

$$u_A = (k\sqrt{\lambda}J_A)^{-1} \tag{334}$$

So, eq (332) is an equation for x'_A, for given t_A. For it to be satisfied, the value x'_O of the first derivative of $x(t)$ at the equatorial plane cannot be chosen arbitrarily (see (6.3.4)).

6.7 ASYMPTOTIC SOLUTION

6.7.1 Asymptotic Bernoulli Equation

Let us take the limit of Bernoulli equation (316) for $t \to \infty$, assuming $x(t)$ and $m(t) \to \infty$, but $x' \to 0$. Then we get, using eqs (312.a) and (311)

$$x^2 - 3 = u^2 + x^2 - 2\lambda + 2\frac{x^2}{m} \tag{335}$$

$$2\lambda - 3 = u^2 + \frac{2x}{kuJ} \tag{336}$$

From this we obtain, if needed, du/dx by differentiation

$$uu' = -\frac{ku^3 J}{ku^3 J - x} \, \frac{x}{kuJ} \left(\frac{x'}{x} + \frac{tx''}{J} \right) \tag{337}$$

6.7.2 Asymptotic Transfield Equation

In the limit m and $t \to \infty$, the transfield equation (330) reduces to

$$\frac{u^2}{2} \, (1 + x'^2) + t\frac{d}{dt}(\frac{u^2}{2}) + u^2 \, x \, x'' + x \, x' \, \frac{d}{dt}(\frac{u^2}{2}) = \lambda - \frac{3}{2} - \frac{3}{4} \frac{x^2}{m} \tag{338}$$

Using the Bernoulli equation (336) and eq (311), this can be written as:

$$\frac{u^2}{2} \, (1 + x'^2) + t \, \frac{d}{dt} \, (\frac{u^2}{2}) + u^2 \, x \, x'' + x \, x' \, \frac{d}{dt} \, (\frac{u^2}{2}) = \frac{u^2}{2} + \frac{x}{4 \, k \, u \, J} \tag{339}$$

For a collimating solution, $x' \to 0$, so that this further simplifies to:

$$t \, u \, u' + u^2 \, x \, x'' + x \, x' \, u \, u' = \frac{x}{4 \, k \, u \, J} \tag{340}$$

u' can be obtained from equation (337) to give for $x(t)$ the equation

$$(t + xx') \frac{x}{kuJ} \left(\frac{x'}{x} + \frac{tx''}{J} \right) + \frac{x}{4kuJ} \left(1 - \frac{x}{ku^3 J} \right) = u^2 \, x \, x'' \left(1 - \frac{x}{ku^3 J} \right) \qquad (341)$$

This can be further simplified by noting that xx'/t, which equals $(1/2) \, d(x^2)/d(t^2)$ goes to zero and that the x'' terms on the l.h.s and r.h.s are in a ratio such that $u^2 xx'' \ll (t^2 xx'')/(kuJ^2)$. This is because $ku^3 J^2$ is proportional to x^2, which is much less than t^2. Thus eq (341) further simplifies to:

$$tx(\frac{x'}{x} + \frac{tx''}{J}) + \frac{x}{4} - \frac{x^2}{4ku^3 J} = 0 \qquad (342)$$

Finally the asymptotic structure is determined by the set of equations coupling the simplified Bernoulli and transfield equations (J being given by eq (308)):

$$2\lambda - 3 = u^2 + \frac{2x}{kuJ} \qquad (343.a)$$

$$t^2 xx'' + t \, x' \, J + \frac{xJ}{4} - \frac{x^2}{4ku^3} = 0 \qquad (343.b)$$

6.7.3 Asymptotic Solution when the Fast Point is at Finite Distance

The fast mode Mach number, M_f for a cold flow is given by

$$M_f^2 \equiv \frac{v^2}{c_f^2} = \frac{\mu_0 \rho v^2}{B_P^2 + B_\theta^2} \qquad (344)$$

This can be expressed in terms of u, x and J, using eqs (289), (291), (307), (310), (312.b), and taking into account the relevant asymptotic properties. The result is

$$M_f^2 \approx \frac{ku^3 J}{x} \qquad (345)$$

Suppose first that the fast mode speed is passed at a point situated at a finite distance, and that the fast mode Mach number becomes very large at infinity. This corresponds to $lim \, (ku^3) \gg 1$. In that case the asymptotic transfield equation looses its last term (eq (343.b)), and becomes

$$t^2 xx'' + (x - tx') \, tx' + \frac{1}{4}x(x - tx') = 0 \qquad (346)$$

It can be solved by the change of variables $z = \ln x$ and $y = \ln t$. The solution is

$$x(t) = K \, t \, e^{-at^{1/4}} \qquad (347)$$

where a and K are integration constants. We find infinite focusing, since $x \to 0$ as $t \to \infty$. Obviously, this is because the pressure has been neglected. Otherwise, the hoop stresses would not be able to indefinitely crunch the flow to the polar axis. Note however that the solution is inconsistent with the ordering ($x \to \infty$) used to derive it!

6.7.4 Asymptotic Solution when the Fast Point is at Infinite Distance

If the fast critical point recedes to infinity, then $M_f \to 1$ as $t \to \infty$. Still assuming axial focusing, $x' \to 0$ and $J \approx x$, we see, from eq (344), that $ku^3 \to 1$. For the fast critical point to be rejected at infinity, the velocity must be such that the Bernoulli equation has a double root at infinity. So, u is fixed by the condition that equation (336) be satisfied at infinity, as well as its derivative with respect to u:

$$2u - \frac{2x}{ku^2J} = 0 \tag{348}$$

$$u^3 = \lim_{t \to \infty} \left(\frac{x}{k(x - tx')} \right) \tag{349}$$

Substituting this expression for ku^3 in the transfield equation (343.b), we obtain:

$$t^2 \frac{x''}{x} - t^2 \frac{x'^2}{x^2} + t \frac{x'}{x} = 0 \tag{350}$$

This equation can be solved by the same change of variables as in (6.7.3), with the result that

$$x = Kt^\alpha \tag{351}$$

where K and α are integration constants. The value of α, however, is constrained by equation (349), which implies that u^3 should equal

$$u^3 = \frac{1}{k(1 - \alpha)} \tag{352}$$

The Bernoulli equation (343.a) then becomes

$$\lambda - \frac{3}{2} = \frac{3}{2} \left(\frac{1}{k(1 - \alpha)} \right)^{2/3} \tag{353}$$

This gives the value of the exponent in terms of the parameters , k and λ, of the model:

$$\alpha = 1 - \frac{3^{3/2}}{k(2\lambda - 3)^{3/2}} \tag{354}$$

The solution found in this case is more self consistent. It shows parabolic focusing, with a power that can be derived from the parameters of the model.

References

Biermann, L., 1957, Observatory 77, 107

Blandford, R.D., Payne, D.G., 1982, Mon. Not. Roy. Astr. Soc. 199, 883

Chan,K.L., Henriksen, R.H., 1980, Astrophys.J. 241, 534

Contopoulos, J., Lovelace, R.V.E., 1994, Astrophys.J. 429, 139

Gringauz K.I.,Bezruvkikh, V.V., Ozerov, V.D., Rybchinskii, R.E., 1960, Soviet Physics Doklady 5, 361

Heyvaerts, J., 1989, in "Basic Plasma processes on the sun", IAU Symp 142, E.R Priest and V. Krishan, ed., Dordrecht, Kluwer Ac. Press, p. 207

Heyvaerts, J., Norman, C.A., 1989, Astrophys.J. 347, 1055
Lovelace,R.V.E., Berk, H. L.,Contopoulos, J.,1991, Astrophys.J. 379, 696
Parker, E.N.,1958, Astrophys. J. 123, 664
Parker, E.N., 1960,Astrophys.J. 132, 821
Sauty, C., Tsinganos, K., 1994, Astron. Astrophys. 287, 893
Tsinganos, K., Sauty, C.,1992, Astron. Astrophys., 257, 790
Weber,D.J., Davis, L.,Jr, 1967 Astrophys. J. 148, 217

Particle Acceleration

Jan Kuijpers[1,2,3]

[1] Astronomical Institute, Utrecht University, P.O. Box 80 000, 3508 TA Utrecht, The Netherlands
[2] Faculty of Science, University of Nijmegen, Toernooiveld 1, 6525 ED Nijmegen
[3] CHEAF (Center for High Energy Astrophysics), P.O. Box 41882, 1009 DB Amsterdam

Abstract. It is the purpose of this series of lectures to give an overview of our understanding of (electromagnetic) particle acceleration processes in astrophysics. For each process I emphasize the basic physics and point out differences and correspondences with other mechanisms. Remaining problems are summarized with references to the recent literature. For instructive reasons I first discuss a number of fundamental aspects which are common to several acceleration processes. Thereafter I have grouped the various processes for particle acceleration into three chapters according to their underlying physical mechanism: plasma turbulence, shock waves, and direct electric fields.

Keywords. Stochastic acceleration, plasma turbulence acceleration, shock wave acceleration, direct acceleration

1 Introduction

Acceleration of particles is a universal phenomenon occuring near a variety of objects, ranging from planetary magnetospheres to solar flares, the heliospheric terminal shock, flare stars, close binaries, protostars and their jets, radio pulsars, accreting white dwarfs, binary neutron stars and black holes and their jets, supernova remnants, and nuclei and jets in active galaxies. Much effort has gone into finding a universal acceleration process. The most promising candidate for such a universal process is *diffusive shock acceleration*, a first-order Fermi-type acceleration process in shock waves. This is because shocks occur in various violent events observed to be associated with particle acceleration and because strong non-relativistic shocks produce a distribution of accelerated particles with a power law dependence on energy $N(\epsilon) \propto \epsilon^{-2}$ under steady state conditions, independent of local physical conditions. Such a spectrum is close to, but somewhat flatter than is observed:

Table 1. Differential energy spectra of high-energy particles

Object	Interval	$N(\epsilon) \propto$	Species	Observation
Extragal.	$0.1 < \epsilon < 10$ GeV	$\epsilon^{-2.0/-3.8}$	electrons	radio [1]
Galaxy	≤ 0.3 GeV	$\epsilon^{-2.3}$	electrons	diffuse γ [2]
Galaxy	$1 \leq \epsilon \leq 10$ GeV [3]	$\epsilon^{-2.2}$	electrons[4]	≤ 1.4 GHz [2]
Galaxy	$4.5 \leq \epsilon \leq 62.5$ GeV	$\epsilon^{-3.15\pm0.20}$	cosmic rays	in situ [5 6]
Galaxy	$10 \leq \epsilon \leq 10^4$ GeV	$\epsilon^{-2.7}$	baryonic CR	in situ [6]
Galaxy	$3 \cdot 10^4 \leq \epsilon \leq 3 \cdot 10^6$ GeV	'bump'[7]	baryonic CR	in situ [6]
Galaxy	up to 10^{10} GeV	$\epsilon^{-3.0/-3.1}$	baryonic CR	in situ [6]
Galaxy	energy per nucl. 10^{20} eV	flattening[8]	protons?	in situ [6]

Apart from local detections fast particles are indirectly observed by their electromagnetic signature, from gamma rays to the radio domain by processes such as bremsstrahlung, inverse Compton emission, (gyro)synchrotron emission, cyclotron emission, and a variety of plasma masers (cyclotron maser, maser at the plasma frequency, linear accelerator emission). These observations make it clear that, although strong shock waves probably form the most important acceleration process, other acceleration mechanisms become important when free

---

[1] Miley 1990; assuming equipartition.

[2] Reynolds 1992

[3] Below a few GeV the cosmic ray electrons are strongly modulated by the solar wind.

[4] Electrons are observed up to 1000 GeV; the energy density in electrons (positrons) is only $6 \cdot 10^3$ eV m^{-3} compared to $5 \cdot 10^5$ eV m^{-3} for nuclei.

[5] Bremsstrahlung below 150 MeV from electrons with energies below 0.3 GeV.

[6] Axford 1992; Sokolsky 1992; Stanev 1992

[7] The 'knee' in the spectrum.

[8] The 'ankle' in the spectrum; expected cut-off due to photo-pion production.

energy is stored in magnetized atmospheres. Moreover a number of fundamental concepts appear over and again in a variety of acceleration processes, such as Fermi-type processes, stochastic acceleration, particle scattering, and plasma turbulence. I have therefore ordered the material as follows: A number of fundamental processes are treated in this Introduction. An introduction to plasma turbulence and its acceleration potential forms the subject of Section 2. Shock acceleration, which requires the presence of plasma turbulence, is treated in Section 3. Finally in Section 4 I discuss direct, in contrast to stochastic, acceleration processes.

1.1 Fundamentals

A distinction is made between heating and acceleration: *Heating* is used for a temperature increase of the gas or, more loosely, for an increase in random energy of the bulk of the particles in a specified volume. *Acceleration* is used either for the increase in energy of a subset of particles in a given volume or for the increase of the net momentum of a gas, resulting in a particle distribution which differs markedly from a Maxwellian. Of course this distinction is not strict: for instance local plasma heating can lead to the appearance of accelerated particle beams at neighbouring sites. Here I will not discuss gravitational acceleration and acceleration by radiation fields, but concentrate on three groups of mechanisms to accelerate *electrically charged* particles:

- Acceleration by plasma waves;
- Acceleration by shock waves;
- Acceleration by unidirectional electric fields.

Each of these processes operates *selectively* on specific particles:

- Waves can accelerate particles, usually when a resonance condition is fulfilled;
- Shocks can only accelerate particles above a certain injection energy;
- Electric fields in a collisonless plasma can cause runaway acceleration, either of a small fraction or of the entire population.

An acceleration process is stochastic or regular. *Stochastic* acceleration is characterized by an increase of the average energy per particle with a simultaneous spread in momenta of the accelerated particles. Examples of stochastic acceleration are: acceleration by *plasma turbulence* (Section 2), *second-order Fermi* acceleration (Section 1.3) and *magnetic pumping* or *betatron* acceleration (Section 1.5). Physically stochastic acceleration results from the tendency of an ensemble of particles that interact with a population of energetic entities via random 'kicks', to distribute the free energy equally over the number of degrees of freedom. This thermodynamic argument leads to *energy equipartition* per scattering entity and per particle. In a collision dominated plasma this leads to a temperature increase; in an astrophysical 'collisionless' plasma, where plasma waves have ample time to develop, this leads to acceleration of a subset of the particle distribution long before thermalization ('dissipation') takes place.

Examples of *regular* acceleration are *first-order Fermi* acceleration (Section 1.2), 'diffusive' acceleration in *quasi-parallel shocks* (shock direction parallel to the magnetic field, Section 3.1), drift acceleration in *quasi-perpendicular shocks* (Section 3.2), and acceleration by *unidirectional electric fields*, e.g. in a current circuit (Section 4.1) or in an electrostatic Double Layer (Section 4.2). Again the distinction is not always strict: for example particle acceleration by magnetic *reconnection* can obtain a stochastic character (Section 4.3).

Recent reviews on acceleration by weak plasma turbulence are: Melrose 1995; and on shock acceleration: Achterberg 1990; Kirk 1995. Further recent conference proceedings on particle acceleration can be found in IAU Coll. **142** (1994), in Zank and Gaisser (1992), and in van Paradijs *et al.* (1991).

1.2 First-Order Fermi Acceleration

First-order Fermi acceleration (Fermi 1949, 1954) operates when a particle reflects elastically between two approaching, massive, perfect mirrors (Fig. 1). After each collision the Lorentz factor of the particle $\gamma = (1 - v^2/c^2)^{-1/2}$ changes according to

$$\gamma'' = \gamma \gamma_c^2 (1 + 2v_\parallel v_c/c^2 + v_c^2/c^2), \tag{1}$$

where $\gamma_c = (1 - v_c^2/c^2)^{-1/2}$ is the Lorentz factor of the mirror or 'cloud' and v_c is the absolute velocity of each cloud in the laboratory frame. If one assumes that the mirrors move at a non-relativistic speed ($v_c \ll c$) while the particles move relativistically ($v_\parallel \approx c$) the relative energy change upon each collision is

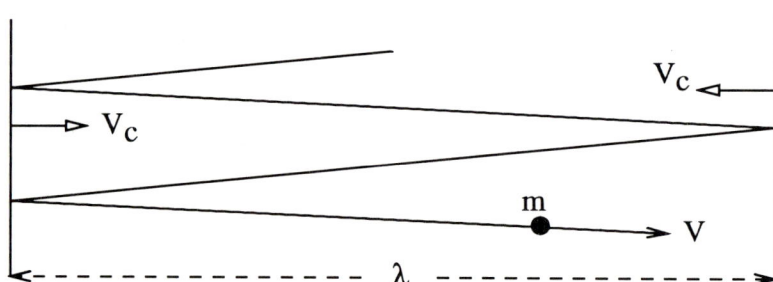

Fig. 1. First-order Fermi acceleration.

$\Delta\epsilon/\epsilon = 2v_c/c$. Further, assuming that the collision time $\Delta t = \lambda/v_\parallel \approx \lambda/c$ is time independent (constant mirror separation λ) one finds that the particle energy increases exponentially in time:

$$\epsilon = \epsilon_0 e^{t/t_{acc}}, \tag{2}$$

where $t_{acc} \approx \lambda(2v_c)^{-1}$ is the acceleration time scale. If one now assumes that the escape time of a particle out of the acceleration region (the actual escape out of the galaxy or its destruction by a nuclear collision) is independent of its energy, $t_{esc}(\epsilon) = t_{esc}$, the probability of finding a particle after time t the differential energy spectrum $f(\epsilon)$ is proportional to (use eq.(2))

$$f(\epsilon)d\epsilon = P(t)dt = e^{-t/t_{esc}}dt/t_{esc} = \frac{d\epsilon}{\epsilon^{1+t_{acc}/t_{esc}}}\frac{t_{acc}}{t_{esc}}\epsilon_0^{t_{acc}/t_{esc}}, \tag{3}$$

and therefore has a *power-law* dependence as is observed for cosmic rays, $f(\epsilon) \propto \epsilon^{-2.7}$ (see Table 1).

It was assumed that the approaching mirrors are approaching magnetized interstellar clouds. A serious shortcoming of the model is that it does not explain why the ratio of acceleration time to escape time should be of order unity, as is required to explain the observed power-law index $2 - 3$ in the cosmic ray spectrum. As we will see below the great success of diffusive shock wave acceleration (Section 3.1) is that, in its most simple form, this mechanism is first-order Fermi acceleration with a *fixed power-law index*, of value 2, independent of local conditions as long as the shock is strong and moves in a non-relativistic hydrogen plasma. In such a shock the system of approaching mirrors is formed by the relative approach of the bulk of the unshocked and of the shocked gas on both sides of the shock.

1.3 Second-Order Fermi Acceleration

In interstellar space magnetized clouds have a random motion and therefore, some 'mirrors' are approaching while others are increasing their separation (Fig. 2). Yet, on average an energetic particle gains energy, both by a relativistic effect (the second-order term $\propto (v_c/c)^2$ in eq. (1) does not vanish) and by the excess of head-on collisions over head-tail collisions with a factor

$$\frac{v_\parallel + v_c}{v_\parallel - v_c}.$$

Using eq. (1) one finds on average per collison

$$\Delta\gamma = 4\gamma\gamma_c^2 v_c^2/c^2 \approx 4\gamma v_c^2/c^2, \tag{4}$$

where the non-relativistic effect alone would lead to a factor 2 instead of 4. Let λ be the average distance between the clouds, so that the collision time is $\Delta t = \lambda/v_\parallel$, then the acceleration rate is given by

$$\frac{d\epsilon}{dt} = \frac{4v_c^2 v_\parallel}{c^2\lambda}\epsilon. \tag{5}$$

As the acceleration rate is proportional to the square of the random velocity v_c this process is called *second-order Fermi* acceleration.

If the particle starts off at a relativistic speed ($v_\parallel \approx c$) the acceleration is again exponential in time and leads, under similar conditions as in the previous section, to a power-law, now however with an increased acceleration time $t_{acc} = \lambda c(4v_c^2)^{-1} = t_{acc}^{(1)}c(2v_c)^{-1}$.

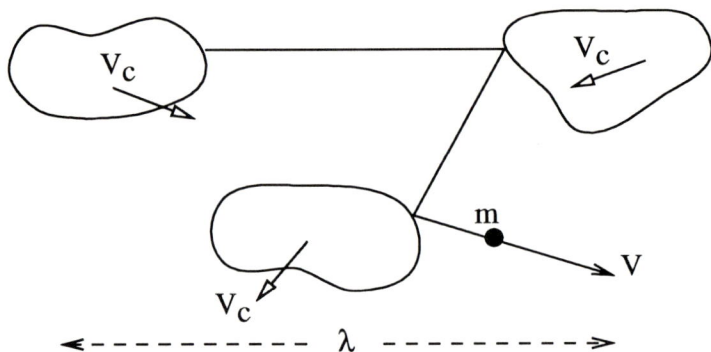

Fig. 2. Second-order Fermi acceleration.

Rewriting eq. (5) in terms of the kinetic energy one finds that the total time required for a particle to be accelerated from a (non-relativistic) speed v to a Lorentz factor γ is

$$t_{acc}(v \to \gamma) \approx \frac{c\lambda}{4v_c^2}(1 + \ell n\gamma). \tag{6}$$

Most of the time is therefore spent in accelerating a particle out of the non-relativistic regime.

Note that second-order Fermi acceleration is nothing else than the tendency of a collection of colliding particles to reach energy equipartition: the kinetic energy of a cloud of mass 10^4 $M_\odot$ and a speed of 10 km/s is $6.2 \cdot 10^{60}$ eV, from which follows a formal temperature of the 'cloud gas' of 10^{64} K. Although this seems promising enough there is a severe injection problem because of the large cloud separation. Due to Coulomb losses electrons are accelerated only if they are 'injected' at energy larger than

$$\epsilon_{inj} \approx 30 \left(\frac{\lambda}{1\,\mathrm{pc}} \frac{n}{10^2\,\mathrm{m}^{-3}}\right)^{1/2} \left(\frac{v}{100\mathrm{km\,s}^{-1}}\right)^{-1} \mathrm{keV}. \tag{7}$$

1.4 Pitch Angle Scattering

Interstellar space is permeated by a magnetic field that guides the majority of cosmic rays: the cyclotron radius of a particle with perpendicular momentum $p_\perp = \gamma m_o v_\perp$ and charge $q = ze$ ($p_\perp/q$ is called the particle *rigidity*) is

$$r_c = \frac{p_\perp}{qB} = \frac{108}{z} \frac{\epsilon}{10^{20}\mathrm{eV}} \frac{10^{-10}\mathrm{T}}{B} \mathrm{kpc}, \tag{8}$$

and is comparable to galactic dimensions only for the largest energies per particle observed (10^{20} eV). Most particles satisfy both $r_c \ll L_B$ and $\omega_c T_B \gg 1$ (L_B is the characteristic magnetic scale length and T_B the period of its variation, $\omega_c = qB(\gamma m)^{-1} = 1.76 \cdot 10^{11} Z B(T) \gamma^{-1}(m_e/m_o)$ c/s is the cyclotron frequency). They are therefore magnetized and their guiding centers drift with a speed (Alfvén 1981)

$$v_D = \frac{\mathbf{F} \times \mathbf{B}}{qB^2} \qquad (9)$$

under a force $\mathbf{F}$ while preserving their first adiabatic moment,

$$\frac{p_\perp^2}{B} \text{ is invariant.} \qquad (10)$$

Therefore the picture of isolated magnetized clouds on which charged particles reflect is inaccurate. Fermi proposed two realizations of the 'clouds': magnetic bottles connecting dense interstellar clouds (Fig. 3a) and magnetic loops with a relative motion between vertex and footpoints (Fig. 3b). In the first case, because of eq. (10), particles only gain energy when they are reflected, which occurs if their pitch angle θ with respect to the magnetic field is large enough: $sin^2\theta > B_{min}/B_{max}$. However, as the energy increase is in the direction along the field the particle pitch angle decreases as its energy increases until the particle escapes from the bottle and the acceleration process stops. Pitch angle *scattering*

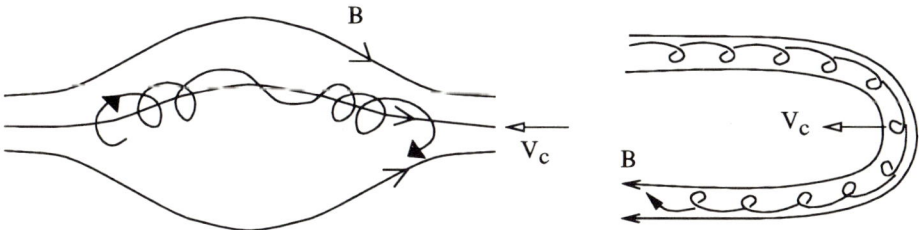

Fig. 3. Two realizations of Fermi acceleration in interstellar space: on the left the mirror points of magnetic 'bottles' are in relative motion; on the right flux tubes have relative motions in transverse directions.

is therefore required in the first case for prolonged acceleration, on a time scale comparable to the acceleration time scale $t_{scat} \le t_{acc} \approx c\lambda(4v_c^2)^{-1}$. Such scattering can occur on magnetic fluctuations with wave numbers $k \ge r_c^{-1} = \omega_c/v$ or frequencies $\omega \ge \omega_c$ when the quantity $p_\perp^2/B$ is not an adiabatic invariant. In the second case (Fig. 3b) pitch angle scattering is not required.

A lower limit to the rate of pitch angle scattering is set by Coulomb scattering (see next section). Under astrophysical (dilute) conditions, however, scattering by external field fluctuations or by self-excited plasma turbulence (see Section 3.1) is often more important.

1.5 Magnetic Pumping

Consider a homogeneous magnetoplasma in a spatially uniform and slowly oscillating magnetic field and a number of sufficiently fast test particles (momentum

p, speed v) which can be considered collisionless:

$$\omega_c^{-1}(p) \equiv \frac{\gamma m_o}{e\,B} \ll t_d(v). \tag{11}$$

As the field changes on a time scale long compared to the cyclotron period of the test particles the first adiabatic invariant $p_\perp^2/B$ (eq. (10)) is conserved and, averaged over several cycles, the test particle energy does not change. Now, however, 'add' some scattering centers so that the fast particles tend to isotropize. Then the average fast particle energy does increase, essentially because the particle has extra degrees of freedom (here one extra spatial dimension) which remains neutral with respect to changes in the field. A simple estimate of the acceleration efficiency is obtained by alternatingly applying a discrete field change $B \to B + \Delta B$, subsequent complete isotropization (so that $< p_\perp^2 >= 2 < p_\parallel^2 >$), a reverse field change, isotropization etc. The result is

$$\frac{\Delta p^2}{p^2} = \frac{2}{9}\left(\frac{\Delta B}{B}\right)^2\left(1 + \frac{\Delta B}{B}\right)^{-1}. \tag{12}$$

This is the process of magnetic pumping or betatron acceleration (Swann 1933; Schlüter 1957; Parker 1958; Melrose 1969, 1983).

(For the general case of M 'active' dimensions on a total of N internal degrees of freedom one finds $M/N(1-M/N)$ instead of 2/9 for the coefficient in eq. (12).) The acceleration is second-order in the (small) field amplitude. As is expected

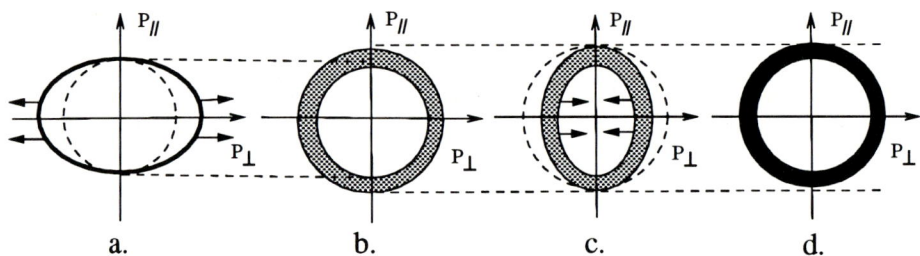

Fig. 4. Sketch of magnetic pumping in momentum space: alternatingly apply a slow field compression (a), relaxation towards isotropy (b), field rarefication (c), isotropization (d), etc.

from these simple arguments, Monte Carlo simulations (Fig 4, Fletcher 1995) show that for varying ratios of t_B/t_{scat} the acceleration maximizes for $t_B/t_{scatt} \approx 1$ when the coefficient becomes 0.05 (four times less than our simplistic estimate).

Magnetic pumping has been proposed for particle acceleration in strong MHD turbulence in the solar corona (Melrose 1983) and for oscillating magnetospheres of rapidly spinning compact objects (AE Aqr, Kuijpers *et al.* 1995).

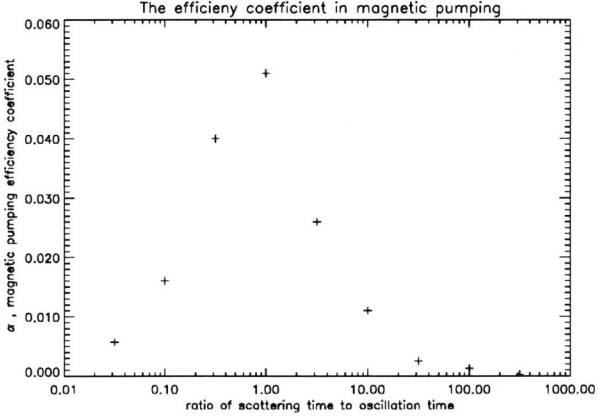

Fig. 5. Dependence of the acceleration efficiency of magnetic pumping as a function of t_B/t_D (Fletcher 1995).

1.6 Physical Similarities

A number of similarities exist between these acceleration processes. Both in Fermi acceleration and in magnetic pumping two effects are at play on different spatial, respectively temporal, scales: differential background motion on a relatively large (compared to the cyclotron radius) or long (compared to the cyclotron period) scale, causing the particle energy to adjust adiabatically, and small-scale perturbations (comparable to either cyclotron radius or period) scattering the fast particles to maintain near-isotropy. Stochastic magnetic pumping (Hall and Sturrock 1967) therefore is the temporal analogue of Fermi acceleration. Further acceleration by weak compressional MHD turbulence (Kulsrud and Ferrari 1971), which occurs when the resonance condition $\omega - k_\parallel v_\parallel = 0$ is satisfied, can be considered as a form of Fermi acceleration, as is also true for transit time damping (Fisk 1976, Achterberg 1981a).

Fermi acceleration reappears in a disguised form in acceleration by strong Alfvénic turbulence where finite-amplitude effects cause particle mirroring and the appearance of a 'Landau' resonance $\omega - k_\parallel v_\parallel = 0$.

1.7 Universal Spectrum

It has been noted by Ginzburg and Syrovatskii (1964) that the cosmic ray energy density in the galactic disk near the sun is comparable to that of the galactic magnetic field and to the characteristic thermal gas energy density 10^6 eV m^{-3}. They proposed a thermodynamic argument for the observed power-law of cosmic rays:

Suppose that inside local regions of acceleration equipartition exists between the total energies in fast particles W_{cr}, turbulent motions W_{turb} and magnetic fields W_b:

$$W_{cr} = W_{turb} = W_b = \frac{1}{3}W. \tag{13}$$

Further assume that energy leaves a particular acceleration site only in the form of 'leakage' of fast particles on a sufficiently large time scale (so that the equilibrium eq. (13) is maintained). If the total number of fast particles in the source N_0 can at any time be characterized by one energy ϵ the leakage is described by:

$$d(3\,N_0\,\epsilon) = \epsilon\,d\,N_0$$

and it follows that $N_0 \propto \epsilon^{-1.5}$. For the differential energy spectrum of particles leaving the source one finds

$$N(\epsilon)d\epsilon = -dN_0 = \kappa\,\epsilon^{-2.5}\,d\epsilon. \tag{14}$$

Note that this result hinges on the assumed exlusive dissipation of turbulent and magnetic energy into cosmic ray energy. For an assumed equilibrium $W_{cr} = (W - W_{cr})\delta$ instead of eq. (14) the spectrum becomes $N(\epsilon)\alpha\,\epsilon^{-(2+\delta)}$. Nowadays the explanation of a universal spectrum – if it exists at all – is thought to be given by shock wave acceleration (Section 3).

1.8 Gains Versus Losses

A particular acceleration process is only effective when the energy gain of the particle exceeds the losses. The energy dependence of a specific loss mechanism determines if it sets a lower, *injection*, limit for particle acceleration or instead a *cut-off* in the particle spectrum at high energies. The dependences of some important loss mechanisms (mainly for electrons) on total particle energy ϵ are compared to the gain by Fermi acceleration (rate $(t_{acc}^F)^{-1}$ independent of γ) in Fig. 6. As can be seen Coulomb losses set a lower injection limit (eq. (7)) for a specific Fermi-type process while radiative losses determine a cut-off at high energies.

Coulomb Losses At the lowest energies suprathermal particles with speed v loose energy by collisions with particles of the thermal background. As the energy gain in a Fermi process is in one spatial direction only it is the *deflection* time ($\propto v^{-3}$) rather than the energy loss time ($\propto v^{-5}$) which is important. For an isothermal plasma with cosmic abundances the deflection rate of both suprathermal electrons and, respectively, protons is mainly determined by the thermal electrons, with density n_e (Spitzer 1962):

$$t_d^{-1} = 3.2 \cdot 10^7\,n_e v^{-3}\,\frac{m_e}{m_{e,p}}\,\frac{ln\Lambda}{20}\,\text{Hz}, \tag{15}$$

where $ln\Lambda$ is the Coulomb logarithm, and $m_{e,p} = m_e$, and respectively $= m_p$, for the electron (respectively proton) deflection rate.

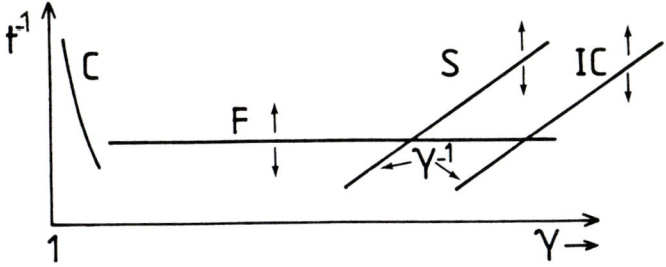

Fig. 6. Energy dependence of a number of loss processes: C for Coulomb losses at suprathermal velocities (electrons and ions); S for synchrotron losses; IC for inverse Compton losses; S and IC apply to electrons only. Also shown is the energy dependence of Fermi acceleration: F.

For electrons with relativistic energies radiative losses of electrons due to bremsstrahlung dominate over collisonal losses with a factor of roughly $\gamma\alpha$ where $\alpha = e^2(4\pi\epsilon_o\hbar c)^{-1} = 1/137$ is the fine structure constant (Rybicki and Lightman 1979) and γ is the Lorentz factor of the electron.

Radiative Losses The power radiated by a relativistic particle of charge q, rest-mass m_o, momentum $\mathbf{p}$ and energy ϵ is

$$P = \frac{1}{6\pi\epsilon_o}\frac{\gamma^2 q^2}{m^2 c^3}\left\{\left|\frac{d\mathbf{p}}{dt}\right|^2 - \frac{1}{c^2}\left(\frac{d\epsilon}{dt}\right)^2\right\}. \tag{16}$$

Because of the mass dependence radiative losses by electrons dominate except at extremely high energies where only nuclei are observed. Also it follows from eq. (16) that incoherent (single particle) radiation processes are most efficient if $d\mathbf{p}/dt \perp \mathbf{p}$. Indeed this is the case for synchrotron radiation, inverse Compton emission and bremsstrahlung.

The energy dependences of the loss rates are given by (Blumenthal and Gould 1970)

Bremsstrahlung

$$t_{ff}^{-1} \approx \frac{16}{3}n_i z^2\left(\frac{e^2}{4\pi\epsilon_o\hbar}\right)\left(\frac{e^2}{4\pi\epsilon_o m_e c^2}\right)^2 \ell n\gamma, \tag{17}$$

where n_i is the density of ions with charge z.

Synchrotron radiation

$$t_S^{-1} = \frac{4\gamma\beta^2\sigma_T}{3m_e c}\frac{B^2}{2\mu_o},$$ (18)

where

$$\sigma_T = e^4(6\pi\epsilon_o^2 m_e^2 c^4)^{-1} = 6.65 \cdot 10^{-29}\mathrm{m}^2$$ (19)

is the Thomson cross section and $\beta = v/c$. The spectral break sometimes observed in the radio spectra of old supernova remnants and Active Galactic Nuclei is commonly interpreted as synchrotron aging of energetic electrons. A different application is the Crab nebula: X-rays from the nebula are probably synchrotron radiation; given the nebular magnetic field strength the synchrotron loss time turns out to be less than the light travel time from the pulsar, forming strong evidence for in situ acceleration of the emitting particles in the nebula.

Inverse Compton radiation

$$t_{IC}^{-1} = \frac{4\gamma\beta^2\sigma_T U}{m_e c}\ \mathrm{for}\ \gamma h\nu \ll m_e c^2,$$

$$= \frac{1}{\gamma}\ell n\left(\frac{2\gamma h\nu}{m_e c^2}\right)\frac{3}{8}\frac{\sigma_T cU}{\gamma^2 h\nu}\ \mathrm{for}\ \gamma h\nu \gg m_e c^2,$$ (20)

where U is the ambient (isotropic) photon energy density. This effect combined with X-ray observations sets limits to the energetic particle content of Active Galactic Nuclei.

Inelastic Photoproduction Protons of sufficiently high energies loose their energy by inverse Compton-like interactions on the cosmic 2.726 K radiation and on ambient protons. For protons above 10^{20} eV secondary pions, protons and neutrons are produced and subsequent decay of pions and muons into gamma rays and neutrinos. The mean free path of such high-energy protons is 6 Mpc and if they are coming from distances larger than 100 Mpc a spectral cut-off below 100 EeV (1 EeV = 10^{18} eV) would exist, in the 'ankle' of the cosmic ray spectrum (Greisen 1966; Zatsepin and Kuzmin 1966).

Adiabatic Losses In an expanding medium of size R energetic particles loose their energy adiabatically, $\epsilon \propto R^{-1}$ for relativistic particles. Such an effect occurs for instance in ejected expanding plasmoids (expanding cloud model or synchrotron bubble, van der Laan 1966; Ball 1994). It forms evidence for in situ particle acceleration in extragalactic jets far out from the parent galaxy.

1.9 Fokker-Planck Equation

During stochastic acceleration a particle population undergoes many random interactions, which are often small in the sense that each interaction produces only a small perturbation of the unperturbed particle orbit. This is the *quasilinear* assumption. Then the evolution of the single-particle particle distribution function can be described by a Fokker-Planck equation (van Kampen 1981; Sturrock 1994). Here I will give a heuristic derivation of this equation.

If the members of an arbitrary ensemble are conserved but continuously change in some of their properties (e.g. position, momentum) a differential conservation equation can be written down which describes the temporal evolution of the distribution u of the members in the ensemble over these continuous properties

$$\frac{\partial u}{\partial t} + \boldsymbol{\nabla} \cdot \mathbf{S} = 0, \tag{21}$$

where $\mathbf{S}$ is the flux of the members in the space corresponding to the changing properties and $\boldsymbol{\nabla} \cdot$ the corresponding divergence. In the case of particle acceleration we assume that the particles are conserved so that we can write down a conservation equation for the single-particle distribution function $f(\mathbf{r}, \mathbf{p}, t)$ in 6-D phase space $\{\mathbf{r}, \mathbf{p}\}$

$$0 = \frac{\partial f}{\partial t} + \boldsymbol{\nabla}_{\mathbf{r}, \mathbf{p}} \cdot \mathbf{S} =$$

$$= \frac{\partial f}{\partial t} + \boldsymbol{\nabla}_{\mathbf{r}} \cdot (f\mathbf{v}) + \boldsymbol{\nabla}_{\mathbf{p}} \cdot (f\mathbf{F}) + \boldsymbol{\nabla}_{\mathbf{r}, \mathbf{p}} \cdot (\mathbf{S}_{\mathbf{r}, \mathbf{p}}^{stoch}), \tag{22}$$

where $\mathbf{F}$ is the force on an individual particle, we have split up the total flux $\mathbf{S}$ in 6-D phase space in a 'regular' flux $\{f\mathbf{v}, f\mathbf{F}\}$ and a 'stochastic' average contribution $\mathbf{S}_{\mathbf{r}, \mathbf{p}}^{stoch}$ due to the cumulative effect of many small jumps $\{\Delta \mathbf{r}, \Delta \mathbf{p}\}$ each of duration Δt. Note that there are seven *independent* variables $\{\mathbf{r}, \mathbf{p}, t\}$. Let us suppose that the stochastic effects cause a particle to make a random walk in phase space. Then a separation $\delta \mathbf{r}$ (respectively $\delta \mathbf{p}$) will be crossed on average after a time $\delta t = (\delta \mathbf{r}/\Delta \mathbf{r})^2 \Delta t$ (respectively $\delta t = (\delta \mathbf{p}/\Delta \mathbf{p})^2 \Delta t$) by an effective number of particles $-(\partial f/\partial \mathbf{r}) \cdot \delta \mathbf{r}/2$ (respectively $-(\partial f/\partial \mathbf{p}) \cdot \delta \mathbf{p}/2$). The stochastic flux is therefore

$$\mathbf{S}_{\mathbf{r}}^{stoch} = -\frac{\partial f}{\partial \mathbf{r}} \cdot \frac{\delta \mathbf{r}}{2} \frac{\delta \mathbf{r}}{\delta t} = -\frac{< \Delta \mathbf{r} \Delta \mathbf{r} >}{2 \Delta t} \cdot \frac{\partial f}{\partial \mathbf{r}},$$

(respectively $\mathbf{S}_{\mathbf{p}}^{stoch} = -\frac{\Delta \mathbf{p} \Delta \mathbf{p}}{2 \Delta t} \cdot \frac{\partial f}{\partial \mathbf{p}}$). Finally substituting the Lorentz force $\mathbf{F} = q(\mathbf{E} + \mathbf{v} \times \mathbf{B})$ with $\mathbf{E}, \mathbf{B}$ the averaged electromagnetic field, we can write eq. (22) as a Fokker-Planck equation

$$\frac{\partial f}{\partial t} + (\mathbf{v} \cdot \boldsymbol{\nabla}) f + q(\mathbf{E} + \mathbf{v} \times \mathbf{B}) \cdot \frac{\partial f}{\partial \mathbf{p}}$$

$$= \boldsymbol{\nabla} \cdot \left(\frac{< \Delta \mathbf{r} \Delta \mathbf{r} >}{2 \Delta t} \cdot \boldsymbol{\nabla} f \right) + \frac{\partial}{\partial \mathbf{p}} \cdot \left(\frac{< \Delta \mathbf{p} \Delta \mathbf{p} >}{2 \Delta t} \cdot \frac{\partial f}{\partial \mathbf{p}} \right). \tag{23}$$

The left-hand side alone would form the usual Liouville equation in the absence of stochastic interactions. The right-hand side describes diffusion in space and, respectively, momentum space. The diffusion tensors

$$\overset{\leftrightarrow}{\mathbf{D}}_{\mathbf{r}} \equiv \frac{< \Delta \mathbf{r} \Delta \mathbf{r} >}{2 \Delta t}, \tag{24}$$

$$\overset{\leftrightarrow}{\mathbf{D}}_{\mathbf{p}} \equiv \frac{< \Delta \mathbf{p} \Delta \mathbf{p} >}{2 \Delta t} \tag{25}$$

have to be calculated for the specific process underlying the stochastic changes. In general they are (*inversely*) related as fast diffusion in momentum space is related to scattering and therefore often leads to pitch angle scattering and slow diffusive propagation in real space.

The right-hand part of eq. (23) contains a systematic and a diffusive part as can be seen by writing explicitly

$$\nabla\nabla : \left(\overset{\leftrightarrow}{\mathbf{D_r}} f\right) - \nabla \cdot \left\{\left(\nabla \cdot \overset{\leftrightarrow}{\mathbf{D_r}}\right) f\right\},$$

and similarly for the momentum term. The last term describes *dynamical friction* whereas the first term is *proper diffusion*. It therefore follows that stochastic particle acceleration – the dynamical friction term causing $< \mathbf{r} >$ (respectively $< \mathbf{p} >$) to change – is always associated with broadening of the distribution, i.e.increase of $< |\mathbf{r} - < \mathbf{r} > |^2 >$ (respectively $< |\mathbf{p} - < \mathbf{p} > |^2 >$) – the diffusion term (see Fig. 7).

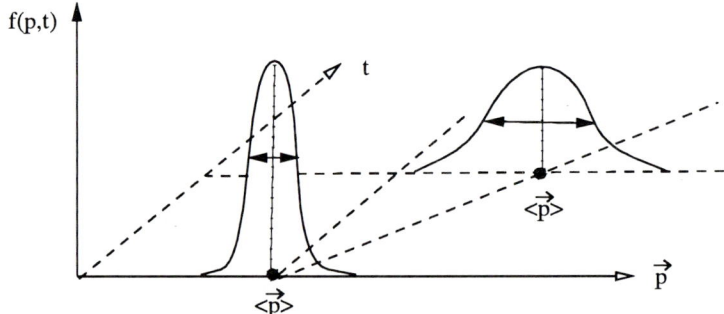

Fig. 7. Stochastic acceleration of a particle distribution is characterized by an increase in the width of the distribution in momentum space simultaneously with an increase of the average particle energy.

For numerical purpose the Fokker-Planck equation can be replaced by a set of stochastic *ordinary* differential equations, which are easier to implement (Krülls and Achterberg 1994). For numerical methods see also Earl (1994), Giacalone and Jokipii (1994) and Ruffolo (1995).

1.10 Diffusion-Convection Equation

Often it is assumed that the distribution function of the energetic particles is nearly isotropic in a local rest frame defined by the (as yet unspecified) scattering centres (speed $\mathbf{v}_S(\mathbf{r})$ which may differ from the average plasma flow velocity $\mathbf{v}_P(\mathbf{r})$). A small anisotropy implies a small spatial inhomogeneity. The Fokker-Planck equation can then be reduced to the *diffusion-convection* equation ('confusion equation') (Jones 1990; Jones & Ellison 1991; Schlickeiser 1994).

It describes changes in the particle distribution both due to advection and due to scattering. The advection terms consist of free streaming in real space caused by inhomogeneity and of 'adiabatic' acceleration (free streaming in momentum space) caused by smooth 'large-scale' electromagnetic fields. The terms due to scattering result in diffusion in real space as well as angular diffusion in momentum space. The equation describes acceleration processes where a particle does not change its energy in the rest frame of the scattering centres in a *single* scattering. Examples are diffusive acceleration by (parallel as well as oblique) shocks and *drift* acceleration of magnetized particles by gradient and curvature drifts (both in shocks and in other inhomogeneous conditions).

2 Plasma Turbulence and Acceleration

The magnetized atmospheres of stars and accretion disks often deviate considerably from local thermodynamic equilibrium. This is because in these dilute and hot *coronae* the collisonal relaxation times are long compared to the characteristic frequencies of plasma waves. For instance the *deflection time* of a thermal electron from encounters with ions in a plasma of temperature T is of order (Spitzer 1965)

$$t_d \simeq \frac{2\pi}{\ell n \Lambda} \frac{N_D}{\omega_{pe}} = 1.8 \cdot 10^3 \frac{T^{3/2}}{n_e} \frac{20}{\ell n \; \Lambda} \, \text{s}, \tag{26}$$

where the *Debye number* is the number of electrons in a sphere of radius λ_D

$$N_D \equiv n_e \lambda_D^3 \approx 3.29 \cdot 10^5 \frac{T^{3/2}}{n_e^{1/2}}, \tag{27}$$

the Coulomb logarithm is defined by $\ell n \; \Lambda \equiv \ell n \; (12\pi n \lambda_D^3)$, n_e is the electron density (m^{-3}), T is the temperature, ω_{pe} the electron plasma frequency defined by

$$\omega_{pe} \equiv \left(\frac{n_e e^2}{m \epsilon_o} \right)^{1/2} = 56.4 \, n^{1/2} \text{c/s}, \tag{28}$$

and the *Debye length* or electrostatic screening distance is given by

$$\lambda_D \equiv \left(\frac{\epsilon_o K \; T}{\sum\limits_{\alpha} n_\alpha (z_\alpha e)^2} \right)^{1/2} \approx 69 \left(\frac{T}{n_e} \right)^{1/2} \text{m}, \tag{29}$$

where the sum is over the particle species α, e is the electron charge (absolute value), m_e is the electron rest mass and K is Boltzmann's constant. Binary collisions become less and less important if the Debye number is large. The Debye number is the number of electrons with which a given particle interacts simultaneously.

As a result waves, once excited, are damped by collisions only after many oscillations. As these plasmas are the sites of magnetic flares or other strong MHD disturbances, they are often in a state known as plasma turbulence (Tsytovich

1970), where many wave modes are present at suprathermal levels. The energy in these waves can be transferred to selected, *resonant*, particles and thereby lead to acceleration. Before describing acceleration by plasma waves (Section 2.2 and further) I will first give an introduction to plasma turbulence.

2.1 Plasma Turbulence

In contrast with hydrodynamical turbulence plasma turbulence is often weak in the sense that growth rates of waves (imaginary part of the wave frequency or of the product of group speed and wave number) are much less than the real parts of the frequencies: $|\gamma| \ll |\Re \omega|$, whereas for hydrodynamical turbulence $|\gamma| \approx \omega$, where $\omega^{-1} \approx \lambda/v_\lambda$ is the eddy turnover time. This makes plasma turbulence a relatively well understood subject.

Waves The plasma is considered to be an ideal gas, *i.e.* a collection of point particles that undergo small and brief interactions or 'collisions' when the characteristic interaction energy is much less than the average particle energy

$$\frac{e^2 n_e^{1/3}}{4\pi\epsilon_o} \ll KT \text{ or } \ll E_F, \tag{30}$$

where we have used $n_e^{-1/3}$ as an estimate for the typical interparticle distance, and $E_F = \hbar^2 n_e^{2/3}/m_e$ is the Fermi energy. We shall neglect quantum effects (degeneracy, recoil) and assume that

$$K\,T \gg (E_F, \hbar\omega_{pe}, \hbar\omega_{ce}), \tag{31}$$

with

$$\omega_{ce} \equiv \frac{eB}{m_e} \approx 1.76 \cdot 10^{11} \text{ c/s}, \tag{32}$$

the electron Larmor frequency (the gyration frequency of a non-relativistic electron in a magnetic field of strength B (units Tesla). These approximations are well satisfied in most cosmic plasmas. Exceptions are the degenerate interiors of white dwarfs and neutron stars, and the strongly magnetized atmospheres of neutron stars (Galeev and Sudan 1983). Further we consider completely ionized hydrogen plasmas ($n_e = n_i$ on average).

As the Debye number (Eq. (27) is large in stellar and disk atmospheres we can use the small parameter

$$\alpha_0 \equiv \frac{1}{N_D} \ll 1. \tag{33}$$

As the Debye number is much larger than unity *collective* effects are much more important than binary interactions or 'collisions' (eq. (26)). In such a *collisionless* plasma the equation of motion is the *Vlasov equation*:

$$\frac{Df(\mathbf{r}, \mathbf{p}, t)}{Dt} = 0,$$

$$\frac{\partial f}{\partial t} + \mathbf{v} \cdot \frac{\partial f}{\partial \mathbf{r}} + q(\mathbf{E} + \mathbf{v} \times \mathbf{B}) \cdot \frac{\partial f}{\partial \mathbf{p}} = 0. \tag{34}$$

Here $\frac{D}{Dt} = \frac{\partial}{\partial t} + \mathbf{v} \cdot \frac{\partial}{\partial \mathbf{r}} + \dot{\mathbf{p}} \cdot \frac{\partial}{\partial \mathbf{p}}$ is the comoving derivative in six-dimensional phase space $\{\mathbf{r}, \mathbf{p}\}$, and $\mathbf{E}$ and $\mathbf{B}$ are the 'smoothed' electromagnetic fields.

The Vlasov equation expresses conservation of the particle distribution function in phase space on a *mesoscopic* scale (large compared to the average interparticle distance). It is obtained from the exact microscopic Klimontovich equation for point-like particles by splitting the electromagnetic fields into a smooth part averaged over a distance large compared to the interparticle distance and the remaining fast fluctuating part caused by the discrete nature of the charges (Ichimaru 1973). Averaging over this mesoscopic length scale (larger than the interparticle distance but still small compared to the Debye length) then leads to the Boltzmann equation, of which the left-hand side is the Vlasov equation and the right-hand side contains the effect of fluctuations or collisions. As the right-hand side is of order $\frac{f}{t_d} \approx \frac{\omega_{pe} f}{N_D}$ it can therefore be neglected if $1 \ll \omega t_d$ or if $1 \ll k v t_d$ (wavelength much less than the electron mean free path), where ω is the characteristic frequency and k the characteristic wave number of the phenomenon under investigation.

Vacuum electromagnetic waves are modified by a plasma by the reaction of the charged particles to the wave fields, in particular at frequencies near the electron plasma frequency or lower. But also new wave modes appear such as electrostatic waves created by charge fluctuations in the plasma. To find the various kinds of *electromagnetic* waves or *eigen modes* of infinitesimally small amplitude which are sustained by a collisionless plasma, one must know the *response* of the plasma to electromagnetic perturbations. This response consists of induced electric current and charge densities. It can be obtained by complementing the Vlasov equation (34) with Maxwell's equations and *linearizing* the equations in the perturbing electromagnetic wave amplitude around the unperturbed equilibrium. Maxwell's equations in a plasma can be written in terms of the macroscopic (smoothed) electric field $\mathbf{E}$ and magnetic induction $\mathbf{B}$ as

$$\nabla \times \mathbf{E}(\mathbf{r}, t) = -\frac{\partial \mathbf{B}(\mathbf{r}, t)}{\partial t}, \tag{35}$$

$$\frac{1}{\mu_o} \nabla \times \mathbf{B}(\mathbf{r}, t) = \mathbf{j}(\mathbf{r}, t) + \epsilon_o \frac{\partial \mathbf{E}(\mathbf{r}, t)}{\partial t}, \tag{36}$$

$$\epsilon_o \nabla \cdot \mathbf{E}(\mathbf{r}, t) = \tau(\mathbf{r}, t), \tag{37}$$

$$\nabla \cdot \mathbf{B}(\mathbf{r}, t) = 0. \tag{38}$$

Here τ is the electric charge density and $\mathbf{j}$ the electric current density. Further we have used that in a strongly ionized plasma electric charge and current densities consist nearly entirely of free charges. The polarization of the material and the magnetization can therefore be neglected so that the relative dielectric constant and the relative dielectric permeability can be put equal to unity and $\mathbf{B} = \mu_o \mathbf{H}$ and $\epsilon_o \mathbf{E} = \mathbf{D}$ ($\mathbf{H}$ the magnetic field strength and $\mathbf{D}$ is the electric flux density or displacement).

In a homogeneous and stationary plasma in equilibrium with a vanishing electric field and a uniform magnetic field $\mathbf{B}_0$ we now find plane wave solutions by Fourier transforming eqs. (35 – 38) (we use the convention $A(\mathbf{r}, t) \propto exp(i[\mathbf{k} \cdot \mathbf{r} - \omega t])$ for the amplitudes A):

$$k \times \mathbf{E}(\mathbf{k}, \omega) = \omega \mathbf{B}(\mathbf{k}, \omega), \tag{39}$$

$$\frac{1}{\mu_o} \mathbf{k} \times \mathbf{B}(\mathbf{k}, \omega) = - i \, \mathbf{j}(\mathbf{k}, \omega) - \epsilon_o \omega \mathbf{E}(\mathbf{k}, \omega), \tag{40}$$

$$\epsilon_o \mathbf{k} \cdot \mathbf{E}(\mathbf{k}, \omega) = -i\tau(\mathbf{k}, \omega), \tag{41}$$

$$\mathbf{k} \cdot \mathbf{B}(\mathbf{k}, \omega) = 0. \tag{42}$$

We now substitute $\mathbf{B}$ from eq. (39) into eq. (40) to find an equation for $\mathbf{E}$ and $\mathbf{j}$

$$k \times [\mathbf{k} \times \mathbf{E}(\mathbf{k}, \omega)] + \frac{\omega^2}{c^2} \mathbf{E}(\mathbf{k}, \omega) = -\mu_o i \omega \mathbf{j}(\mathbf{k}, \omega). \tag{43}$$

Note that eq. (39) will be used as a definition of $\mathbf{B}(\mathbf{k}, \omega)$, eq. (41) as a definition of $\tau(\mathbf{k}, \omega)$, and that eq. (42) is not independent but follows from eq. (39), essentially because in Fourier transforming we have lost the (arbitrary) uniform field $\mathbf{B}_0$. To proceed we note that the current density $\mathbf{j}$ in eq. (43) fulfills a dual function: it acts as a source term for the electromagnetic field, which expresses the excitation of electromagnetic fields by a prescribed current distribution; it can, however, also constitute the current induced in the plasma by an eigen mode, or electromagnetic wave, of the system. We therefore split up the current density as

$$\mathbf{j}(\mathbf{r}, t) = \mathbf{j}^{ind}(\mathbf{r}, t) + \mathbf{j}^{ext}(\mathbf{r}, t), \tag{44}$$

where $\mathbf{j}^{ind}$ is the response by the plasma to the wave field and where $\mathbf{j}^{ext}$ is the externally imposed current.

To derive the eigen modes of the plasma we therefore first consider the part $\mathbf{j}^{ind}$. The induced part can be easily obtained for small electromagnetic field perturbations because the response will be linear in the perturbing field and can therefore be written as

$$\mathbf{j}^{ind}(\mathbf{r}, t) = \int_V dV \int_{-\infty}^{t} dt' \boldsymbol{\sigma}(\mathbf{r}, \mathbf{r}'; t, t') \cdot \mathbf{E}(\mathbf{r}', t'). \tag{45}$$

Here $\boldsymbol{\sigma}$ is called the conductivity tensor. If the equilibrium plasma is homogeneous and stationary, the conductivity tensor depends on its arguments only through the combinations $\mathbf{r} - \mathbf{r}'$ and $t - t'$. Fourier transformation in space and time then leads to

$$j_i^{ind}(\mathbf{k}, \omega) = \sigma_{ij}(\mathbf{k}, \omega) E_j(\mathbf{k}, \omega). \tag{46}$$

Note that the conductive properties can be described by a multiplicative operation in the Fourier domain only, not in real space. Substituting eq. (46) into eq. (43) and allowing for an externally prescribed part $\mathbf{j}^{ext}$ one finds the wave equation

$$\Lambda_{ij}(\mathbf{k}, \omega) E_j(\mathbf{k}, \omega) = -\frac{i}{\epsilon_o \omega} j_i^{ext}(\mathbf{k}, \omega), \tag{47}$$

where

$$\Lambda_{ij}(\mathbf{k}, \omega) \equiv \frac{k_i k_j c^2}{\omega^2} - \left(\frac{k^2 c^2}{\omega^2} - 1\right) \delta_{ij} + \frac{i}{\epsilon_o \omega} \sigma_{ij} \mathbf{k}, \omega$$

$$= \frac{k^2 c^2}{\omega^2}(\kappa_i \kappa_j - \delta_{ij}) + \epsilon_{ij}(\mathbf{k}, \omega), \tag{48}$$

with the dielectric tensor $\overset{\leftrightarrow}{\epsilon}$ defined by

$$\epsilon_{ij}(\mathbf{k}, \omega) \equiv \delta_{ij} + \frac{i}{\epsilon_o \omega} \sigma_{ij}(\mathbf{k}, \omega), \tag{49}$$

and $\kappa \equiv \mathbf{k}/|\mathbf{k}|$.

To find the kinds of plane waves sustained by the plasma as specified by its conductivity tensor (or, equivalently, its dielectric tensor), one puts the external current to zero. The solutions of the *homogeneous* wave equation

$$\Lambda_{ij}(\mathbf{k}, \omega) E_j(\mathbf{k}, \omega) = 0 \tag{50}$$

are determined by

$$\Lambda(\mathbf{k}, \omega) \equiv \det\{\Lambda_{ij}(\mathbf{k}, \omega)\} = 0. \tag{51}$$

The resulting algebraic expressions between wave vector $\mathbf{k}$ and wave frequency ω are the so-called *dispersion relations* $\omega^\sigma(\mathbf{k})$ for the various kinds of wave modes σ (André 1985). In general the relations have complex and real parts. For a particular wave mode and a real $\mathbf{k}$ the resulting ω has a real part and a complex part. The latter describes damping ($\Im\omega < 0$) or growth ($\Im\omega > 0$). If the wave is growing, energy is transferred from the plasma into the eigen mode: an *instability* arises as the equilibrium state of the plasma is not stable to a small perturbation of the right kind.

The conductivity tensor can be found from the material properties of the gas. Linearizing and Fourier transforming the equation of motion eq. (34), assuming for simplicity $\mathbf{B}_0 = 0$

$$\mathbf{E}_0 = \mathbf{B}_0 = \mathbf{j}_0 = \tau_0 = 0,$$

$$n_i = n_e = n_0,$$

$$f_e(\mathbf{r}, \mathbf{v}, t) = n_0 f^0(\mathbf{v}) + f^1(\mathbf{r}, \mathbf{v}, t), \tag{52}$$

one finds the first-order perturbation of the electron (charge $q = -e$) distribution function

$$f^1(\mathbf{k}, \omega, \mathbf{v}) = \frac{\frac{n_0 e}{m_e} \frac{\partial f^o}{\partial \mathbf{v}} \cdot \left\{\mathbf{I}\left(1 - \frac{\mathbf{k} \cdot \mathbf{v}}{\omega}\right) + \frac{\mathbf{k}\mathbf{v}}{\omega}\right\} \cdot \mathbf{E}(\mathbf{k}, \omega)}{i(\mathbf{k} \cdot \mathbf{v} - \omega - i\eta)}, \tag{53}$$

where **I** is the identity operator. Substitution of the result into the expression for the linear current density (here assuming that the ions remain fixed)

$$\mathbf{j}(\mathbf{k}, \omega) = \sum_\alpha q_\alpha \int \mathbf{v} f_\alpha(\mathbf{k}, \omega, \mathbf{v}) d^3\mathbf{v} = -e \int d^3\mathbf{v} \mathbf{v} f^1(\mathbf{k}, \omega, \mathbf{v}) \qquad (54)$$

and comparison with eq. (46) then directly gives the Fourier transform $\boldsymbol{\sigma}(\mathbf{k}, \omega)$.

In the integration the poles are handled by using the *Landau prescription* or the *causal condition* $\eta \to 0^+$. This prescription arises from proper use of a *Laplace transformation* instead of a Fourier transformation in time: The temporal function $F(t)$ is transformed as

$$F(s) = \int\limits_0^\infty F(t)e^{-st}dt, \ \Re s > 0, F(t) = \frac{1}{2\pi i} \int\limits_{-i\infty+\sigma}^{i\infty+\sigma} F(s)e^{st}ds, \qquad (55)$$

where in the first transformation $\Re s > 0$ is sufficiently large to ensure existence of the integral in case of a growing disturbance. Note that causality requires that the function cannot have been infinite in the infinite past. As a result the real part σ of the contour integral in the reverse transformation has to lie *to the right* of any poles in $F(s)$. The correspondence with the not entirely correct but 'easier' Fourier transform is found by replacing $\omega \to \omega + i\eta$ with $\Re s = \eta$ and indenting the Fourier integration path along the real axis below any singularities in $F(s)$, which amounts to choosing the analytic continuation.

The simplest plasma without any damping or growth, is that of a non-streaming cold electron plasma with a neutralizing immobile ion background. It is easy to verify that in this case the conductivity tensor is the identity tensor multiplied by

$$\sigma(\mathbf{k}, \omega) = \frac{i}{4\pi} \frac{\omega_{pe}^2}{\omega}. \qquad (56)$$

Instabilities and Resonances Instabilities are classified according to their physical origin: *Configuration instabilities* arise from spatial inhomogeneities (In this case one cannot in general make a Fourier transformation in space but has to solve an ordinary differential equation to find the spatial part of the eigen modes); *Velocity space or micro instabilities* include instabilities in homogeneous plasmas and are caused either by an energy inversion or a velocity anisotropy of the particle distribution functions. Depending on the offset of the resonance $|i\Im\omega + \Re\omega - k_\| v_\| - N\omega_c|$ in comparison to the width $< k\Delta v >$ (Δv is the spread in the particle distribution) one speaks of *reactive or hydrodynamic* (small spread) and *resistive or kinetic* (large thermal spread) instabilities. Usually a particular physical instability occurs under both forms depending on the extent of the deviation of the particle distributions from thermodynamic equilibrium.

In the *kinetic* case for a given wave mode and a given wave vector a clear distinction can be made between two groups of particles: the resonant particles, corresponding to the near-poles in eq. (53), and the other, non-resonant, particles. The reaction of both groups to the wave is markedly different.

Most of the particles interact *adiabatically* with the electromagnetic wave field. Energy is transferred from the particles to the fields in a periodic, reversible manner. On average there is no energy transfer between these particles and the electromagnetic wave. This is the bulk of the plasma which sustains the particular wave mode. In fact the wave consists partially of an electromagnetic field and partially of kinetic energy of the oscillating particles.

For some particles however the projection of their velocity in the direction of the wave vector equals the phase speed of the wave. These particles feel a constant electromagnetic field and are accelerated (or retarded). They contribute to damping (or growth) of the particular wave. The general form of these wave–particle *resonances* can be found as follows: The force on a particle of speed $\mathbf{v}$ and orbit $\mathbf{r}(t)$ exerted by a plane wave $(\mathbf{k}, \omega)$ is proportional to $exp[i(\mathbf{k} \cdot \mathbf{r}(t) - \omega t)]$. In a magnetic field the particle orbit is given by

$$\mathbf{r}(t) = \left(v_{\|}t, \frac{v_{\perp}}{\omega_{cq}}sin(\omega_{cq}t), \frac{v_{\perp}}{\omega_{cq}}cos(\omega_{cq}t) \right), \tag{57}$$

where

$$\omega_{cq} = \frac{qB}{\gamma m_o} \tag{58}$$

is the cyclotron frequency of a particle with charge q, rest mass m_o and Lorentz factor γ. Calculating the dependence of the force along the particle orbit and making use of the Bessel identities then results in the following spatio-temporal dependence of the force

$$e^{i(k_{\|}v_{\|}-\omega)t+i\frac{k_{\perp}v_{\perp}}{\omega_{cq}}\sin(\omega_{cq}t)} = \sum_{n=-\infty}^{+\infty} (-1)^n J_n \left(\frac{k_{\perp}v_{\perp}}{\omega_{cq}} \right) e^{-i(\omega-k_{\|}v_{\|}+N\omega_{cq})t}.$$

It follows that a resonance between a particle and a wave exists provided

$$\omega - k_{\|}v_{\|} - N\omega_{cq} = 0, \tag{59}$$

for an integer value of N. In fact the resonances appear as *poles* in the induced current similar to the unmagnetized case (eq. (53) and eq. (54)). These resonant particles are responsible for damping (stable plasma) or growth of the wave (plasma instability), depending or their distribution in phase space. For $N > 0$ the resonance is called normal cyclotron resonance, as in this case the direction of gyration of the particle in the frame moving along the ambient magnetic field with the speed $\omega/k_{\|}$ is in the same sense as in the lab frame. For $N < 0$ one has an anomalous cyclotron resonance as the sense of gyration of the particle in the wave frame at constant wave phase along the magnetic field is reversed with respect to the sense in the lab frame. For $N = 0$ one has a Čerenkov or Landau resonance as a special case of the resonance condition

$$\omega - \mathbf{k} \cdot \mathbf{v} = 0 \tag{60}$$

for unmagnetized ($r_c k_{\perp} \ll 1$, eq. (8)) particles.

Most of the kinetic instabilities are particular cases of an *inversion* in momentum space (Fig. 8), such as a beam, a bump-in-tail, a relative drift, an electric current, or an *anisotropy* in momentum space, such as a loss-cone (one- or two-sided) or a ring distribution. Such anisotropic distributions are present

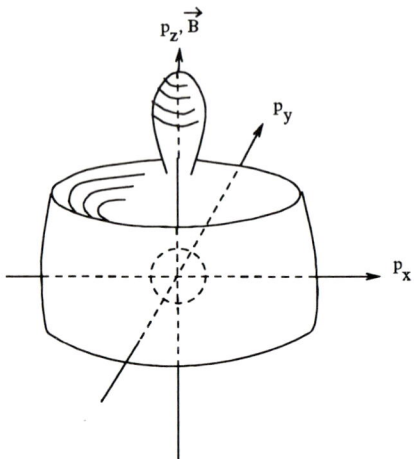

Fig. 8. Examples of unstable particle distributions in momentum space indicated by a representative level surface: a beam parallel to the magnetic field, and an anisotropy across the field. Also drawn is a characteristic surface of the 'thermal' background population near the origin.

in magnetic traps by mirroring or in collisionless shocks. In a magnetic reconnection event such as a flare, many instabilities can occur as sketched in Fig. 9.

The growth rate in such a case is in fact determined by the integrated (slopes) of the distribution function across the resonant surfaces determined by eq. (59). The resonant surface is in general an ellipsoid (Fig. 10) in momentum (velocity) space because of the dependence of the cyclotron frequency eq. (58) on particle energy

$$\left(\frac{p_\perp}{mc}\right)^2 + \left(\frac{p_\parallel}{mc} - \frac{N\omega_{co}k_\parallel c}{\omega^2 - k_\parallel^2 c^2}\right)^2 \left(1 - \frac{k_\parallel^2 c^2}{\omega^2}\right) = \frac{N^2\omega_{co}^2 - \omega^2 + k_\parallel^2 c^2}{\omega^2 - k_\parallel^2 c^2}, \qquad (61)$$

where $\omega_{co} = qB/m$ is the Larmor frequency of the rest mass m_o. The unstable frequencies depend sensitively on the precise shape of the distribution function as can be seen in the example of Fig. 10. On the left a Maxwellian distribution with a superimposed ring (or a Guest-Dory-Harris distribution) is unstable to a value $k_\parallel = 0$ (resonance curve centered on the origin) and therefore (eq. (61)) to a frequency below a cyclotron harmonic. On the right a proper loss-cone distri-

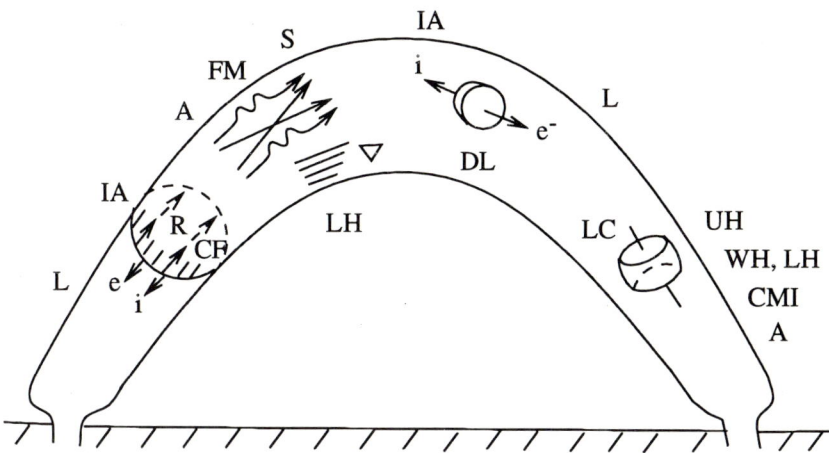

Fig. 9. Reconnection inside a magnetic flux tube excites growth of various kinds of waves indicated by abbreviations along the edge of the magnetic tube: MHD waves (A: Alfvén, FM: fast magnetosonic, S: shocks and sound waves) are excited by sudden reconnection; electrostatic and electromagnetic waves by loss-cone distributions (LC) of the fast particles mirroring in the trap (UH: upper-hybrid, WH: whistler, LH: lower-hybrid, CMI: cyclotron maser instability, A: Alfvén); electrostatic waves by beams running ahead of conduction fronts (CF) or reverse currents inside them (L: Langmuir waves, IA: ion-acoustic waves) or by beams produced inside double layers (DL); lower-hybrid waves from inhomogeneity drifts (∇).

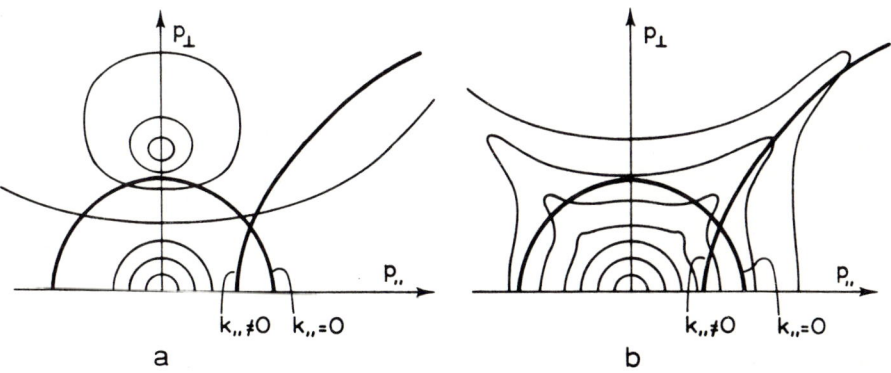

Fig. 10. The instability is very sensitive to the shape of the distribution function. The distribution on the left is unstable to parallel propagating waves while the one on the right is not.

bution is drawn where the instability appears for $k_\parallel \neq 0$ (the ellipse off-axis) and therefore (eq. (59)) at a frequency above a harmonic. It also follows from eq. (61) that in the case $k_\parallel = 0$ the dependence of the cyclotron frequency on particle energy cannot be neglected, even at non-relativistic energies (Zheleznyakov and Zlotnik 1975).

Waves at frequencies much below the relevant cyclotron frequency are very effective in scattering resonant particles in pitch angle and are therefore easily excited by an anisotropic particle distribution (as in shock waves, see below). This follows from comparing the energy loss of a particle when emitting a wave (Kennel and Engelmann 1966) $\Delta \epsilon = m_o v \Delta v = -\hbar \omega_{\mathbf{k}}$ to its loss in energy parallel to the ambient field $\Delta \epsilon_\parallel = m_o v_\parallel \Delta v_\parallel = -\hbar k_\parallel v_\parallel$, causing a change in pitch angle α (note that $v \sin \alpha \Delta \alpha = \cos \alpha \Delta v - \Delta v_\parallel$)

$$\Delta \alpha = \frac{\Delta v}{v \sin \alpha \cos \alpha} \left(\cos^2 \alpha - \frac{k_\parallel v_\parallel}{\omega} \right) = \frac{\Delta v}{v} \frac{N \omega_{cq}/\omega - \sin^2 \alpha}{\sin \alpha \cos \alpha}, \tag{62}$$

where we have used the resonance condition eq. (59). It follows that $\Delta \alpha$ increases quickly as ω falls below ω_c. The effect can also be understood from the relative unimportance of the electric field with respect to the magnetic field field in a wave much below the cyclotron frequency. From Faraday's law one has $E_{\mathbf{k}}/(B_{\mathbf{k}} c) = \omega/k_\perp < \omega r_c = (v/c)\omega/\omega_{cq} \ll 1$.

On the other hand high-frequency waves (large ratio of energy to momentum ω/k) are primarily excited by a particle distribution inverted in energy.

Effective Temperature The energy density of plasma waves of kind σ and wave vector $\mathbf{k}$ can be characterized by an *effective temperature*

$$KT^\sigma(\mathbf{k}) \equiv W^\sigma(\mathbf{k}), \tag{63}$$

where $W^\sigma(\mathbf{k})$, the energy density of wave mode σ per unit volume of space and per unit volume of wave vector space, is related to W^σ, the energy density of wave mode σ per unit volume of space, by

$$W^\sigma \equiv \int W^\sigma(\mathbf{k}) d^3\mathbf{k} (2\pi)^{-3}. \tag{64}$$

The definition is chosen so that in thermodynamic equilibrium for electromagnetic waves with $\hbar\omega \ll KT$ the wave temperature (distinguish between two polarizations) coincides with the temperature of the plasma (Tsytovich 1973; Melrose 1986). Note that the 'Rayleigh-Jeans' definition eq. (63) only makes sense for non-quantum plasmas (eq. (31)).

In a collisionless plasma (Debye number much larger than unity, $N_D \gg 1$) the effective temperature of plasma waves can be much larger than the gas temperature (or the characteristic particle kinetic energy divided by Boltzmann's constant). For instance for Langmuir waves of effective temperature T^ℓ the energy density per unit volume is of order $W^\ell \approx KT^\ell (N_D 18\pi^2)^{-1} W_{gas}$ and is smaller than unity up to an effective temperature $T^\ell \approx N_D T$, where $W_{gas} = 3nKT$

is the kinetic energy density in the gas. Further the effective temperature of a given wave mode can be high only for a small range of wave vectors $\mathbf{k}$ so that the integrated energy density still remains small. Such a suprathermal wave level can occur when the particle distribution functions deviate from Maxwellians.

As electric charge is discrete even a quiescent plasma has a noise level in its eigen modes. For undamped waves in a thermal plasma of tempertaure T_0 the energy density of the noise level per unit volume per unit volume of momentum space is just KT_0. For example Langmuir waves (around the plasma frequency) are practically undamped down to the Debye wavelength. Their integrated energy density per unit volume is

$$W_0^L \equiv \int W_0^L(\mathbf{k}) \frac{d^3\mathbf{k}}{(2\pi)^3} \simeq \frac{KT}{6\pi^2 \lambda_D^2} = 0,06 \frac{3nKT}{N_D}. \tag{65}$$

Weak Turbulence The weak turbulence approximation is based on a perturbation expansion of the combined Maxwell and Vlasov equations in the electric field (Melrose 1986, Ch. VI). One uses the relatively small wave energy densities as given by the expansion parameter (Davidson 1972)

$$\alpha_1 \equiv \frac{W^\sigma}{nKT} \ll 1. \tag{66}$$

We shall briefly summarize the general procedure. The first part is identical to the derivation of the wave equation. Again the Maxwell and Vlasov equations are Fourier transformed in space and time and the time-dependent magnetic field is expressed in the electric field. The difference is that the equation of motion is not linearized but higher-order terms in the perturbing electric field are kept. If one uses the Vlasov equation its Fourier transform is solved by a perturbation expansion of the distribution function. The distribution function consists of a zero-order part, the homogeneous background, and higher order terms (in E or W) induced by the electric field. One writes

$$f(\mathbf{r}, \mathbf{p}, t) = \sum_{n=0}^{\infty} f^{(n)}(\mathbf{r}, \mathbf{p}, t), \tag{67}$$

where $f^o(\mathbf{r}, \mathbf{p}, t) = n^o f^o(p)$ is the stationary and homogeneous unperturbed equilibrium solution, and $f^{(n)}$ is proportional to the $n-$th power of the electric field. Since the acceleration term in the Vlasov equation is proportional to the electric field, a formal solution for the distribution function can readily be written down in the form of a hierarchy of contributions Eq. (67), each of which is expressed recursively as a convolution of the next lower-order contribution and the electric field. These successively higher–order (in E) terms of the distribution function contribute to successively higher–order terms in the electric current density induced in the plasma by the electric field. By straightforward multiplication of $f^{(n)}$ with the particle charge and the velocity, subsequent integration over momentum space and summing over different species of charged particles one finds:

$$j_i(\mathbf{k}, \omega) = \sigma_{ij}(\mathbf{k}, \omega) E_j(\mathbf{k}, \omega) +$$
$$+ \int d\lambda^{(2)} \sigma_{ij\ell}(\mathbf{k}, \omega, \mathbf{k}_1, \omega_1, \mathbf{k}_2, \omega_2) E_j(\mathbf{k}_1, \omega_1) E_\ell(\mathbf{k}_2, \omega_2) +$$
$$+ \int d\lambda^{(3)} \sigma_{ij\ell m}(\mathbf{k}, \omega, \mathbf{k}_1, \omega_1, \mathbf{k}_2, \omega_2, \mathbf{k}_3, \omega_3) E_j(\mathbf{k}_1, \omega_1) E_\ell(\mathbf{k}_2, \omega_2) E_m(\mathbf{k}_3, \omega_3)$$
$$+ \dots, \tag{68}$$

where

$$d\lambda^{(n)} \equiv \frac{d^3\mathbf{k}_1 d\omega_1}{(2\pi)^4} \dots \frac{d^3\mathbf{k}_n d\omega_n}{(2\pi)^4} (2\pi)^4 \delta^3(\mathbf{k} - \mathbf{k}_1 - \dots - \mathbf{k}_n) \delta(\omega - \omega_1 - \dots - \omega_n).$$

Substituting Eq.(68) for the induced current density into Ampère's law, the wave equation for the electric field can now be derived in the standard manner from Maxwell's equations. Collecting the terms which are linear in the electric field on the left-hand side of the equation and the rest on the other side, one arrives at the wave equation in the form

$$\Lambda_{ij}(\mathbf{k}, \omega) E_j(\mathbf{k}, \omega) = -\frac{i}{\epsilon_o \omega} j_i^{nonl}(\mathbf{k}, \omega), \tag{69}$$

which differs from Eq. (47) in that the nonlinear induced current contributions (second and higher order in the electric field) are kept and placed on the right-hand side as an effective 'external' current density. For a given nonlinear current density Eq.(68) the corresponding electric field can be found by inverting Eq.(69). In words: the nonlinear properties of the plasma (through the action of electric fields on the particle motion) create electric perturbations at beats and harmonics (in Fourier space) of primary electric fields.

The *power per unit volume radiated* by this nonlinear current is given by

$$P = -\frac{lim}{T, V \to \infty} \frac{1}{TV} \int \frac{d^3\mathbf{k} d\omega}{(2\pi)^4} \mathbf{j}^{nonl}(\mathbf{k}, \omega) \cdot \mathbf{E}^*(\mathbf{k}, \omega). \tag{70}$$

Random Phase Approximation Apart from small amplitudes the weak turbulence approximation assumes the validity of the *Random Phase Approximation* (RPA) for the ensemble average of the product of the field Fourier components

$$< E_i(\mathbf{k}, \omega) E_j(\mathbf{k}', \omega') > = \frac{lim}{T, V \to \infty} \frac{(2\pi)^4}{TV} < E_i(\mathbf{k}, \omega) E_j^*(\mathbf{k}, \omega) > \delta^3(\mathbf{k}+\mathbf{k}') \delta(\omega+\omega'). \tag{71}$$

Physically this approximation is valid if the *autocorrelation time* of a wave packet τ_{ac} is much less than the interaction time τ_{int} with the wave field of a characteristic particle

$$\alpha_2 \equiv \frac{\tau_{ac}}{\tau_{int}} \ll 1. \tag{72}$$

Here τ_{ac} is the time during which a resonant particle feels a force exerted by the wave packet

$$\tau_{ac} \equiv \frac{1}{k\Delta\frac{\omega}{k}}. \tag{73}$$

The interaction time is given by

$$\tau_{int} = min(\tau_{tr}, \tau_D, \gamma^{-1}), \tag{74}$$

where the trapping time for a particle of mass m, charge q in waves near wave vector $\mathbf{k}$ and (real) electric field amplitude E is

$$\tau_{tr} = \left(\frac{m}{qkE}\right)^{1/2}, \tag{75}$$

the diffusion time of a resonant particle under the action of the waves is

$$\tau_D = \frac{v^2}{D}, \tag{76}$$

D is the *diffusion coefficient* in velocity space corresponding to the process considered and depends on $\mathbf{v}$ and $W^\sigma(\mathbf{k})$, and finally γ is the growth rate of the waves involved.

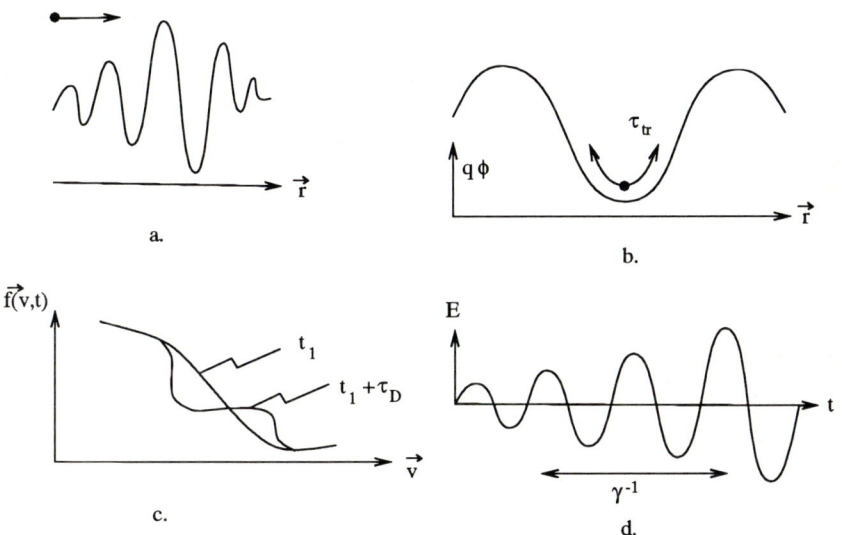

Fig. 11. Illustration of the autocorrelation time (a), the trapping time (b), the diffusion time (c), and the growth time (d).

In a plasma the electric Fourier component at arbitrary argument $(\mathbf{k}, \omega)$ can be written as

$$E_i(\mathbf{k}, \omega) = \sum_\sigma E_i^\sigma(\mathbf{k}, \omega) 2\pi \delta\left(\omega - \omega^\sigma(\mathbf{k})\right), \qquad (77)$$

where the dispersion relations of the various wave modes are indicated by the label σ. Technically for each wave mode in Eq.(77) positive and negative frequencies are involved, and one must extend the summation over "forward" and "backward" waves with separate dispersion relations $\sigma^\pm$ for each mode. Emission of electromagnetic waves of kind t occurs if the integrand in Eq. (58) does not vanish at the argument $(\mathbf{k}, \omega^t(\mathbf{k}))$ where the label t indicates the dispersion relation $\omega = \omega^t(\mathbf{k})$. The radiated power per unit volume of space and per unit volume of wave-vector space is now obtained from Eq. (70) with Eqs.(71) and (77) and integration over frequency

$$Q^t(\mathbf{k}) \equiv \frac{dW^t(\mathbf{k})}{dt} = -\frac{lim}{V \to \infty} \frac{1}{V} \mathbf{j}^{ext}(\mathbf{k}) \cdot \mathbf{E}^{t*}(\mathbf{k}) + C.C. \qquad (78)$$

Here $W^\sigma(\mathbf{k})$ is the energy density of wave mode σ per unit volume of space and wave–vector space

$$W^\sigma(\mathbf{k}) = \frac{\epsilon_o \mid \mathbf{E}^\sigma(\mathbf{k}) \mid^2}{V R_E^\sigma(\mathbf{k})}, \qquad (79)$$

and $R_E^\sigma(\mathbf{k})$ is the ratio of electric to total energy density of mode σ at wave vector $\mathbf{k}$ (Melrose 1986). In eq. (78) and hereafter the frequencies of the wave modes are always positive – unless explicitly stated otherwise –, in contrast with the earlier expressions.

'All' one has to do now is to single out the contributions to the nonlinear current for a specific process and to substitute this for the external current in Eq.(70).

To illustrate the general procedure we again consider the unmagnetized homogeneous equilibrium plasma (eq. (52)). With eq. (67) the nth order terms of the (Fourier transform of the) Vlasov equation give

$$f^{(n)}(\mathbf{k}, \omega, \mathbf{v}) = \frac{1}{\omega - \mathbf{k} \cdot \mathbf{v}} \int d\lambda^{(2)} \frac{q}{m} g_{sj}(\mathbf{k}_1, \omega_1) E_j(\mathbf{k}_1, \omega_1) \frac{\partial f^{(n-1)}(\mathbf{k}_2, \omega_2, \mathbf{v})}{\partial v_s}, \qquad (80)$$

with

$$i g_{sj}(\mathbf{k}, \omega) \equiv (1 - \frac{\mathbf{k} \cdot \mathbf{v}}{\omega}) \delta_{js} + \frac{k_s v_j}{\omega}. \qquad (81)$$

Using

$$j_i^{(n)}(\mathbf{k}, \omega) = \sum_\alpha \int q_\alpha v_i f_\alpha^{(n)}(\mathbf{k}, \omega, \mathbf{v}) d^3\mathbf{v} = \qquad (82)$$

$$\int d\lambda^{(n)} \sigma_{ij\ell t}(\mathbf{k}, \omega, \mathbf{k}_1, \omega_1, \mathbf{k}_2, \omega_2, ..., \mathbf{k}_n, \omega_n) E_j(\mathbf{k}_1, \omega_1) E_\ell(\mathbf{k}_2, \omega_2)...E_t(\mathbf{k}_n, \omega_n),$$

one finds

$$\sigma_{ij}(\mathbf{k}, \omega) = \sum_\alpha \frac{q_\alpha^2}{m_\alpha} n_\alpha \int d^3\mathbf{v} \frac{v_i}{\omega - \mathbf{k} \cdot \mathbf{v}} g_{sj}(\mathbf{k}, \omega) \frac{\partial f_\alpha^o(\mathbf{v})}{\partial v_s},$$

$$\sigma_{ij\ell}(\mathbf{k}, \omega, \mathbf{k}_1, \omega_1, \mathbf{k}_2, \omega_2) = \sum_\alpha \frac{q_\alpha^3}{m_\alpha^2} \int d^3\mathbf{v} \frac{v_i}{\omega - \mathbf{k} \cdot \mathbf{v}} g_{sj}(\mathbf{k}_1, \omega_1) \frac{\partial}{\partial v_s}$$
$$\left[\frac{g_{r\ell}(\mathbf{k}_2, \omega_2)}{\omega_2 - \mathbf{k}_2 \cdot \mathbf{v}} \frac{\partial f_\alpha^o(\mathbf{v})}{\partial v_r} \right],$$

$$\sigma_{ij\ell m}(\mathbf{k}, \omega, \mathbf{k}_1, \omega_1, \mathbf{k}_2, \omega_2, \mathbf{k}_3, \omega_3) = \sum_\alpha \frac{q_\alpha^4}{m_\alpha^3} \int d^3\mathbf{v} \frac{v_i}{\omega - \mathbf{k} \cdot \mathbf{v}} g_{sj}(\mathbf{k}_1, \omega_1) \frac{\partial}{\partial v_s}$$
$$\left[\frac{1}{\omega_2 + \omega_3 - (\mathbf{k}_2 + \mathbf{k}_3) \cdot \mathbf{v}} g_{r\ell}(\mathbf{k}_2, \omega_2) \frac{\partial}{\partial v_r} \left\{ \frac{g_{tm}(\mathbf{k}_3, \omega_3)}{\omega_3 - \mathbf{k}_3 \cdot \mathbf{v}} \frac{\partial f_\alpha^o(\mathbf{v})}{\partial v_t} \right\} \right]. \tag{83}$$

(For a covariant formulation see Melrose and Kuijpers 1984.) To investigate the nature of the higher-order interactions we consider the lowest-order nonlinear radiation. From eq. (78) it is clear that only $Re j^{nonl}(\mathbf{k}) \cdot \mathbf{E}(-\mathbf{k})$ contributes (because of the reality condition $\mathbf{E}^*(\mathbf{k}) = \mathbf{E}(-\mathbf{k})$). Therefore with eq. (77) and the RPA (eq. (71)) it follows that only the real parts of $\sigma_{ij...}(\mathbf{k}, \mathbf{k}_1, ...)$ contribute. To lowest (nonlinear) order the following couplings exist:

Wave coupling (wave fusion or decay): The contribution comes from

$$\mathbf{j}^{(2)}(\mathbf{k}_3, \omega_3) \cdot \mathbf{E}^*(\mathbf{k}_3, \omega_3),$$
$$\mathbf{E}^*(\mathbf{k}_3, \omega_3) \propto \mathbf{j}^{(2)*}(\mathbf{k}_3, \omega_3) \propto \mathbf{E}^{\sigma_1*}(\mathbf{k}_1, \omega_1) \mathbf{E}^{\sigma_2*}(\mathbf{k}_2, \omega_2), \tag{84}$$

where the first proportionality derives from inverting eq. (69). As in this case $\sigma_{ij\ell}(\mathbf{k}_3, \omega_3, \mathbf{k}_1, \omega_1, \mathbf{k}_2, \omega_2)$ is real, the proportionality factor between $\mathbf{E}^*(\mathbf{k}_3, \omega_3)$ $\propto \mathbf{j}^{(2)*}(\mathbf{k}_3, \omega_3)$ has to be real, which in view of eq. (51), requires $1/\Lambda(\mathbf{k}_3, \omega_3)$ to have a pole. As the convolution eq. (83) imposes the condition $(\mathbf{k}_3, \omega_3) = (\mathbf{k}_1, \omega_1) + (\mathbf{k}_2, \omega_2)$ this means that $(\mathbf{k}_3, \omega_3)$ is a solution to the zero-order wave equation and is itself an eigen mode:

$$\mathbf{k}_1 + \mathbf{k}_2 = \mathbf{k}_3,$$
$$\omega^{\sigma_1}(\mathbf{k}_1) + \omega^{\sigma_2}(\mathbf{k}_2) = \omega^{\sigma_3}(\mathbf{k}_3). \tag{85}$$

Clearly waves σ_1 and σ_2 interact to produce a beat which itself is an eigen mode.

Nonlinear scattering and double emission: The contribution comes from

$$\mathbf{j}^{(2)}(\mathbf{k}_2, \omega_2) \cdot \mathbf{E}^{\sigma_2*}(\mathbf{k}_2, \omega_2), \text{ with}$$
$$\mathbf{j}_i^{(2)}(\mathbf{k}_2, \omega_2) \propto \sigma_{ij\ell} E_j^{\sigma_1}(\mathbf{k}_1, \omega_1) E_\ell(\mathbf{k}_2 - \mathbf{k}_1, \omega_2 - \omega_1), \text{ and}$$
$$E_\ell(\mathbf{k}_2 - \mathbf{k}_1, \omega_2 - \omega_1) \propto \mathbf{j}_\ell^{(2)}(\mathbf{k}_2 - \mathbf{k}_1, \omega_2 - \omega_1)$$
$$\propto \sigma_{\ell qr} E_q^{\sigma_1}(-\mathbf{k}_1, -\omega_1) E_r^{\sigma_2}(\mathbf{k}_2, \omega_2), \tag{86}$$

where the first proportionality in the last line derives from inverting eq. (69). As, in general, $(\mathbf{k}_2 - \mathbf{k}_1, \omega_2 - \omega_1)$ is not an eigenmode the inversion eq. (69) only gives an imaginary part. Therefore only an imaginary part in $\sigma_{\ell q r}$ contributes. Inspection of eq. (83) shows that such a contribution would come from

$$\Im \frac{1}{\omega_1 - \omega_2 - (\mathbf{k}_1 - \mathbf{k}_2) \cdot \mathbf{v}} = \pi \delta(\omega_1 - \omega_2 - (\mathbf{k}_1 - \mathbf{k}_2) \cdot \mathbf{v}), \qquad (87)$$

This lead to the condition

$$\omega^{\sigma_1}(\mathbf{k}_1) - \mathbf{k}_1 \cdot \mathbf{v} = \pm\{\omega^{\sigma_2}(\mathbf{k}_2) - \mathbf{k}_2 \cdot \mathbf{v}\}. \qquad (88)$$

The plus sign corresponds to nonlinear scattering; the minus sign to double emission (Melrose 1982). This process of nonlinear scattering should be complemented with the Thomson scattering on free particles. The influence of the plasma therefore drastically influences the scattering on single particles 'in vacuo'. For instance scattering of an electrostatic plasma wave on thermal electrons is inefficient as Thomson scattering on an electron is about equal but in antiphase to the nonlinear scattering on the associated Debye cloud. Scattering of a low-frequency electrostatic wave on thermal ions on the other hand becomes important as Thomson scattering on the heavy ions can be neglected compared to the nonlinear scattering on the Debye cloud of electrons.

We therefore find that higher-order interactions between waves themselves and between waves and 'dressed' particles only exist when certain relations (eq. (85) or eq. (88)) are satisfied. Classically these relations can be interpreted as the requirement that 1. the beat in frequency *and* in wave vector between two eigen modes again is an eigen mode, or that, 2. the Doppler shifted frequency of the incoming wave in the frame of the scattering particle equals the Doppler shifted frequency of the outgoing wave. Semiclassically these relations can be considered as conservation of wave energy and wave momentum in the coupling process.

In principle one could think of contributions to the same order arising from other poles in the expressions for the $\sigma_{ij...}$. However no further possibilities exist (see for the nonexistence of *turbulent bremsstrahlung* Melrose and Kuijpers 1987). Details of the calculation for a particular case (nonlinear scattering of a particle on a static electric field) can be found in Kuijpers (1990).

It is now easy to write down generalizations of eq. (85) and eq. (88) to higher-order nonlinear interactions. As an example we write down second-order nonlinear scattering for magnetized particles with $kv/\omega_{cq} < 1$:

$$\omega^{\sigma_1}(\mathbf{k}_1) - \mathbf{k}_{1\parallel}v_\parallel = \pm\{\omega^{\sigma_2}(\mathbf{k}_2) - \mathbf{k}_{2\parallel}v_\parallel + N\omega_{cq}\}, \qquad (89)$$

where N is integer and the components parallel to the ambient magnetic field are involved.

Strong Turbulence If the RPA approximation eq. (71) is not satisfied the nonlinear interactions become coherent (cf. nonlinear optics) and the turbulence is called *'strong' plasma turbulence*. We refer to Similon and Sudan (1990) and to Hasegawa (1975) for a review on strong turbulence.

A first effect of finite amplitude turbulence is *resonance broadening*. The orbits of the resonant particles (satisfying eq. (59) are perturbed and the resonant diffusion broadens the originally sharp resonance. Phenomenologically the process can be taken into account by replacing a resonance as eq. (60) by

$$\omega - \mathbf{k} \cdot \mathbf{v} + i/\tau_{tr} = 0, \tag{90}$$

where we have added an imaginary part to the frequency because of particle trapping in waves of characteristic electric field strength E at wave number k: $\tau_{tr}^{-1} \approx \left(\frac{ekE}{2m_o}\right)^{-0.5}$. Resonance broadening is responsible for scattering of cosmic rays across the pitch angle interval at 90 degrees with respect to the magnetic field, where Alfvén waves are non-existent.

Ponderomotive Force An important nonlinear effect destroying the homogeneity of turbulence and thereby invalidating the weak turbulence approach is the ponderomotive force. As the name says the force is due to matter in motion. More precisely the ponderomotive force is used for the average net force from a collection of small scale oscillations as soon as the force amplitude varies on a spatial scale much larger than the characteristic displacement of an individual particle in the oscillation. Starting from the relevant equation of motion one splits up the physical quantitities into averaged quantitities (which satisfy the equation of motion in the absence of fluctuations) plus a small oscillating part. Upon averaging the equation of motion all terms linear in the fluctuating quantities vanish. However the terms of second order in the fluctuations do not in general vanish and their summed effect is the ponderomotive force for the oscillations under study.

A simple example (Landau and Lifshitz 1976) is the one-dimensional motion of particles in the force field $f(x,t)$ given by

$$f(x,t) = f(x)cos(\omega t + \phi), \tag{91}$$

where the amplitude $f(x)$ varies on a length scale large with respect to the excursion of a particle from the average flow during one oscillation. As the excursion of a particle depends on its inertial mass the condition will be satisfied for sufficiently large frequencies ω. Expanding the individual particle orbit $x(t) = x^o(t) + x^1(t)$, where $< x^1(t) >= 0$ averaged over the fast oscillations, the equation of motion can be expanded

$$m\frac{d^2x^o(t)}{dt^2} + m\frac{d^2x^1(t)}{dt^2} = f(x^o(t),t) + x^1(t)\frac{\partial f(x^o,t)}{\partial x^o}. \tag{92}$$

To first order the solution is

$$x^1(t) = -\frac{f(x^o,t)}{m\omega^2}. \tag{93}$$

Substituting the result back into eq. (92) and averaging gives

$$m\frac{d^2x^o(t)}{dt^2} = -\frac{1}{m\omega^2}\left\langle f(x^o,t)\frac{\partial f(x^o,t)}{\partial x^o}\right\rangle$$

$$= -\frac{d}{dx^o}\frac{[f(x)]^2}{4m\omega^2} = -\frac{d}{dx^o}\frac{1}{2}m_o\left\langle\left(\frac{dx^1(t)}{dt}\right)^2\right\rangle. \tag{94}$$

The net effect on all particles is therefore an effective potential density equal to the average kinetic oscillation energy density.

Using the properties of the particular waves the force can often be expressed as the gradient of (a fraction of) the wave energy density (and in case of excitation of waves by an energetic particle component, as the pressure gradient of the particular particle component which excites the waves (Achterberg 1981b)).

Some examples: An electromagnetic laser beam in a plasma gives rise to a ponderomotive force density (Bell 1993)

$$\mathbf{F} = -\frac{1}{2}n_e m_e \boldsymbol{\nabla} < |\mathbf{v}(\mathbf{r},t)|^2 >= -\frac{\omega_{pe}^2}{\omega^2}\boldsymbol{\nabla}\left\langle\frac{\epsilon_o}{2}|\mathbf{E}(\mathbf{r},t)|^2\right\rangle, \tag{95}$$

where $\mathbf{v}(\mathbf{r},t)$ and $\mathbf{E}(\mathbf{r},t)$ are the actual velocity fluctuations and, respectively, electric field fluctuations of the radiation. The reduction factor $(\omega_{pe}/\omega)^2$ is due to the reduced mobility of the plasma electrons in the electric field of the electromagnetic waves as the radiation frequency increases from the electron plasma frequency upwards.

Another example is the ponderomotive force density due to Alfvén waves of energy density $W^A = \int\frac{d^3\mathbf{k}}{(2\pi)^3}\left(\frac{|\mathbf{B}^A(\mathbf{k})|^2}{2\mu_o} + \frac{1}{2}\rho|\mathbf{v}(\mathbf{k})|^2\right) = \int\frac{d^3\mathbf{k}}{(2\pi)^3}\frac{|\mathbf{B}^A(\mathbf{k})|^2}{\mu_o}$ (Belcher 1971, Achterberg 1981b):

$$\mathbf{F} = -\frac{1}{2}\rho\boldsymbol{\nabla} < |\mathbf{v}(\mathbf{r},t)|^2 >= -\boldsymbol{\nabla}\frac{< |\mathbf{B}(\mathbf{r},t)|^2 >}{2\mu_o} = -\boldsymbol{\nabla}\frac{1}{2}W^A, \tag{96}$$

Finally for electron oscillations at the plasma frequency (Langmuir waves) with energy density W_L the ponderomotive force density is

$$\mathbf{F} = -\frac{1}{2}\rho\boldsymbol{\nabla} < |\mathbf{v}(\mathbf{r},t)|^2 >= -\boldsymbol{\nabla} < \frac{\epsilon}{2}|\mathbf{E}(\mathbf{r},t)|^2 >= -\boldsymbol{\nabla}\frac{1}{2}W^L. \tag{97}$$

Modulational Instabilities The ponderomotive force plays an important role in destroying the average homogeneity of a weakly turbulent plasma. The resulting modulational instability appears to be a generic feature of a variety of eigen modes. For most eigen modes the frequency $\omega^\sigma(\mathbf{k})$ increases with increasing wave vector $\mathbf{k}$. Let us follow the history of waves of such a mode after they have been excited around wave vector $\mathbf{k}$. Suppose that their linear damping is small. Then the next-order process of nonlinear scattering (eq. (89)) will become important and shape the spectrum. It can be shown that during this process the action density (Lichtenberg and Lieberman 1983)

$$S^\sigma \equiv \int\frac{d^3\mathbf{k}}{(2\pi)^3}\frac{W^\sigma(\mathbf{k})}{\omega^\sigma(\mathbf{k})}, \tag{98}$$

is conserved. Therefore the number of wave 'quanta' $W^\sigma(\hbar\omega^\sigma(\mathbf{k}))^{-1}$ is conserved. As the wave level is supra-thermal, it will not come as a surprise that the wave-particle interactions tend to transfer energy from the waves to the plasma (*nonlinear Landau damping*). Because of the conservation of quanta the waves are scattered to lower frequencies and, because of the shape of the dispersion curve, towards lower wave numbers. This causes a pile-up or *condensation* in wave vector space near zero wave numbers. Thereby the wave temperature $W^\sigma(\mathbf{k})/K$ increases and ultimately higher-order interactions come into play. This highly nonlinear stage has first been described by Zakharov for Langmuir waves (Fig. 12) (see Sitenko (1982) for a self-consistent nonlinear fluctuation theory). These *Zakharov* equations incorporate the ponderomotive force of inhomogeneous turbulence. A four-wave interaction between the $k \approx 0$ Langmuir condensate, two Langmuir waves at finite wave number and one ion-acoustic wave, returns energy to large wave numbers by the effect of the ponderomotive force. Note that the homogeneous weak turbulence regime has been left. The turbulence becomes quickly inhomogeneous and collapses into *cavitons*, finite-amplitude low-frequency ion-acoustic waves filled with high-frequency Langmuir waves. In fact the latter burn out a density depression by their ponderomotive action.

An approximate criterion for onset of the modulational instability and collapse can be derived as follows (Papadopoulos and Freund 1979; Goldman 1984). A packet of intense electrostatic Langmuir waves will, by the ponderomotive

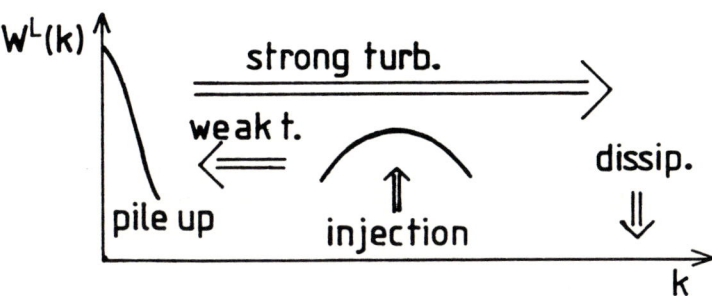

Fig. 12. Sketch of the fate of Langmuir turbulence. After their injection at k_{inj} the waves condensate at zero wave number by weak turbulence scattering. There the ponderomotive force causes a fast transfer of the waves to high wave numbers (collapse in space) where they are dissipated quickly, e.g. by ordinary (linear) Landau damping.

force, exert an excess pressure $\frac{1}{2}W^L$ on the ambient plasma and therefore create a density depression of magnitude $\delta n \approx \frac{1}{2}W^L(KT)^{-1}$. The dispersion relation for Langmuir waves is modified by the density depression as

$$\omega^L(\mathbf{k}) = \omega_{pe}\left(1 + \frac{3}{2}k^2\lambda_D^2 + \frac{1}{2}\frac{\delta n}{n}\right) \approx \omega_{pe}\left(1 + \frac{3}{2}k^2\lambda_D^2 + \frac{W^L}{4nKT}\right). \qquad (99)$$

It is expected that the collapse sets in as soon as the last term becomes comparable to the second term, that is when the characteristic wave number of the weak turbulence k^L has decreased to about

$$k\lambda_D \approx W^L(6nKT)^{-1} = \epsilon_o < |\mathbf{E}(\mathbf{r},t)|^2 > (12nKT)^{-1}, \qquad (100)$$

which happens sooner or later.

2.2 Weak Turbulence Acceleration

Acceleration by weak turbulence occurs mainly through a Čerenkov resonance (Landau damping) eq. (60) or cyclotron resonance eq. (59). Electrostatic waves, which are driven by charge excesses, have a relatively large electric field. It is therefore no surprise that they can accelerate particles quickly. Their effect is limited however to sub-relativistic energies. We will first have a closer look at Langmuir turbulence above the electron plasma frequency (with dispersion relation $\omega^L(\mathbf{k}) = (\omega_{pe}^2 + 3k^2 v_{te}^2)^{0.5}$.)

Electrostatic Turbulence Acceleration We start from the Fokker-Planck equation eq. (23) and first calculate the diffusion coefficient in momentum space $< \Delta p \Delta p > /(2\Delta t)$. This is obtained from the equation of motion of a single particle of charge q and rest mass m_o in the absence of a magnetic field

$$\frac{d\mathbf{p}}{dt} = q\mathbf{E}, \qquad (101)$$

since we are considering electrostatic waves without a magnetic field.

Calculation of Momentum Diffusion To calculate $< \Delta p \Delta p > /(2\Delta t)$ we integrate the momentum equation eq. (101) along the unperturbed particle orbit in the time interval $t, t + \Delta t$. The electric field at the position of the particle at time t is $\mathbf{E}(\mathbf{r},t) = \mathbf{E}(\mathbf{r}(t),t)$ with $\mathbf{r}(t) = \mathbf{r}_0 + \mathbf{v}t$. We find

$$< \Delta p_i \Delta p_j > = q^2 \int_t^{t+\Delta t} dt' \int_t^{t+\Delta t} dt'' < E_i(\mathbf{r}(t'),t')E_j(\mathbf{r}(t''),t'') >$$

$$\overset{=}{\overbrace{s \equiv t'' - t'}} q^2 \int_t^{t+\Delta t} dt' \int_{t-t'}^{t-t'+\Delta t} ds < E_i(\mathbf{r}(t'),t')E_j(\mathbf{r}(t'+s),t'+s) >$$

$$\overset{=}{\overbrace{\Delta t \gg \tau_c}} q^2 \int_t^{t+\Delta t} dt' \int_{-\infty}^{\infty} ds \int \frac{d^3\mathbf{k}\,d\omega}{(2\pi)^4} \int \frac{d^3\mathbf{k}'\,d\omega'}{(2\pi)^4} < E_i(\mathbf{k},\omega)E_j(\mathbf{k}',\omega') >$$

$$exp\{i(\mathbf{k}+\mathbf{k}')\cdot\mathbf{r}_0 + i\mathbf{k}\cdot\mathbf{v}t' + i\mathbf{k}'\cdot\mathbf{v}(t'+s) - i\omega t' - i\omega'(t'+s)\}$$

$$\overset{=}{\overbrace{RPA}}\; \lim_{T,V\to\infty}\frac{q^2}{TV}\int_t^{t+\Delta t}dt'\int_{-\infty}^{\infty}ds\int\frac{d^3k\,d\omega}{(2\pi)^4}<E_i(\mathbf{k},\omega)E_j^*(\mathbf{k},\omega)>e^{-i(\mathbf{k}\cdot\mathbf{v}-\omega)s}$$

$$= \Delta t\;\lim_{T,V\to\infty}\frac{q^2}{TV}\int\frac{d^3k\,d\omega}{(2\pi)^4}<E_i(\mathbf{k},\omega)E_j^*(\mathbf{k},\omega)>2\pi\delta(\omega-\mathbf{k}\cdot\mathbf{v}), \tag{102}$$

where in the last equality we have made use of

$$\int_{-\infty}^{\infty}e^{-i(\mathbf{k}\cdot\mathbf{v}-\omega)s}=\lim_{T\to\infty}\int_{-T}^{T}ds\,e^{i(\omega-\mathbf{k}\cdot\mathbf{v})s}=\lim_{T\to\infty}\frac{2\sin[(\omega-\mathbf{k}\cdot\mathbf{v})T]}{\omega-\mathbf{k}\cdot\mathbf{v}}=2\pi\delta(\omega-\mathbf{k}\cdot\mathbf{v}). \tag{103}$$

The occurrence of the delta-function in eq. (102) expresses that only resonant particles receive *kicks* from the turbulence.

For purely electrostatic plane wave modes the electric field is completely determined by Poisson's equation eq. (37) and is therefore parallel to the wave vector. We can now write ($\hat{\mathbf{k}}\equiv\mathbf{k}/k$ is the unit wave vector)

$$\mathbf{E}(\mathbf{k},\omega)=\sum_\sigma \mathbf{E}^\sigma(\mathbf{k})2\pi\delta(\omega-\omega^\sigma(\mathbf{k}))=\sum_\sigma\hat{\mathbf{k}}E^\sigma(\mathbf{k})2\pi\delta(\omega-\omega^\sigma(\mathbf{k})), \tag{104}$$

$$\frac{\partial f}{\partial t}=\frac{\partial}{\partial\mathbf{p}}\left(\overset{\leftrightarrow}{\mathbf{D}}\cdot\frac{\partial f(\mathbf{p})}{\partial\mathbf{p}}\right), \tag{105}$$

where the diffusion tensor is given by

$$\overset{\leftrightarrow}{\mathbf{D}}=\frac{<\Delta\mathbf{p}\Delta\mathbf{p}>}{2\Delta t}$$

$$=\lim_{T,V\to\infty}\frac{q^2}{TV}\int\frac{d^3k\,d\omega}{(2\pi)^4}<\mathbf{E}(\mathbf{k},\omega)\mathbf{E}^*(\mathbf{k},\omega)\pi\delta(\omega-\mathbf{k}\cdot\mathbf{v})$$

$$=\lim_{V\to\infty}\sum_\sigma q^2\int\frac{d^3k}{(2\pi)^3}\frac{\mathbf{k}\mathbf{k}}{k^2}\frac{|E^\sigma(\mathbf{k})|^2}{V}\pi\delta(\omega^\sigma(\mathbf{k})-\mathbf{k}\cdot\mathbf{v}). \tag{106}$$

For Langmuir waves ($\sigma=L$) half of the energy is in kinetic energy of the oscillating electrons and half in the electric field so that the wave energy density is

$$W^L(\mathbf{k})=\frac{2\epsilon|\mathbf{E}^L(\mathbf{k})|^2}{V}, \tag{107}$$

and eq. (106) in the non-relativistic regime leads to

$$\frac{\partial f}{\partial t}=\frac{\partial}{\partial\mathbf{v}}\cdot\left[\sum_\pm 2\epsilon_0\frac{q^2}{m_o^2}\int\frac{d^3k}{(2\pi)^3}W^L(\mathbf{k})\delta(\omega^L(\mathbf{k})-\mathbf{k}\cdot\mathbf{v})\frac{\mathbf{k}\mathbf{k}}{k^2}\cdot\frac{\partial f}{\partial\mathbf{v}}\right]$$

$$=\frac{\partial}{\partial v}\left[\frac{4\epsilon_0 q^2}{m_o^2}\int\frac{d^3k}{(2\pi)^3}W^L(\mathbf{k})\delta\left(k_\parallel-\frac{\omega^L(\mathbf{k})}{v}\right)\frac{k_\parallel^2}{vk^2}\frac{\partial f}{\partial v}\right], \tag{108}$$

where $k_{\parallel} \equiv \mathbf{k} \cdot \mathbf{v}/v$.

The characteristic acceleration time scale of non-relativistic electrons then is

$$t_{acc}^L = \frac{1}{\omega_{pe}} \left(\frac{W^L}{nKT_e} \right)^{-1} \left(\frac{v}{v_{te}} \right)^3 \frac{\Delta k}{k_D}. \tag{109}$$

In the *subrelativistic* regime acceleration of electrons by Langmuir is therefore very efficient. The waves are however not efficient accelerators at relativistic energies. The reason is that as the particle speed approaches the speed of light the wave energy available for acceleration ($k \approx \omega/v \approx \omega/c$) occupies an ever smaller volume of wave vector space. They do form however a suitable preaccelerator or injector of fast electrons, provided a source of Langmuir waves is present, for other acceleration processes.

Apart from Langmuir waves the *low-frequency* electrostatic lower-hybrid waves are efficient accelerators. The latter can be excited by anisotropic ion distributions (Benz and Smith 1987; Thejappa 1987; Bryant 1993) in contrast with the high-frequency Langmuir waves which are not easily produced out of low-frequency turbulence.

Magnetic Turbulence Acceleration Both electrons and protons can be accelerated to highly *relativistic* energies by *Alfvén* waves, *magnetosonic* waves and *whistler* waves (Miller and Ramaty 1987; Ramaty and Murphy 1987; Smith 1990; Schlickeiser 1989; Steinacker and Schlickeiser 1989; Miller, Guessoum, and Ramaty 1990; Miller 1991). The energy is transferred through a cyclotron resonance Eq. (59) between particle and wave of the right frequency, direction and polarization.

The primary proton spectrum inferred from observations of a solar flare seems to agree better with a Bessel function spectral type than with a power-law (Rieger 1989). Such a fit agrees with second-order Fermi acceleration (whistlers or Alfvén waves) which produce a modified second-order Bessel function in the non-relativisic regime.

Acceleration by Alfvén waves, however, requires the particles already to be very relativistic. This can be seen as follows. As Alfvén waves have no electric field component parallel to the ambient magnetic field a resonance eq. (59) only exists for $N \neq 0$. In view of the dispersion relation for Alfvén waves $\omega^A(\mathbf{k}) = k_{\parallel}v_A$ the lowest energy resonance $\omega = k_{\parallel}v_{\parallel} \pm \omega_c$ requires

$$\omega^A(\mathbf{k}) = \frac{\omega_{co}}{\left| \frac{v}{v_A} - 1 \right| \gamma}. \tag{110}$$

Now as the source of MHD waves comes from large-scale motions or reconnections (lenght-scale L), which per se generate low-frequency waves ($\omega \leq v_A/L$), the primary waves are excited much below the nonrelativistic gyrofrequency ω_{co} and condition (110) therefore requires extremely relativistic cosmic rays $\gamma \gg 1$ to begin with.

It may be possible however that sufficient wave energy is first transferred to higher frequencies by nonlinear coupling (eq. (85), Smith 1990). A different possibility is when the waves propagate down a magnetic field gradient in a magnetized atmosphere and thereby approach the cyclotron frequencies from below (Note that an undamped wave in a static atmosphere with scale height H such that $kH \gg 1$ will preserve its absolute frequency and adiabatically adjust its wave vector to satisfy the local dispersion relation).

For low-frequency magnetosonic waves such a problem does not exist as their parallel electric field component allows a Landau resonance $\omega^{FMS} - k_\parallel v_\parallel = 0$ at sub-relativistic speeds.

For parallel (to the magnetic field) whistler waves, which in fact are the higher frequency continuation of the Alfvén branch with circular polarization in the electron gyration sense, the dispersion relation reads

$$\omega^{WH} = kv_A \left(\frac{m_p}{m_e}\right)^{1/2} \left[\frac{\omega}{\omega_{ce}}\left(1 - \frac{\omega}{\omega_{ce}}\right)\right]^{1/2}. \tag{111}$$

The polarization matching between resonant particle and circularly polarized whistler determines the lowest energy resonance to occur at

$$\omega - k_\parallel v_\parallel = |\omega_{ce}|, \text{ for the electrons,} \tag{112}$$

$$\omega - k_\parallel v_\parallel = -|\omega_{ci}|, \text{ for the ions,} \tag{113}$$

that is resonant electrons are approaching the whistlers (normal cyclotron resonance) whereas positive ions are overtaking the waves (anomalous resonance). For example whistlers at $\omega^{WH} = 0.5\omega_{ce}$; $k^{WH} = \omega_{ce}[v_A(m_p/m_e)^{\frac{1}{2}}]^{-1}$ resonate with electrons with a speed $v_\parallel = v_A 0.5(m_p/m_e)^{\frac{1}{2}}$.

2.3 Strong Turbulence Acceleration

In a non-perturbative approach, valid at sufficiently large amplitudes, the restrictive resonance conditions eq. (59) broaden or vanish altogether. For large amplitude Alfvén waves (de la Beaujardière and Zweibel 1989) cyclotron resonance is not a necesssary condition for particle acceleration. As particles start to mirror between large wave crests (due to the mirror force $-\mu \nabla B$, $\mu = \frac{1}{2}m_o v_\perp^2/B$ is the non-relativistic magnetic moment) charge separation occurs and parallel electric fields arise. Then the Čerenkov condition ($N = 0$ in Eq. (59)) is already sufficient for energy transfer from waves to particles. Further although particles become trapped in the strong waves the nonlinear orbits in different waves overlap in phase space and *chaotic* diffusion (and acceleration) of particles in momentum becomes possible. It is not clear if particles can attain speeds much larger than ten times the ambient Alfvén speed.

Chaotic acceleration also appears in a 'monochromatic' strong wave due to resonance broadening to such an extent that the resonant volumes corresponding to subsequent harmonics N in eq. (59) overlap in momentum space (Menyuk *et al.* 1987; Hizanidis *et al.* 1989; Karimabadi *et al.* 1990). The characteristic wave amplitude required for this acceleration process to work follows from $\omega_{tr} > \omega_{co}/\gamma$ and the expression for the bounce or trapping frequency $\omega_{tr} \approx \tau_{tr}^{-1}$ eq. (75).

Finally, a special case is acceleration of single particles by the extremely strong 'vacuum' electromagnetic wave of a radio pulsar impinging on the surrounding plasma (Ostriker and Gunn 1969), initially with a plasma frequency above the frequency of the electromagnetic wave. As particles at the plasma boundary get accelerated to extreme Lorentz factors the (transverse) plasma frequency decreases ($\omega_p^t \propto \gamma^{-0.5}$) and the wave 'burns' itself into the plasma thereby accelerating the particles. Later work is by (Tsintsadze 1989) and Thielheim (1989).

2.4 Conclusion

The main problem with acceleration by waves is the *origin* of the particular kind of waves invoked. Somehow large-scale kinetic energy must be converted into waves. A good case may be particle acceleration in the solar corona. The observations tell us that here particle acceleration predominantly occurs in solar flares and for instance not in large-scale shocks, which is also supported by calculations (Achterberg and Norman 1980). In a solar flare the stored magnetic energy (created by dynamo action below the photosphere) is explosively released on a time scale not much larger than the Alfvén crossing time of the relaxing magnetic structure. While it is true that the expected sudden changes in coronal field topology during flares must be *abundant radiators of low-frequency MHD waves* these waves primarily have periods of the order of milliseconds or larger, corresponding to the characteristic Alfvén travel time of the reconnecting regions and their frequency is too small to accelerate low-energy particles from typical coronal energies upward. Therefore an *injection problem* exists. Either a cascade of MHD waves to higher frequencies (e.g. by nonlinear wave couplings) ensures efficient acceleration from coronal to subrelativistic energies or preacceleration of the coronal plasma is required by other mechanisms, or some kind of strong turbulence is at play.

Preacceleration by high-frequency electrostatic waves does not really solve the problem if the generation of these waves is not explained. Again a process would be needed to produce such waves out of the low-frequency flare turbulence. More interesting are the *lower-hybrid* waves (Benz and Smith 1987).

The second alternative way for preacceleration is through *shock* wave heating in shocks in the nonsteadily reconnecting flare plasma (Cargill, Goodrich and Vlahos 1988).

Finally the third alternative possibility for acceleration from coronal energies upwards consists of direct electric fields, possibly inside *electrostatic double layers* (Alfvén 1981).

Although mhd waves, shocks and double layers have different acceleration properties and effiencies, in a realistic solar flare they may all be generated by reconnection, and probably in a stochastic fashion. Observational evidence for the existence of a multitude of acceleration centers comes from the spatial (Tapping *et al.* 1983; Kattenberg 1981) and spectral properties of the radio emission (Kuijpers, van der Post and Slottje 1981), from the association of radio spikes with Type III bursts and hard X-rays (Benz 1985; Vlahos 1989) and from X-rays (Schadee *et al.* 1983; Martens, van den Oord and Hoyng 1985; Schadee 1986 *'nanoflares'*; de Jager *et al.* 1987; Lin *et al.* 1984; Athay 1984; Canfield and Metcalf 1987). It appears that the energy release in a solar flare is highly fragmented (Vlahos 1994).

Observations of gamma ray emission in solar flares show that in the first few seconds of the energy release both electrons and ions are accelerated to relativistic energies: electrons in 1 s up to 100 MeV and ions within seconds to GeV energies (Rieger 1989, 1994). At present the best candidates for acceleration in the flare involve a *multitude of nonlinear structures* either in the form of double layers or of shocks or of a high level of mhd waves (Vlahos 1989, Cargill 1991). The action of many such coherent structures in turbulent flows can accelerate particles with a power-law dependence on particle energy (Anastasiades and Vlahos 1993).

The question of the nature of particle acceleration and the role of waves is connected to the problem of the *energy partitioning* in a stellar flare. *What determines the quality of the flare products?* How much of the energy appears as bulk motion, how much as acceleration of a small fraction of the particle population and how much as heating (and perhaps also how much as intense radiation (Melrose, Hewitt and Dulk 1984)). What parameters determine the end products? In some cases (Mätzler and Wiehl 1980) the flare consists of pure heating only. It may however be that the primary energy release is unique, for instance in the form of pure runaway acceleration (Moghaddam-Taaheri and Goertz 1990; Holman and Benka 1992) but that the column density of the overlying gas determines whether the accelerated particles can be observed or whether their energy is completely degraded into heating.

3 Shock Wave Acceleration

3.1 Diffusive Shock Acceleration

Collisonless shocks traveling along an ambient magnetic field can support a first-order Fermi type acceleration process, which is known as *diffusive* or *regular* acceleration (Axford, Leer and Skadron 1977; Krimsky 1977; Bell 1978; Blandford and Ostriker 1978). For sufficiently energetic particles the shock behaves as a discontinuity. While such particles are scattered back and forth across the shock they effectively find themselves in between two approaching mirrors, the role of the mirrors being played by the shocked and the unshocked background plasmas.

In Fig. 13 the shock is standing still; the upstream plasma (index 1) approaches from the right with the shock speed w_1 along the magnetic field of magnitude B_0, while the shocked plasma (index 2) leaves downstream to the left with a speed w_2 with respect to the shock front. Also indicated is the pitch angle ζ of a fast particle with respect to the magnetic field. Shock acceleration works if

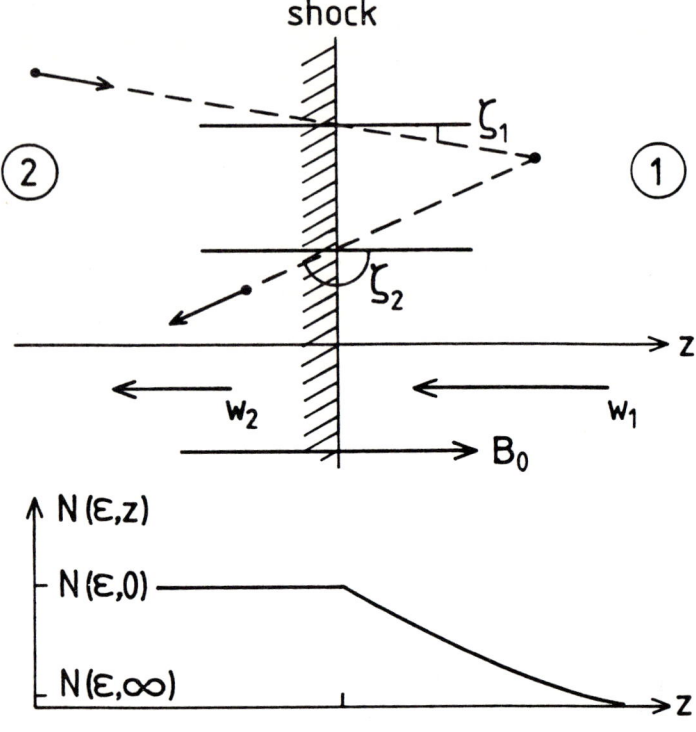

Fig. 13. Top: shock geometry and notation. Below: shape of the steady energy spectrum of accelerated particles.

1. energetic particles are present with a scattering mean free path larger than the width of the collisionless shock, and
2. the energetic particle population is kept isotropic on each side with respect to the respective flow speed of the ambient plasma.

If the acceleration time scale t_{acc} and the escape time scale of the fast particles t_{esc} have the same dependence on energy a differential power-law distribution is established under steady state conditions, which depends only on the strength of the shock and reaches a universal slope 2 ($f(\epsilon) \propto \epsilon^{-2}$) for strong shocks in 'monatomic' plasmas (essentially because in such shocks $t_{acc} = t_{esc}$). This result, together with the generic occurrence of shocks near supernova remnants and active galactic nuclei – believed to be the most important producers of cosmic rays – has focussed most of the efforts in particle acceleration on shock waves. Reviews of diffusive shock acceleration are Blandford and Eichler (1987), Achterberg (1990), Jones and Ellison (1991), and Kirk (1995).

Scattering and Spatial Diffusion In the upstream plasma the fast particles are thought to scatter on self-excited Alfvén waves. Sufficiently energetic particles excite Alfvén waves as long as their systematic velocity with respect to the ambient plasma exceeds the Alfvén speed. Conversely a strong anisotropy of streaming cosmic rays will be reduced by the excitation of and scattering on Alfvén (or on whistler) waves until the anisotropy is reduced to a fraction v_A/w_1. In the downstream plasma appeal is made to 'turbulent fluctuations' created or amplified inside the presumably turbulent collisionless shock to maintain near-isotropy. I will first discuss the scattering in self-excited Alfvén turbulence.

Pitch Angle Scattering We are interested in the particular part of the Fokker-Planck equation eq. (23) which describes scattering in pitch angle

$$\frac{\partial f}{\partial t} = \frac{1}{\sin\zeta}\frac{\partial}{\partial\zeta}\left[\sin\zeta D_\zeta \frac{\partial f}{\partial\zeta}\right], \tag{114}$$

where the pitch angle diffusion coefficient

$$D_\zeta \equiv \frac{<\Delta\zeta\Delta\zeta>}{2\Delta t} \tag{115}$$

has to be calculated.

As we are interested in pitch angle scattering on Alfvén waves we use the magnetostatic approximation in which the effect of the wave electric field is neglected with respect to that of the wave magnetic field. This is satisfied if the phase speed of the waves is much less than the particle speed (which will be true for non-relativistic Alfvén waves):

$$\frac{|\mathbf{E}(\mathbf{k},\omega)|}{|\mathbf{v}\times\mathbf{B}(\mathbf{k},\omega)|} \approx \frac{|\mathbf{E}(\mathbf{k},\omega)|}{|\mathbf{v}\times(\mathbf{k}\times\mathbf{E}(\mathbf{k},\omega)/c)|} \approx \frac{\omega}{kv} \ll 1. \tag{116}$$

In fact (see Section) the assumption implies that we consider only momentum transfer between waves and particles (which arises from the wave magnetic field) but will neglect any energy exchange (which comes from the electric field). As in Section () we approximate the equation of motion to first-order along the unperturbed particle orbit

$$\frac{d\mathbf{p}}{dt} = q(\mathbf{v}\times\mathbf{B}_0 + \mathbf{v}\times\mathbf{B}(\mathbf{r},t)). \tag{117}$$

It follows that the particle energy is conserved:

$$\frac{d\epsilon^2}{dt} = c^2\mathbf{p}\cdot\frac{d\mathbf{p}}{dt} = 0, \tag{118}$$

and that the unperturbed orbit is given by (Cartesian coordinates with $\mathbf{z} \parallel \mathbf{B}_0$)

$$\mathbf{v} = v_\perp\cos(\phi_0 - \omega_c t)\hat{\mathbf{x}} + v_\perp\sin(\phi_0 - \omega_c t)\hat{\mathbf{y}} + v_\parallel\hat{\mathbf{z}},$$

$$\mathbf{r} = \left(x_0 - \frac{v_\perp}{\omega_c}\{\sin\phi_0 - \sin(\phi_0 - \omega_c t)\}\right)\hat{\mathbf{x}}$$

$$+ \left(y_0 + \frac{v_\perp}{\omega_c}\{\cos\phi_0 - \cos(\phi_0 - \omega_c t)\}\right)\hat{\mathbf{y}}$$

$$+ (z_0 + v_\parallel t)\hat{\mathbf{z}}, \tag{119}$$

where the cyclotron frequency $\omega_c = qB(\gamma m_o)^{-1}$ carries the sign of the charge. Let us consider plane waves along B_0, with

$$\mathbf{k} = k\hat{z}, \ \mathbf{B}(\mathbf{r}, t) = B_x(\mathbf{r}, t)\hat{x} + B_y(\mathbf{r}, t)\hat{y}. \tag{120}$$

Defining the pitch angle ζ by

$$\cos\zeta = \frac{\mathbf{p} \cdot \mathbf{B}_0}{pB_0} \tag{121}$$

we find from eq. (117) using eq. (118) (γ is conserved)

$$\frac{d\zeta}{dt} = -\omega_c \left\{ \cos\phi\frac{B_y}{B_0} - \sin\phi\frac{B_x}{B_0} \right\}, \tag{122}$$

with $\phi = \arctan(p_y/p) = \phi_0 - \omega_c t/\gamma$. Defining

$$B_\pm(\mathbf{r}(t), t) \equiv B_x(\mathbf{r}(t), t) \pm iB_y(\mathbf{r}(t), t), \tag{123}$$

we can write eq. (122) as

$$\frac{d\zeta}{dt} = \frac{1}{2}i\omega_c \left(\frac{B_+}{B_0}e^{-i\phi(t)} - \frac{B_-}{B_0}e^{i\phi(t)} \right). \tag{124}$$

Then a calculation completely similar to that for eq. (102) leads to the result

$$< \Delta\zeta\Delta\zeta > = \int\limits_{t}^{t+\Delta t} dt' \int\limits_{-\infty}^{\infty} ds \frac{\omega_c^2}{4B_0^2}$$

$$* [- < B_+(t')B_+(t' + s) > e^{-i(2\phi_0 - \omega_c(2t'+s))}$$
$$+ < B_+(t')B_-(t' + s) > e^{-i\omega_c s} + < B_-(t')B_+(t' + s) > e^{i\omega_c s}$$
$$- < B_-(t')B_-(t' + s) > e^{i(2\phi_0 - \omega_c(2t'+s))}]. \tag{125}$$

As the first and the last term are rapidly varying in t' ($\propto exp(\pm 2i\omega_c(t'))$) their contribution vanishes since we take $\Delta t > \omega_c^{-1}$. Fourier transforming and using the RPA,

$$\mathbf{B}(\mathbf{r}, t) = \int \frac{dK\, d\omega}{(2\pi)^2} \mathbf{B}(K, \omega)e^{iKz - i\omega t},$$

$$< B_i(K, \omega)B_j(K', \omega') > = \lim_{T, L \to \infty} \frac{B_i(K, \omega)B_j^*(K, \omega)}{TL}(2\pi)^2\delta(K + K')\delta(\omega + \omega'),$$

$$B_i(K, \omega) = B_i^*(-K, -\omega), \tag{126}$$

we arrive at

$$\frac{< \Delta\zeta\Delta\zeta >}{\Delta t} = \lim_{T, L \to \infty} \frac{1}{TL}\frac{\omega_c^2}{4} \int \frac{dK\, d\omega}{(2\pi)^2} \tag{127}$$

$$\left(\frac{< |B_+(K, \omega)|^2 >}{B_0^2}2\pi\delta(\omega - Kv_\parallel - \omega_c) + \frac{< |B_-(K, \omega)|^2 >}{B_0^2}2\pi\delta(\omega - Kv_\parallel + \omega_c) \right),$$

where $B_\pm \equiv iB_x \pm B_y$, $v_\parallel \equiv v\cos\zeta$. Note that the appearance of $\pm\omega_c$ in the delta function does not arise from the phase of the wave as seen by the particle $((\omega - Kv_\parallel)t)$ but from the velocity dependence of the Lorentz force (see eq. (117) and eq. (124)).

To find the scattering efficiency by *Alfvén waves* we decompose the Alfvén waves in opposite circular polarizations: $B_x(K)/B_y(K) = \pm i$ for L, respectively R, polarization, and write

$$\mathbf{B}_L(K,\omega) = \frac{1}{2}B_L(K)(\hat{\mathbf{x}} - i\hat{\mathbf{y}})2\pi\delta(\omega - Kv_A),$$

$$\mathbf{B}_R(K,\omega) = \frac{1}{2}B_R(K)(\hat{\mathbf{x}} + i\hat{\mathbf{y}})2\pi\delta(\omega - Kv_A), \qquad (128)$$

so that $|\mathbf{B}_\pm|^2 = 2|\mathbf{B}_{L,R}|^2$. Assuming $|Kv_A| \ll |\omega_c| + Kv_\parallel$ and remembering that $\omega_c = \omega_{co}/\gamma$ it follows from eq. (128)

$$D_\zeta = \frac{<\Delta\zeta\Delta\zeta>}{2\Delta t}$$

$$= \lim_{L\to\infty} \left(\frac{|\omega_{co}K||B_L(K)|^2}{8\gamma B_0^2 L}\Big|_{K=-\frac{\omega_{co}}{v_\parallel\gamma}} + \frac{|\omega_{co}K||B_R(K)|^2}{8\gamma B_0^2 L}\Big|_{K=+\frac{\omega_{co}}{v_\parallel\gamma}} \right). (129)$$

Spatial Diffusion Particles which are scattered in pitch angle reduce their anisotropy, as we will show first. The isotropic part of the distribution is

$$\bar{f}(p) = \frac{1}{2}\int_0^\pi d\zeta \sin\zeta f(p,\zeta), \qquad (130)$$

and a measure of the anisotropy is

$$I = \frac{1}{2}\int_0^\pi d\zeta \sin\zeta [f(p,\zeta) - \bar{f}(p)]^2 > 0. \qquad (131)$$

It follows from eq. (114)

$$\frac{\partial I}{\partial t} = -\int_0^\pi d\zeta \sin\zeta D_\zeta \left(\frac{\partial f}{\partial\zeta}\right)^2 < 0. \qquad (132)$$

Because of the scattering, particles are inhibited in their propagation along the ambient magnetic field. Both diffusion in pitch angle and diffusion along the magnetic field depend therefore on the level of fluctuations, be it *inversely !* In a steady state, the inhomogeneity and the anisotropy are coupled, as follows from calculating the effective propagation speed of a (monoenergetic) particle distribution along the ambient magnetic field:

$$<v_\parallel> = \frac{1}{2}\int_0^\pi d\zeta \sin\zeta v\cos\zeta f(p,\zeta) = -\frac{1}{2}v\int_0^\pi \sin^2\zeta \frac{\partial f(p,\zeta)}{\partial\zeta}d\zeta. \qquad (133)$$

We will now derive diffusion along the ambient magnetic field as a function of the scattering rate. In the case of strong scattering we can allow for a slight inhomogeneity by complementing the diffusion equation eq. (114) with a free-streaming term

$$\frac{\partial f}{\partial t} + v\mu \frac{\partial f}{\partial z} = \frac{\partial}{\partial \mu}\left(\frac{1-\mu^2}{2}\nu(\mu)\frac{\partial f}{\partial \mu}\right), \tag{134}$$

where we have changed variables $\zeta \rightarrow \mu \equiv \cos\zeta$ and we have introduced the scattering rate

$$\nu(\mu) \equiv \frac{<\Delta\zeta\Delta\zeta>}{\Delta t} = 2D_\zeta. \tag{135}$$

Suppose ν is large enough so that the particles are nearly isotropic. We develop f in the small parameter $\epsilon \propto \nu^{-1} \ll 1$

$$f(p,\mu) = f_0 + f_1 + O(\epsilon^2). \tag{136}$$

Then to order ϵ^{-1} eq. (134) gives

$$\frac{\partial}{\partial \mu}\left(\frac{1-\mu^2}{2}\nu\frac{\partial f}{\partial \mu}\right) = 0, \tag{137}$$

so that $f_0(p,\zeta) = f_0(p)$ is isotropic. Substituting this result into eq. (134) and averaging over μ proves the stationarity of f_0:

$$\frac{1}{2}\int_{-1}^{1}\frac{\partial f_0}{\partial t}d\mu = \frac{\partial f_0}{\partial t} = 0. \tag{138}$$

To order ϵ^0 eq. (134) gives

$$v\mu\frac{\partial f_0}{\partial z} = \frac{\partial}{\partial \mu}\left(\frac{1-\mu^2}{2}\nu(\mu)\frac{\partial f_1}{\partial \mu}\right), \tag{139}$$

and, after integrating twice,

$$f_1(p,\mu) - f_1(p,-1) = -\int_{-1}^{\mu}d\mu'\frac{v}{\nu(\mu')}\frac{\partial f_0}{\partial z}. \tag{140}$$

To order ϵ we find from eq. (134)

$$\frac{\partial f_1}{\partial t} + v\mu\frac{\partial f_1}{\partial z} = \frac{\partial}{\partial \mu}\left(\frac{1-\mu^2}{2}\nu(\mu)\frac{\partial f_2}{\partial \mu}\right). \tag{141}$$

Substituting eq. (140) into eq. (141), and averaging over μ, results in

$$\frac{\partial \bar{f_1}}{\partial t} - \frac{1}{2}\frac{\partial}{\partial z}\int_{-1}^{1}d\mu v\mu\int_{-1}^{\mu}d\mu'\frac{v}{\nu(\mu')}\frac{\partial f_0}{\partial z} = 0, \tag{142}$$

which after partial integration and the notation $\bar{f} = f_0 + f_1$ results in the transport equation (correct up to order ϵ^2)

$$\frac{\partial \bar{f}}{\partial t} = \frac{\partial}{\partial z} D_\parallel \frac{\partial \bar{f}}{\partial z},\tag{143}$$

where

$$D_\parallel \equiv \frac{1}{2} \int\limits_{-1}^{1} d\mu v^2 \frac{1 - \mu^2}{2\nu(\mu)} = \frac{v^2}{3\nu_\zeta},\tag{144}$$

where the last equality only holds when the scattering rate is independent of angle, $\nu(\mu) = \nu_\zeta$. Note that this result can be derived in a heuristic way by recalling that a spatial diffusion coefficient can be written in terms of the mean free path λ and the speed v as

$$D_\parallel = \frac{1}{3}\lambda v = \frac{v^2}{3\nu_\zeta} = \frac{v^2}{6D_\zeta}.\tag{145}$$

For magnetic fluctuations $\delta \mathbf{B}$ one has (see eq. (128), eq. (129) or eq. (135))

$$\nu_\zeta \approx \frac{|\omega_{co}|}{2\gamma} \frac{|\delta \mathbf{B}|^2}{|\mathbf{B}|^2}.\tag{146}$$

We are now in a position to derive the steady spectrum of accelerated particles.

Particle Spectrum in Parallel Shocks We first determine the average energy gain per fast particle when it has crossed the shock back and forth once. Referring for the notation to Fig. 13 and applying one Lorentz transformation for a particle which crosses the shock from the downstream to the upstream region 1, we find for its energy ϵ and parallel momentum $p_\parallel = p_z$ for a parallel (to $\mathbf{B}_0$) shock

$$\epsilon_1 = \gamma_s \{\epsilon + (w_1 - w_2)p \cos \zeta_1\},$$
$$p_{1z} = \gamma_s \{p_z + \gamma m_o(w_1 - w_2)\},\tag{147}$$

where the Lorentz factor of the particle is $\gamma = (1 - v^2/c^2)^{-0.5}$ and of the relative fluid motion $\gamma_s = \{1 - (w_1 - w_2)^2/c^2\} \approx 1$ for a non-relativistic shock. If the scattering is elastic with respect to the upstream plasma (magnetostatic fluctuations) the particle energy upon its return to the downstream side is

$$\epsilon_2 = \gamma_s \{\epsilon_1 - (w_1 - w_2)p_1 \cos \zeta_2\},\tag{148}$$

and the energy change for a non-relativistic shock and a fast particle (with $p \gg \gamma m_o |w_1 - w_2|$)

$$\Delta \epsilon = \epsilon_2 - \epsilon \approx (w_1 - w_2)p(-\cos \zeta_2 + \cos \zeta_1).\tag{149}$$

If we assume isotropy of the fast particles in the upstream frame and of those in the downstream frame we find the average change of energy per crossing back and forth by multiplying eq. (149) with the particle flux ($\propto 2\pi \cos \zeta \sin \zeta d\zeta$) and integrating over the corresponding half solid angle (from $0 - \pi/2$ for ζ_1 and from $\pi/2 - \pi$ for ζ_2)

$$< \Delta\epsilon > = \frac{4}{3}(w_1 - w_2)p = \alpha\frac{v}{c}\epsilon,$$

$$\alpha \equiv \frac{4}{3}\frac{w_1 - w_2}{c}. \tag{150}$$

Note that this result is first order in $w_1 - w_2$ and corresponds to first-order Fermi acceleration only for fast particles with a mean free path $\lambda = v/\nu_\zeta$ much exceeding the shock thickness $L_s \ll \lambda$.

Steady Spectrum As the fast particles are isotropic with respect to the fluid rest frame in which they satisfy eq. (143), their evolution in the shock frame is found by allowing for a free streaming. Changing variables according to

$$N(\epsilon)d\epsilon = 4\pi p^2 f(p)dp, \tag{151}$$

the approximate evolution of the energy spectrum in the shock frame is

$$\frac{\partial N(\epsilon, z, t)}{\partial t} + \frac{\partial}{\partial z}\left\{-wN(\epsilon, z, t) - D_\parallel\frac{\partial N(\epsilon, z, t)}{\partial z}\right\} = 0, \tag{152}$$

where, as before, $D_\parallel = v^2(3\nu_\zeta)^{-1}$. The stationary solution of eq. (152) is of the form

$$N(\epsilon, z) = A + B\exp\left(-\int_0^z dz\frac{w}{D_\parallel}\right). \tag{153}$$

The upstream solution for the differential density $N(\epsilon, z)$ is

$$N_1(\epsilon, z) = N(\epsilon, \infty) + \{N(\epsilon, 0) - N(\epsilon, \infty)\}\exp\left(-\int_0^z dz\frac{w}{D_\parallel}\right), \tag{154}$$

and downstream:

$$N_2(\epsilon, z) = N(\epsilon, 0). \tag{155}$$

while the downstream flux $F(\epsilon)$ is

$$F_2(\epsilon) = -w_2 N(\epsilon, 0). \tag{156}$$

The average chance of escape of a fast particle towards the downstream region whence it will never return, is determined by comparing the flux F_2 to the flux $F(2 \to 1)$. The latter is approximately

$$F(2 \to 1) \approx \int_0^{\pi/2} 2\pi v\cos\zeta\frac{N(\epsilon)}{4\pi}\sin\zeta d\zeta = N(\epsilon, 0)v/4, \tag{157}$$

assuming $w \ll v$. Finally the chance of escape is

$$\eta = \frac{|F_2|}{|F(2 \to 1)|} = \frac{4w_2}{v} \ll 1. \tag{158}$$

As the acceleration time relates to the escape time according to

$$\frac{t_{acc}}{t_{esc}} = \frac{\eta}{\alpha}, \tag{159}$$

we can immediately write down the first-order Fermi spectrum (Section 1.2)

$$N(\epsilon) \propto \left(\frac{\epsilon}{\epsilon_0}\right)^{-\left(\frac{\eta}{\alpha}+1\right)} = \left(\frac{\epsilon}{\epsilon_0}\right)^{-\frac{2w_2+w_1}{w_1-w_2}} \propto \epsilon^{-2}, \tag{160}$$

where the last equality is valid only for a strong non-relativistic shock in a gas with three degrees of freedom. Note that it is assumed that there are two populations of particles: the cosmic rays for which we have derived the steady spectrum, and the ambient gas which is shocked and which obeys the Rankine-Hugoniot shock relations based on conservation of mass, energy and momentum. These adiabatic shock relations imply a maximum density contrast for a strong shock (Mach number much larger than unity)

$$\frac{\rho_2}{\rho_1} = \frac{w_1}{w_2} = \frac{\gamma_P + 1}{\gamma_P - 1}, \tag{161}$$

where γ_P is the adiabatic gas (Poisson) index (5/3 for a non-relativistic and 4/3 for a relativistic, monatomic gas).

Injection Energy The excitation of Alfvén waves by the cosmic rays puts a lower limit on the particles to be accelerated. They should at least satisfy the lowest cyclotron resonance

$$\omega - k_\| v_\| \pm \omega_{co}/\gamma = 0. \tag{162}$$

Using $\omega^A(\mathbf{k}) = k_\| v_A$ for Alfvén waves and $v_\| \gg v_A$ for the cosmic rays it follows that $|v_\|| \approx v_A |\omega_{co}(\gamma\omega)^{-1}|$. For a particle of species α and Larmor frequency $\omega_{c\alpha} = q_\alpha B_0/m_\alpha$ we find an injection energy ϵ_α and an injection momentum p_α

$$p_\alpha = \epsilon_\alpha \frac{v}{c^2} > \left|\frac{\omega_{c\alpha}}{\omega}\right| m_\alpha v_A \geq 2m_p v_A, \tag{163}$$

where we have used that Alfvén waves have frequencies much below the proton Larmor frequency $\omega \leq \omega_{cp}/2$ and the index p stands for proton. Electrons require a higher starting energy than protons (with a factor m_p/m_e) and are therefore less easy to accelerate.

As electrons of lower energies can also excite the higher-frequency whistler waves, their injection energy may be lower than given by eq. (163). This has been worked out for shock acceleration with a thermal subshock (Levinson 1994).

Beyond the Classical Case We have given the derivation of the steady spectrum of cosmic rays for the classical case of a strong, non-relativistic shock moving along the (uniform) magnetic field direction in the test particle limit, for a low-level of (magnetostatic) fluctuations $\delta B/B \ll 1$ and in the absence of losses. Each of these conditions can be relaxed and a pictorial overview of the resulting changes (Ballard and Heavens 1992) can be found in Longair (1994). The upshot is that diffusive shock acceleration is a very robust process in astrophysics. In particular it is not very sensitive to alignment of the shock direction with the magnetic field as long as the shock is not moving nearly perpendicular.

A point to consider is the magnetic field amplitude of the scattering fluctuations. As $\delta B/B \to 1$ a perturbation approach becomes invalid (see Giacalone *et al.* 1993).

In a realistic shock the cosmic ray component may not be a small perturbation in the energy budget of the shock. In particular the gradient of the cosmic ray pressure works on the incoming gas and retards it. As a result the shock proper is smeared out. How this affects the efficiency is treated by Zank *et al.* 1993. Diffusive shock acceleration can be very efficient, eg. Jun *et al.* 1994 with a two-fluid model of a piston-driven shock attain a conversion efficiency of kinetic shock energy into cosmic rays up to 90 %.

If the magnetostatic assumption about the turbulence is dropped not only scattering in pitch angle but also diffusion in energy takes place (Section 2.1). This second-order Fermi type acceleration can be incorporated and sometimes leads to important modifications (Schlickeiser 1994).

3.2 Shock Drift Acceleration

A different acceleration process occurs in a shock moving perpendicular to the magnetic field. A charged particle experiences a 'grad B' drift when it crosses the shock front. As the drift is along the shock front and parallel to the electric force, it gains energy. This process is called *shock drift acceleration* (Pesses 1981; Pesses, Decker and Armstrong 1982; Webb, Axford and Terasawa 1983; Holman and Pesses 1983; Decker 1988).

Perpendicular Non-relativistic Shocks Let me illustrate the mechanism for the simplest perpendicular shock (in which not only the magnetic field is perpendicular to the shock normal but also the upstream speed is parallel to the shock normal). Fig. 14 is in the frame of the shock; the sketch on the left contains the field lines while the sketch on the right is perpendicular to the magnetic field. As the flow in this frame is stationary only a uniform electric field $\mathbf{E} = -\mathbf{w_1} \times \mathbf{B_1} = -\mathbf{w_2} \times \mathbf{B_2}$ exists. (That these expressions are consistent with continuity of the transverse electric field across the shock as required by Maxwell's laws follows from mass-conservation and magnetic flux-freezing applied to the transverse magnetic component.) Now a charged particle experiences a grad B force in an inhomogeneous magnetic field $\mathbf{F}_B = -\mu \boldsymbol{\nabla} B$, where

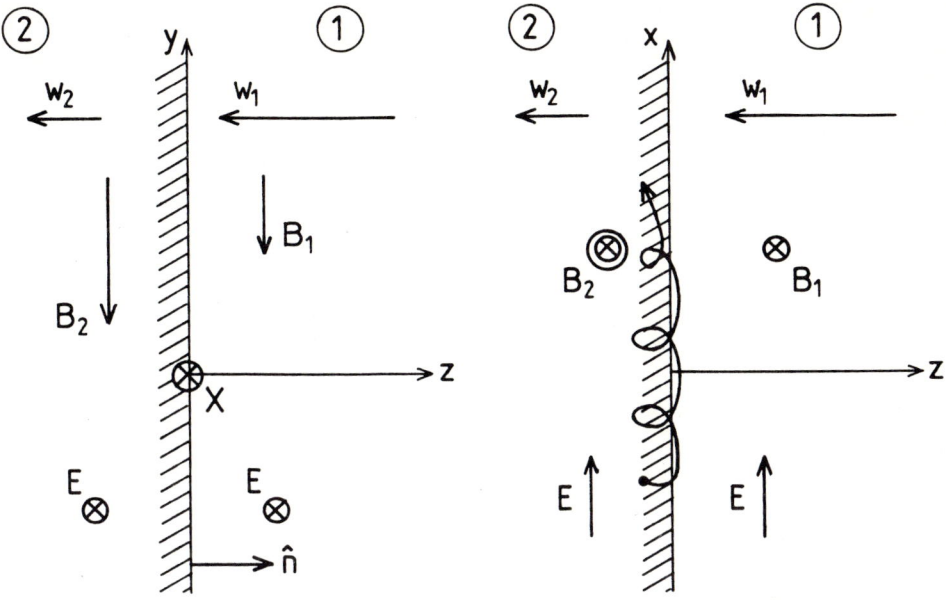

Fig. 14. Perpendicular shock. A positive ion is accelerated along the electric field in the x-direction.

the magnetic moment is

$$\mu = \frac{1}{2} \int |\mathbf{r} \times \mathbf{j}| d^3\mathbf{r} = \frac{\gamma m_o v_\perp^2}{2B} = \frac{p_\perp^2}{2m_o \gamma B}, \tag{164}$$

and undergoes therefore a drift eq. (9)

$$\mathbf{v}_D = -\mu \frac{\nabla B \times \mathbf{B}}{qB^2}. \tag{165}$$

It follows that this drift is in the same direction as the electric force $q\mathbf{E} = -q\mathbf{w} \times \mathbf{B}$ for both positive and negative charges. Therefore charges are accelerated during their passage of the shock. Moreover in a perpendicular shock all particles from the upstream region that hit the shock, also cross it eventually by the $\mathbf{E} \times \mathbf{B}$ drift.

To find the average energy gain of the particles we first consider the particles with a radius of gyration smaller than the thickness of the shock front. They form the majority of the plasma and only experience adiabatic changes as they cross the shock. For these the quantity $p_\perp^2/B = 2m_o\gamma\mu$ is conserved.

For particles with a cyclotron radius equal to or larger than the shock front $p_\perp^2$ is not conserved and these particles may experience larger gains than the bulk. However numerical calculations show that for distributions that are isotropic in gyrophase to start with, again their average $< \gamma\mu >$ is conserved. It has been pointed out by Pesses (1981) that this result can be easily understood if one considers the angular momentum density of the particles due to their gyration around the magnetic field. For isotropic (in gyrophase) distributions the average (per species) angular momentum per particle is

$$< \mathbf{L} >= \frac{-q\mathbf{B}}{|q\mathbf{B}|}\gamma m_o v_\perp r_c = -\frac{\gamma^2 m_o^2 v_\perp^2 q\mathbf{B}}{|q\mathbf{B}|^2}, \tag{166}$$

and the average flux of angular momentum in the upstream ($j = 1$), respectively downstream ($j = 2$), region

$$<\overleftrightarrow{\mathbf{M}}_j >= n_j \mathbf{w}_j < \mathbf{L}_j >= \rho_j w_{jz}\frac{\gamma_j^2 m v_{j\perp}^2}{q B_j}\hat{\mathbf{z}}\hat{\mathbf{y}}, \tag{167}$$

where $\mathbf{y} \parallel \mathbf{B}$ and $\mathbf{z} \parallel \mathbf{w}_j$. In a steady state and in the absence of collisions this flux is conserved across the shock. However, using particle conservation $\rho_1 w_{1z} = \rho_2 w_{2z}$ eq. (167) then immediately implies that on average $< p_\perp^2/B >$ is conserved also for these 'nonadiabatic' particles. The result is that particle drift acceleration in a perpendicular shock cannot be much more than in adiabatic compression: particles can not gain in energy by more than a factor of a 'few' only.

The General Case: Oblique Shocks In the shock frame (any frame in which the shock front does not move) of an oblique shock the downstream gas flow in general makes an angle with respect to the upstream flow (see Fig. 15) because at the shock front the Lorentz force can have a component along the shock front in the y-direction so that $w_{2y} \neq 0$. However as before Maxwell's laws for a steady state require continuity of the transverse electric field and of the parallel magnetic field

$$E_{1x} = B_{1y}w_1 = B_{2y}w_{2z} = E_{2x},$$
$$B_{1z} = B_{2z}. \tag{168}$$

Superluminal and Subluminal Shocks The physical distinction between a perpendicular and a parallel shock is in fact determined by the ratio $|w_1(c\tan\alpha_1)^{-1}|$ as compared to unity. As can be seen from Fig 15 $w_1/\tan\alpha_1$ is the velocity with which the kinks in the field lines at the shock front slide along the front. If this speed is sub-relativistic a Lorentz transformation can be made (with a boost $\mathbf{v}_L = \hat{\mathbf{y}}w_1/\tan\alpha_1$) to a frame in which the field pattern is static. This frame is called the de Hoffmann-Teller frame (de Hoffmann and Teller 1950). In this frame the flow is along the magnetic field lines. In the upstream region this can be seen from $|\tan\alpha_1| = |B_{1z}/B_{1y}| = |w_1|/|\mathbf{v}_L|$. As a result the upstream

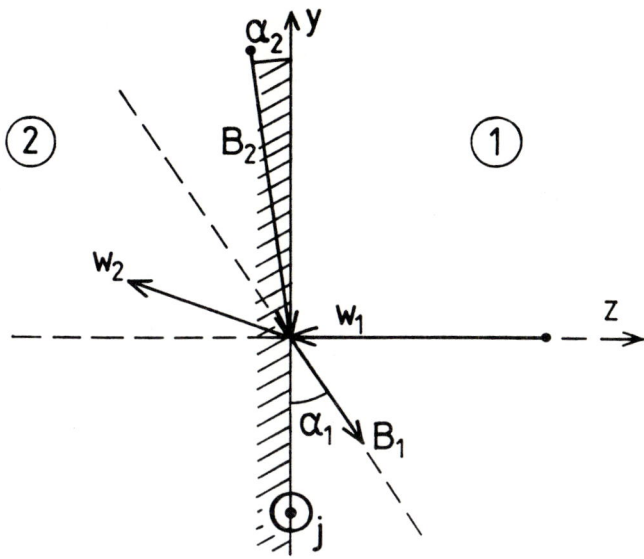

Fig. 15. Oblique shock.

electric field vanishes $\mathbf{E}'_1 = 0$ and, because of continuity of the tangential electric field, also the downstream tangential electric field must vanish. Further a downstream electric field component in the z-direction would be partly along the shocked magnetic field which itself by assumption (oblique shock) makes an angle with the shock front. As this would imply an infinite current in the ideal MHD picture this is not permitted. Therefore finally $\mathbf{E}'_2 = 0$. (Note that a vanishing $\mathbf{E}_1$ also follows generally from the Lorentz transformation $E'_{1y} = E_{1y} = 0$, $\mathbf{E}'_{1\perp} = \gamma_L(\mathbf{E}_{1\perp} + \mathbf{v}_L \times \mathbf{B}_1) = \gamma(w_1 B_1 \cos\alpha_1 + |\mathbf{v}_L| B_1 \sin\alpha_1)\hat{\mathbf{x}} = 0.)$

Superluminal shocks are shocks for which a de Hoffmann-Teller frame does not exist. In any shock frame they are characterized by $|w_1/\tan\alpha_1| > c$. In the laboratory frame, in which the upstream plasma is at rest and the shock moves at a speed $-w_1\hat{\mathbf{z}}$ this corresponds to the condition $|w_1/\sin\alpha_1^{lab}| > c$ where α_1^{lab} is the angle between the unshocked magnetic field and the shock frontal plane. The perpendicular shock itself belongs to this class. Physically important is that the electric field can not be transformed away everywhere. In this case, however, the perpendicular component of the magnetic field can be transformed away by a Lorentz boost along the shock front downwards with a speed $\mathbf{v}_L = -\hat{\mathbf{y}}c^2\tan\alpha_1|w_1|^{-1}$. A superluminal shock can therefore always be transformed to a *perpendicular* shock, similar to Fig. 14, in which the magnetic field is aligned with the shock front but where the upstream inflow speed is not necessarily perpendicular to the front (see for an extensive discussion Kirk 1995).

For *subluminal* shocks a de Hoffmann-Teller frame exists as these shocks satisfy $|w_1/\tan\alpha_1| > c$ in (any) shock frame. Physically important is that the electric field can be transformed away.

Now, returning to our discussion of shock drift acceleration in oblique shocks, let us consider subluminal shocks first. As a de Hoffmann-Teller frame exists in which the electric field vanishes completely it would seem that no particles are accelerated at all and that the previous raesoning on the conservation of $p_\perp^2/B$ certainly applies in this case. However, what was assumed there but is not true anymore now, is that all upstream particles cross the shock. Instead some fast particles could be reflected and gain energy (in the laboratory system). Nevertheless, calculations (Decker 1988) show that again $p_\perp^2/B$ does not vary greatly. This, moderate, form of drift acceleration or reflection on the shock, has been used to explain electron beams ahead of the earth's bow shock (Wu 1984), and electrons radiating Type II solar radio bursts from coronal shocks (Holman and Pesses 1983; Benz and Thejappa 1988). For the effect of multiple encounters see Decker (1990), Kirk and Wassmann (1992).

On the other hand for relativistic superluminal shocks numerical calculations show that substantial particle acceleration can occur (Begelman and Kirk 1990; Lieu and Quenby 1990). An important difference with the non-relativistic perpendicular shock is that the drift speed can approach the particle speed and that the particle orbit at the shock front is very sensitive to its initial gyrophase. It is precisely because such relativistic superluminal shocks occur near relativistic outflows from rapidly spinning neutron stars, e.g. the Crab Nebula (Kennel and Coroniti 1984; Gallant and Arons 1994) or black holes, that shock drift acceleration is an important process. The maximum energy a particle can gain in shock drift acceleration is set by the total potential jump along the (finite) shock front $\int |(\mathbf{v} \times \mathbf{B}) \cdot d\mathbf{s}|$.

3.3 Combined Diffusive/Drift Shock Acceleration, Conclusion

Diffusive acceleration can be extended to oblique shocks and to perpendicular shocks. Such a combination has been proposed by Jokipii (1982, 1987), Lieu et al. (1994) and Achterberg and Ball (1994). This is a natural thing to do as pitch angle scattering implies inhibited transport along the ambient magnetic field but increased transport across. Suppose that upon every scattering with mean free path λ along the field, the particle makes a perpendicular excursion of a cyclotron radius r_c, than the spatial diffusion coefficients $D_{1\parallel}$, $D_{1\perp}$ along, respectively across the field in the upstream region are (using $\lambda = v/\nu_\zeta$ and eq. (145))

$$D_{1\parallel} = \frac{v^2}{3\nu_\zeta},$$

$$D_{1\perp} = \frac{D_{1\parallel}}{1 + \left(\frac{\lambda}{r_c}\right)^2} \approx D_{1\parallel} \left(\frac{r_c}{\lambda}\right)^2 = \frac{r_c^2 \nu_\zeta}{3}. \tag{169}$$

Provided perpendicular diffusion is fast enough diffusive acceleration also works in superluminal shocks.

Of course the maximum energy a particle gains is limited to the total potential jump along the shock front.

Finally, although we started off by claiming that shock wave acceleration is a first-order Fermi mechanism it turns out to be second-order (Jones 1994). This can be seen by calculating the *average* energy gain per unit time for a particle of momentum p, speed v, which is scattered across the shock repeatedly, and has a mean free path λ_1 in the upstream region, and λ_2 in the downstream region. The flow speed ahead is w_1, downstream w_2, and the shock compression ratio is called C so that $w_2 = w_1/C$ and $\Delta w = (1 - C^{-1})w_1$. Then the time spent by the particle upstream is on average $\Delta t_1 = \lambda_1/w_1$; the time spent downstream is $\Delta t_2 = \lambda_2/w_2 = C\lambda_2/w_1$; and the cycle time is $\Delta t \approx (\lambda_1 + C\lambda_2)/w_1$. As the momentum gain per cycle is $\Delta p = p4\Delta w(3v)^{-1}$, the rate of change of momentum is

$$< \frac{dp}{dt} > \approx \frac{4(C-1)w_1^2}{3vC(\lambda_1 + C\lambda_2)}p = \nu_{eff}\frac{w_1^2}{v^2}p, \tag{170}$$

where the last term defines an effective scattering rate. Clearly the process is *second-order* in the relative speed w_1, as an ordinary second-order Fermi-type process. However the important differences are its magnitude (the speed w_1 is much larger than the Alfvén speed v_A which is usually the maximum for the ordinary second-order Fermi process), and, secondly, the fixed power-law, which is moreover of the correct observed magnitude.

4 Direct Electric Field Acceleration

In contrast to stochastic acceleration processes particle acceleration can also occur in electric fields that change slowly in the frame moving with the particle. Such electric fields are excited near moving magnetized objects. In astrophysics the dimensions L are relatively large and the electric resistance is small, $\propto L^{-1}$ for a conductor of cross-section L^2 and length L. As a result it can often be assumed that the object is well-conducting and has a zero electric field in the comoving (with the object) frame. In the laboratory frame the electric field is then found from a Lorentz transformation $\mathbf{E} = -\mathbf{v} \times \mathbf{B}$ (the ideal MHD approximation). If, in the next-order description, some particles can cross this potential jump

$$V \equiv \left| \int (\mathbf{v} \times \mathbf{B}) \cdot d\mathbf{s} \right| \approx vBL, \tag{171}$$

they are heated if their mean free path is short but accelerated otherwise. Many examples of objects generating such potential jumps exist (Möbius 1994), such as the solar wind across the earth (9.6 kV, *convection* electric field $E_{con} \approx \alpha B_{sw}v_{sw} = 5 \cdot 10^{-4}$ V/m, $\alpha \approx 0.1$ is the coupling to the magnetosphere, B_{sw} the fieldstrength of the solar wind and v_{sw} its speed), rotating planets in particular Jupiter (11 kV, *corotation* electric field $E_{cor} = R\omega B(R) = 4.2 \cdot 10^{-3}$ V/m), rotating magnetized stars in particular white dwarfs and neutron stars, magnetized accretion disks, accretion onto magnetized objects, and magnetized plasma near rotating black holes in Active Galactic Nuclei.

Shock drift acceleration, discussed before, is another example. In fact more generally, any drift of a charged particle across a magnetic field in the direction of the potential jump established in a moving plasma belongs to the class discussed here. The drift can be a grad B drift (as in eq. (165)) or a curvature drift due particles following curved magnetic field lines, or any other drift.

4.1 Parallel Electric Fields

In a differentially moving plasma the voltage source eq. (171) can give rise to potential jumps *along* some magnetic field lines and this alleviates the difficulty of transporting particles in the right direction. In the MHD context of moving plasmas a finite $\int E_\parallel ds$ can arise as soon as non-ideal effects occur (Schindler *et al.* 1991).

Stellar Flares Also an electric current system can be set up by such dynamo action (unipolar inductor) if a suitable closed circuit exists and if sufficient charge carriers are available. The driving voltage is

$$V = \oint |(\mathbf{v} \times \mathbf{B}) \cdot d\mathbf{s}|\,, \tag{172}$$

If the particle density is large enough the electric field heats the plasma ohmically although a certain fraction can be accelerated in *runaway acceleration* (see below).

Further as under astrophysical conditions the resistance is often small while the allowed current has an upper bound, the circuit is usually not steady but evolves. The upper bound derives from the force-free nature of a part of the electric circuit $\mathbf{j} \times \mathbf{B} = 0$ (Aly 1985; Low 1986; Kuijpers 1992 for a detailed explanation). At some stage the 3-D evolution leads to very small length scales, and a kinetic regime is entered in which magnetic fields reconnect quickly. In these reconnection regions strong parallel electric fields are thought to appear, either because of flux annihilation (Bulanov and Cap 1988) or because of 'short circuiting' effects (Haerendel 1989, 1994; Winglee *et al.* 1991). For applications to compact objects see Aly and Kuijpers (1990) and Hamilton *et al.* (1994).

Particles can also be accelerated by electric fields at reconnection sites in stellar flares (Kaastra 1985; Martens 1988; Zarro *et al.* 1995). Modelling of acceleration by the electric fields (partially from temporally varying magnetic fields, partially from convective electric fields) has been done by Kliem (1994). In this case the acceleration acquires stochastic properties as the particle orbits are episodically demagnetized near magnetic islands in current sheets.

Rotating magnetized neutron stars. If the star is rotating in vacuo, inside the star the electric field satisfies $\mathbf{E} = -(\mathbf{\Omega} \times \mathbf{r}) \times \mathbf{B}$ and in the surrounding vacuum a strong electric field is set up with a potential jump *along* the external magnetic field (Deutsch 1955; Goldreich and Julian 1969) of magnitude eq. (171)

$$V = \frac{\Omega \Phi}{2\pi} = \frac{\Omega^2 B R_*^3}{2c} = 6.58 \cdot 10^{12} P^{-2} B_8 \text{Volt}, \tag{173}$$

where $\Omega = 2\pi/P$ is the stellar rotation frequency, P the period in seconds, Φ is the 'open' magnetic flux through the pole cap crossing the light cylinder at distance $r_{lc} = c/\Omega$, we have used an aligned rotating magnetic dipole ($\sin^2\theta/r$ constant on a fieldline), and B_8 the strength of the stellar field at the surface in units of 10^8 Tesla. This electric field is so strong at the stellar surface ($9.09 \cdot 10^{10}$ V/m) that the atmosphere does not remain a vacuum but is filled up by a pair cascade. However as the (electron-positron) plasma is expelled by centrifugal slingshot on the open field lines an electric potential gap of some magnitude remains at the polar cap. Clearly this potential jump can accelerate particles to high energies be it not to what is needed for the highest-energy cosmic rays. Similar electric *gaps* may occur further out (Cheng and Ruderman 1990; Cheung and Cheng 1994). The Crab nebula is an established source of TeV γ-rays (up to 8 TeV, 1 TeV $= 10^{12}$ eV; Vacanti *et al.* 1991; Baillion *et al.* 1993).

It is of interest to compare eq. (173) with the potential jump along a shock in the relativistic wind far from the pulsar. Assuming the magnetic field to have a dipolar dependence from the star to the light cylinder, and further out to fall off inversely with distance (conservation of the transverse flux in a stellar wind) until the shock (at un unknown distance r), further assuming the wind speed to remain relativistic $v \approx c$ till the shock, and finally estimating the maximum transverse extent of the shock at $L \approx r$, I find a potential jump which is of the same magnitude as eq. (173). It follows that shock acceleration far out does not perform better as far as the maximum particle energy is concerned!

4.2 Runaway Acceleration

In a thermal plasma a small and constant electric field ($\mathbf{E} \parallel \mathbf{B}$) generates a current, carried mostly by the light electrons. The average drift speed is determined by momentum balance between acceleration and Coulomb friction

$$0 = < \frac{dv}{dt} > = < -\frac{e\mathbf{E}}{m} - \frac{\mathbf{v}}{t_{ei}} > \tag{174}$$

where t_{ei} is the collision time of thermal electrons on ions (eq. (26)). The average drift speed is

$$v_D \simeq \frac{eEt_{ei}}{m} \tag{175}$$

For the suprathermal electrons with $v > v_{te}$

$$\frac{d\mathbf{v}}{dt} = -\frac{e\mathbf{E}}{m} - \frac{\mathbf{v}}{t_D(v)} \tag{176}$$

where (eq. (15))

$$t_D(v) = t_{ei} \left(\frac{v}{v_{te}} \right)^3 . \tag{177}$$

Therefore, particles with

$$v_c = \left(\frac{E_c}{E} \right)^{\frac{1}{2}} v_{te} \tag{178}$$

will *run away* (Dreicer 1959, 1960; Kaastra 1982). Here the critical field strength at which the entire plasma runs away is

$$E_c \simeq 0,215 E_D \qquad (179)$$

and the *Dreicer field* is defined as

$$E_D \equiv \frac{e \; \ell n \; \Lambda}{\lambda_D^2}. \qquad (180)$$

Since the particle speed is limited by the speed of light it follows that the minimum required field strength for runaway particles to appear is

$$E > \frac{KT}{mc^2} E_c. \qquad (181)$$

For an application of runaway acceleration to solar flares see Kaastra (1985).

4.3 Double Layer Acceleration

It has been suggested that currents dissipate their energy in electrostatic double layers (DLs) or thin structures of parallel electric fields (Alfvén 1958; Jacobsen and Carlqvist 1964; Alfvén and Carlqvist 1967; Alfvén 1981; Raadu 1989). Inside a DL the electric field is much stronger than outside. The total charge of the DL is nearly zero. In fact the DL is a large-amplitude solution of the combined collisionless Vlasov equation and Poisson's equation. It is thin, $10 - 1000\lambda_D$, and collisions are therefore unimportant. The DL is called strong if

$$q\Phi \gg KT, \qquad (182)$$

where Φ is the potential jump across the DL and T the ambient plasma temperature. Such strong DLs have so far only been directly observed in the laboratory (Torvén and Lindberg 1980; Volwerk 1993). In the terrestrial magnetosphere weak DLs have been observed (Boström 1991). The electrostatic gaps which on theoretical grounds are thought to occur in pulsar magnetospheres, e.g. above the polar caps or in the outer gaps further away, are the clearest examples of strong DLs in astrophysics (Beskin *et al.* 1993; Chiang and Romani 1994). For the purpose of particle acceleration a DL has the following properties:

1. In contrast with acceleration by mhd waves or shock waves DLs accelerate both electrons and ions (in opposite directions) and to the same energy $q\Phi$. Moreover in a relativistic and stationary DL the acceleration rate of electrons equals that of protons;
2. Being part of a circuit the DL has an (inertial) resistivity and a strong potential drop along the magnetic field. It is therefore the location where ideal MHD is violated (Schindler *et al.* 1991) and where a twisted magnetic field tube can unwind. Magnetic energy stored in the circuit is converted into kinetic energy by the DL. In fact the DL is a source of Alfvénic motions.

These properties plus the fact that electric fields in astrophysical plasmas tend to concentrate in small structures, make DLs interesting for particle acceleration. In a solar flare it may be that the electric current circuit dissipates by a multitude of transient electric double layers (Hénoux 1987; Kuijpers 1990; Moghaddam–Taaheri and Goertz 1990; van Oss and van den Oord 1995).

Note that also from a theoretical point of view the acceleration in one large double layer (as in the original Alfvén–Carlqvist model (Alfvén and Carlqvist 1967)) is unlikely: If in a solar flare the entire electric current were to dissipate in one strong double layer on the short timescale observed in the actual flare ≈ 1000 s, primarily extremely relativistic particles would be produced (Raadu 1989). This follows from putting the observed rate of dissipation in a strong solar flare equal to the rate of dissipation in an electric circuit $IV = 10^{22}$ Watt, estimating the current from the liberated energy $0.5LI^2 = 10^{25}$ J, with the circuit inductance approximately given by $L = \mu_o \ell (4\pi)^{-1}$ and $\ell \geq 10^4$ km, the length of the circuit. If a solar flare would dissipate its energy in one DL only, mainly relativistic particles would be produced of a typical energy $\epsilon \geq 2.24 \cdot 10^9$ eV, much larger than is observed.

It has been claimed that DLs, being electrostatic structures, cannot accelerate particles (Bryant 1993). This is not true: an active DL (in contrast to a contact DL separating two different plasmas) forms part of a current carrying circuit. The DL is maintained by the external current and dissipates its energy similar to Ohmic dissipation, now however in the form of accelerated (neutral) beams (van den Oord 1990; van Oss and van den Oord 1995) and Alfv'enic motions instead of heating.

Acknowledgements. I would like to thank Claudio Chiuderi and Giorgio Einaudi for their excellent organization of this Plasma Astrophysics School and the participants for their stimulating reactions. Part of this work was performed under EC-twinning Contract No. SCI*-CT91-0727 on Coherent Radiation and Particle Acceleration in Magnetized Plasmas.

References

Achterberg, A. 1981a *Astron. Astrophys.* **97**, 259

Achterberg, A. 1981b *Astron. Astrophys.* **98**, 195

Achterberg, A. 1990 in *Physical Processes in Hot Cosmic Plasmas,* NATO ASI, eds. W. Brinkmann, A.C. Fabian and F. Giovanelli, Kluwer Acad. Publ., p. 67

Achterberg, A. and Ball, L. 1994 *Astron. Astrophys.* **285**, 687

Achterberg, A. and Norman, C.A. 1980, *Astron. Astrophys.* **89**, 353.

Alfvén, H. 1958, *Tellus,* **10**, 104.

Alfvén, H. 1981, *Cosmic Plasma,* Reidel Publ. Cy, Dordrecht, Holland.

Alfvén, H., and Carlqvist, P. 1967, *Solar Phys.* **1**, 220.

Aly, J.J. 1985 *Astron. Astrophys.* **143**, 19

Aly, J.J. and Kuijpers, J. 1990 *Astron. Astrophys.* **227**, 473

Anastasiades, A., Vlahos, L. 1993 *Astron. Astrophys.* **275**, 427

André, M. 1985 *J. Plasma Phys.* **33**, 1

Athay, R.G. 1984, *Solar Phys.* **93**, 123.

Axford, W.I. 1992, in *Particle Acceleration in Cosmic Plasmas,*, AIP Conf. Proc. 264, eds. G.P. Zank and T.K. Gaisser, American Institute of Physics, New York, p. 45.

Axford, W.I., Leer, E., and Skadron, G. 1977, *EOS,* **57**, 780, Proc. 15th Int. Cosmic Ray Conf., Plovdiv.

Baillon, P. *et al.* 1993 *Astroparticle Phys.,* **1**, 341

Ball, L. 1994 *Astrophys. J. Suppl. Ser.* **90**, 889

Ballard, K.R. and Heavens, A.F. 1992 *Monthly Notices Roy. Astron. Soc.* **259**, 89

Begelman, M.C. and Kirk, J.G. 1990 *Astrophys. J.* **353**, 66

Belcher, J.W. 1971 *Astrophys. J.* **168**, 509

Bell, A.R. 1978 *Monthly Notices Roy. Astron. Soc.* **182**, 147

Bell, A.R. 1993, in *Plasma Physics,* ed. R. Dendy, Cambridge Univ. Press, p. 319

Benz, A.O. 1985, *Solar Phys.* **96**, 357

Benz, A.O. 1994, in *Fragmented Energy Release in Sun and Stars,* ed. G.H.J. van den Oord, Kluwer Acad. Publ., Dordrecht, p. 135

Benz, A.O. and Smith, D.F. 1987, *Solar Phys.* **107**, 299

Benz, A.O. and Thejappa, G. 1988 *Astron. Astrophys.* **202**, 267

Beskin, V.S., Gurevich, A.V., Istomin, Ya.N. 1993 *Physics of the Pulsar Magnetosphere,* Cambridge Univ. Press, Cambridge, U.K.

Blandford, R.D. and Eichler, D. 1987, *Physics Reports,* **154**, 1

Blandford, R.D. and Ostriker, J.P. 1978 *Astrophys. J. (Letters)* **221**, L29

Blumenthal, G.R. and Gould, R.J. 1970 *Rev. Modern Phys.* **42** 237

Boström, R. 1991, in *EPS-8, Trends in Physics,* ed. J. Kaczér, Prometheus, Prague, p. 845

Bryant, D.A. 1993, in *Plasma Physics,* ed. R. Dendy, Cambridge Univ. Press, p. 209

Bulanov, S.V. and Cap, F. 1988, *Soviet Astron.,* **32**, 436

Canfield, R.C. and Metcalf, T.R. 1987, *Astrophys. J.* **321**, 586

Cargill, P. 1991, *Astrophys. J.* **376**, 771

Cargill, R.J., Goodrich, C.C. and Vlahos, L. 1988, *Astron. Astrophys.* **189**, 254.

Cheng, K., Ruderman, M. 1990 *Astrophys. J.* **235**, 576

Cheung, W.H., Cheng, K.S. 1994 *Astrophys. J. Suppl. Ser.* **90**, 827

Chiang, J. and Romani, R.W. 1994 *Astrophys. J.* **436**, 754

Davidson, R.C. 1972, *Methods in Nonlinear Plasma Theory,* Academic Press, New York

Decker, R.B. 1988 *Space Sci. Rev.,* **48**, 195

Decker, R.B. 1990 in *Particle Acceleration in Cosmic Plasmas*,, AIP Conf. Proc. 264, eds. G.P. Zank and T.K. Gaisser, American Institute of Physics, New York, p. 183

de Hoffmann, F., Teller, E., 1950 *Phys. Rev.* **80**, 692

de Jager, C., Kuijpers, J., Correia, E., and Kaufmann, P. 1987, *Solar Phys.* **110**, 317

de la Beaujardière, J.F. and Zweibel, E.G. 1989, *Astrophys. J.* **336**, 1059.

Deutsch, A.J. 1955 *Ann. d'Astrophysique,* **18**, 1

Dreicer, H. 1959 *Phys. Rev.,* **115**, 238

Dreicer, H. 1960 *Phys. Rev.,* **117**, 329

Earl, J.A. 1994 *Astrophys. J.* **425**, 331

Fermi, E. 1949 *Phys. Rev.* **75**, 1169

Fermi, E. 1954 *Astrophys. J.* **119**, 1

Fisk, L.A. 1976 *J. Geophys. res.,* **81**, 4643

Fletcher, L. 1995 *On Magnetic Pumping in Astrophysical Sources,* in preparation

Galeev, A.A., and Sudan, R.N., eds., 1983, *Handbook of Plasma Physics,* Basic Plasma Physics I, II, North Holland Publ. Cy., Amsterdam.

Gallant, Y. and Arons, J. 1994 *Astrophys. J.* **435**, 230

Giacalone, J., Burgess, D., Schwartz, S.J., Ellison, D.C. 1993 *Astrophys. J.* **402**, 550

Giacalone, J. and Jokipii, J.R. 1994 *Astrophys. J. (Letters)* **430**, L137

Ginzburg, V.L. and Syrovatskii, S.I. 1964 *The Origin of Cosmic Rays,* Pergamon Press, Oxford

Goldman, M.V. 1984, *Rev. Modern Physics,* **56**, 709

Goldreich, P., Julian, W.H. 1969 *Astrophys. J.* **157**, 869

Greisen, K. 1966 *Phys. Rev. Lett.* **16**, 748

Haerendel, G. 1989, in *Plasma Astrophysics,* ESA SP-285, vol. I, eds. T.D. Guyenne and J.J. Hunt, p. 37

Haerendel, G. 1994 *Astrophys. J. Suppl. Ser.* **90**, 765

Hall, D.E. and Sturrock, P.A. 1967 *Phys. Fluids* **10**, 1593

Hamilton, R.J., Lamb, F.K., Miller, M.C. 1994 *Astrophys. J. Suppl. Ser.* **90**, 837

Hasegawa, A. 1975 *Plasma Instabilities and Nonlinear Effects,* Springer-Verlag, Berlin

Hénoux, J.C. 1987, in *Solar Maximum Analysis,* eds. V.E. Stepanov and V.N. Obridko, VNU Science Press, Utrecht, The Netherlands, p. 105

Hizanidis, K., Vlahos, L., Polymilis, C. 1989, *Phys. Fluids B* **1**, 682

Holman, G.D., and Benka, S. 1992 *Astrophys. J. (Letters)* **400**, L79

Holman, G.D. and Pesses, M.E. 1983 *Astrophys. J.* **267**, 837

IAU Coll. **142**, 1994 *Particle Acceleration Phenomena in Astrophysical Plasmas,* ApJ Suppl. Ser. **90**, 511 – 983

Ichimaru, S. 1973 *Basic Principles of Plasma Physics,* W.A. Benjamin Enc., London

Jacobsen, C. and Carlqvist, P. 1964, *Icarus,* **3**, 270

Jokipii, J.R. 1982, *Astrophys. J.* **255**, 716

Jokipii, J.R. 1987, *Astrophys. J.* **313**, 842

Jones, F.C. 1990 *Astrophys. J.* **361**, 162

Jones, F.C. 1994 *Astrophys. J. Suppl. Ser.* **90**, 561

Jones, F.C., Ellison, D.C. 1991 *Space Sci. Rev.* **58**, 259

Jun, B.-I., Clerke, D.A., Norman, M.L. 1994 *Astrophys. J.* **429**, 748

Kaastra, J.S. 1982, *J. Plasma Phys.*, **29**, 287

Kaastra, J.S. 1985, *Solar Flares, An Electrodynamic Model*, Ph.D. Thesis, Utrech University, The Netherlands, Chs.4 and 5

Karimabadi, H., Akimoto, K., Omidi, N., and Menyuk, C.R. 1990, *Phys. Fluid B* **2**, 606.

Kattenberg, A. 1981, *Solar Radio Bursts and their Relation to Coronal Magneti Structures*, PhD Thesis, Utrecht University.

Kennel, C.F., Coroniti, F.V. 1984 *Astrophys. J.* **283**, 694

Kennel, C.F. and Engelmann, F. 1966 *Phys. Fluids,* **9**, 2377

Kirk, J.G. 1995, in *Plasma Astrophysics*, 24th "Saas Fee" Advanced Course o the Swiss Society of Astrophysics and Astronomy, ed. A.O. Benz

Kirk, J.G., Wassmann, M. 1992 *Astron. Astrophys.* **260**, 49

Krimsky, G.E. 1977, *Doklady Akad. Nauk. SSR* **242**, 1306

Krülls, W.M. and Achterberg, A. 1994 *Astron. Astrophys.* **286**, 314

Kuijpers, J. 1980, in *Radio Physics of the Sun*, IAU Coll. No. 86, eds. M.R Kundu and T.E. Gergely, Reidel, Dordrecht, p. 341

Kuijpers, J. 1990, in *Plasma Phenomena in the Solar Atmosphere,* 1989 Cargése Workshop, eds. M.A. Dubois, F. Bély-Dubau and D. Grésillon, Les Editions de Physique, Les Ulis, France, p.17.

Kuijpers, J. 1992 in *The Sun, A Laboratory for Astrophysics,* eds. J.T. Schmelz and J.C. Brown, Kluwer Acad. Publ., pp. 535–597

Kuijpers, J., Fletcher, L., Abada-Simon, M., Horne, K., Raadu, M.A., Ramsay G., Steeghs, D. 1995 'Magnetic Pumping in the Cataclysmic Variable AE Aqr', in preparation

Kuijpers, J., van der Post, P., and Slottje, C. 1981, *Astron. Astrophys.* **103**, 331

Kulsrud, R.M. and Ferrari, A. 1971 *Astrophys. Space Sci.* **12**, 302

Landau, L.D. and Lifshitz, E.M. 1976 *Mechanics,* Pergamon Press, Oxford, par 30

Levinson, A. 1994 *Astrophys. J.* **426**, 327

Lichtemberg, A.J. and Lieberman, M.A. 1983 *Regular and Stochastic Motion,* Springer-Verlag, Berlin

Lieu, R. and Quenby, J.J. 1990 *Astrophys. J.* **350**, 692

Lieu, R., Quenby, J.J., Drolias, B., Naidu, K. 1994 *Astrophys. J.* **421**, 211

Lin, R.P., Schwartz, R.A., Kane, S.R., Pelling, R.M. and Hurley, K.C. 1984 *Astrophys. J.* **283**, 421

Longair, M.S. 1994 *High Energy Astrophysics* II, Cambridge Univ. Press, Cambridge, U.K.

Low, B.C. 1986 *Astrophys. J.* **307**, 205

Mätzler, C. and Wiehl, H.J. 1980, in *Radio Physics of the Sun*, IAU Coll. No.86, eds. M.R. Kundu and T.E. Gergely, Reidel, Dordrecht, p. 177.

Martens, P.C.H. 1988, *Astrophys. J. (Letters)* **330**, L131

Martens, P.C.H., van den Oord, G.H.J. and Hoyng, P. 1985, *Solar Phys.* **96**, 253

Melrose, D.B. 1969 *Astrophys. & Space Sci.* **4**, 143

Melrose, D.B. 1982 *Austr. J. Phys.* **35**, 67

Melrose, D.B. 1983 *Solar Phys.* **89**, 149

Melrose, D.B. 1986, *Instabilities in Space and Laboratory Plasmas*, Cambridge Univ. Press, UK.

Melrose, D.B. 1995, in *Plasma Astrophysics*, 24th "Saas Fee" Advanced Course of the Swiss Society of Astrophysics and Astronomy, ed. A.O. Benz

Melrose, D.B., Hewitt, R.G. and Dulk, G.A. 1984, *J. Geophys. Res.* **89**, 897.

Melrose, D.B. and Kuijpers J. 1984 *J. Plasma Phys.* **32**, 239

Melrose, D.B. and Kuijpers, J. 1987, *Astrophys. J.* **323**, 338

Menyuk, C.R., Drobot, A.T., Papadopoulos, K., Karimabadi, H. 1987 *Phys. Rev. Lett.* **58**, 2071

Miley, G. 1980 *Ann. Rev. Astron. Astrophys.* **18**, 165

Miller, J.A. 1991, *Astrophys. J.* **376**, 342

Miller, J.A. and Ramaty, R. 1987, *Solar Phys.* **113**, , 195

Miller, J.A., Guessoum, N., and Ramaty, R. 1990, *Astrophys. J.* **361**, 701

Moghaddam-Taaheri, E. and Goertz, C. 1990, *Astrophys. J.* **352**, 361

Möbius, E. 1994 *Astrophys. J. Suppl. Ser.* **90**, 521

Ostriker, J.P. and Gunn, J.E. 1969 *Astrophys. J.* **157**, , 1395

Papadopoulos, K. and Freund, H.P. 1979 *Space Sci. Rev.* **24**, 511

Parker, E.N. 1958 *Phys. Rev.* **109**, 1328

Pesses, M.E. 1981, *J. Geophys. Res.,* **86**, 150

Pesses, M.E., Decker, R.B., and Armstrong, T.P. 1982, *Space Science Rev.,* **32**, 185

Raadu, M.A. 1989, *Physics Reports* **178**, 25

Ramaty, R. and Murphy, R.J. 1987 *Space Science Rev.,* **45**, 213

Reynolds, S.P. 1992, in *Particle Acceleration in Cosmic Plasmas,*, AIP Conf. Proc. 264, eds. G.P. Zank and T.K. Gaisser, American Institute of Physics, New York, p. 409

Rieger, E. 1989, *Solar Phys.* **121**, 323

Rieger, E. 1994 *Astrophys. J. Suppl. Ser.* **90**, 645

Ruffolo, D. 1995 *Astrophys. J.* **442**, 861

Rybicki, G.B. and Lightman, A.P. 1979 *Radiative Processes in astrophysics,* John Wiley & Sons, New York

Schadee, A. 1986, *Adv. Space Res.* **6**, No. 6, 41

Schadee, A., de Jager, C. and Svestka, Z. 1983, *Solar Phys.* **89**, 287

Schindler, K., Hesse, M., and Birn, J. 1991, *Astrophys. J.* **380**, 293

Schlickeiser, R. 1989, *Astrophys. J.* **336**, 264

Schlickeiser, R. 1994, *Astrophys. J. Suppl. Ser.* **90**, 929

Schlüter, A. 1957 *Zeitschr. Naturforsch.* **12a**, 822

Similon, P.L. and Sudan, R.N. 1990, *Annual Review Fluid Mech.,* **22**, 317

Sitenko, A.G. 1982 *Physica Scripta* **T2/1**, 67

Smith, D.F. 1990, in *Basic Plasma Processes on the Sun*, IAU Symp. 142, eds. E.R. Priest and V. Krishan, p. 375

Sokolsky, P. 1992, in *Particle Acceleration in Cosmic Plasmas,*, AIP Conf. Proc. 264, eds. G.P. Zank and T.K. Gaisser, American Institute of Physics, New York, p. 373

Spitzer, L. 1965, *Physics of Fully Ionized Gases,* Interscience Publ., New York.

Stanev, T. 1992, in *Particle Acceleration in Cosmic Plasmas,*, AIP Conf. Proc. 264, eds. G.P. Zank and T.K. Gaisser, American Institute of Physics, New York, p. 379

Steinacker, J. and Schlickeiser, R. 1989, *Astron. Astrophys.* **224**, 259

Sturrock, P.A. 1994, *Plasma Physics,* Cambridge Univ. Press, UK, Ch. 10, 18.

Swann, W.F.G. 1933 *Phys. Rev.* **43**, 217

Tapping, K.F., Kuijpers, J., Kaastra, J.S., van Nieuwkoop,J., Graham, D., and Slottje, C. 1983, *Astron. Astrophys.* **122**, 177

Thejappa, G. 1987, *Solar Phys.* **111**, 45

Thielheim, K.O. 1989 *Fundamentals of Cosmic Phys.,* **13**, 357

Torvén, S. and Lindberg, L. 1980 *J. Phys. D.: Appl. Phys.* **13**, 2258

Tsintsadze, N.L. 1989, in *Laboratory and Space Plasmas,* ed. H. Kikuchi, Springer-Verlag, Berlin, p. 164

Tsytovich , V.N. 1970, *Nonlinear Effects in Plasma,* Plenum Press, New York

Tsytovich, V.N. 1973, *Ann. Rev. Astron. Astrophys.* **11**, 363

Vacanti, *et al.* G. 1991 *Astrophys. J.* **377**, 467

van den Oord, G.H.J. 1990 *Astron. Astrophys.* **234**, 496

van der Laan, H. 1966 *Nature* **211**, 1131

van Kampen, N.G. 1981, *Stochastic Processes in Physics and Chemistry,* North-Holland Publ. Cy.

van Oss, R.F. and van den Oord, G.H.J. 1995 *Astron. Astrophys.* in press

van Paradijs, J. van der Klis, M., Achterberg, A. (eds.) 1991 *Particle Acceleration near Accreting Compact Objects,* Royal Netherlands Academy of Arts and Sciences, 1000 GC Amsterdam

Vlahos, L. 1989 *Solar Phys.* **121**, 431

Vlahos, L. 1994, in *Fragmented Energy Release in Sun and Stars,* ed. G.H.J. van den Oord, Kluwer Acad. Publ., Dordrecht, p. 39

Volwerk, M. 1993 *J. Phys. D.: Appl. Phys.* **26**, 1192

Webb, G.M., Axford, W.I., and Terasawa, T. 1983 *Astrophys. J.* **270**, 537

Winglee, R.M., Dulk, G.A., Bornmann, P.L., and Brown, J.C. 1991, *Astrophys. J.* **375**, 382

Wu, C.S. 1984, *J. Geophys. Res.* **89**, 8857

Zank, G.P., Gaisser, T.K. 1992 *Particle Acceleration in Cosmic Plasmas,* AIP Conf. Proc. 264, American Inst. of Physics, New York

Zarro, D.M., Mariska, J.T., Dennis, B.R. 1995 *Astrophys. J.* **440**, 880

Zatsepin, G.T., Kuzmin, V.A. 1966 *JETPh Letters* **4**, 78

Zheleznyakov, V.V. and Zlotnik, E.Y. 1975, *Solar Phys.* **43**, 431

Turbulence, Statistics and Structures: an Introduction

A. Pouquet

Laboratoire Cassini, CNRS URA 1362, Observatoire de la Côte d'Azur, BP 229, 06304 Nice Cedex 4, France

1. Weak turbulence and waves

1.1 Introduction

Could turbulence possibly be seen as a superposition of waves, in which case the paraphernalia of linear physics, with some proper modifications, could be applied and the problem presumably solved in simple cases ? Or are they intrinsic effects of nonlinearities that shape the flow in an irreversible way (shocks being one such example) ? And what is meant by waves ? Turbulence, nowadays, is taken to be associated with strong nonlinearities in the evolution equation of the problem, and not simply (as originally meant) rotating vortex (turbine) motions. These nonlinearities imply mode–coupling and thus will lead to a Fourier spectrum of modes which is decreasing as a power–law rather than exponentially decreasing. Such power laws in general can be recovered by simple phenomenological arguments.

Weakly interacting waves, as we study in the first part of this chapter, lead to power–law spectra as well. Then, what else can characterize a fully turbulent flow ? Departures from simple power laws derived from phenomenology, as data in a variety of contexts (wind–tunnel experiments, atmospheric motions, solar wind *in situ* measurements, and the interstellar medium) seem to suggest ?

If so, where do such corrections come from ? Non–Gaussian statistics, as seen for example in the wings of the probability distribution functions of velocity distributions, in the laboratory as well as in interstellar clouds (Falgarone *et al.*, 1994) ? The development and persistence of coherent structures, such as vortex filaments ? Or rather the formation of shocks, and in general fronts that dominate – in some sense – the dynamics, at least at small scale ? A theory of turbulent flows will have to take into account all such phenomena, and in particular the strong mode–coupling and ensuing enhanced transport, the power–law solutions, the small–scale intermittency and the presence of well–defined structures. A first successful attempt at encompassing both the structural and statistical aspects of turbulent flows can be found for example in the spiral model of Lundgren (1982), and its generalization by Pullin and Saffman (1992).

Fluids in astrophysics are encountered everywhere (see *e.g.* Shu, 1992). Even though densities are often quite low (*e.g.* 10 particles per cm^3 or less in the interstellar medium (ISM), outside the molecular clouds), a fluid approach can still be taken because the typical scale of the object (a kiloparsec for the ISM at large) is larger than the mean free path of particles, $\sim 10^{14}$cm for the above density). The fluid is in fact a plasma made up of charged (and neutral) particles, with a degree of ionization $\chi = \rho_i/(\rho_i + \rho_n)$ where ρ_i (*resp.* ρ_n) is the density of ions (*resp.* neutrals) that varies from fully ionized to weakly ionized ($\chi = 10^{-6}$ in the ISM at large). This implies that electromagnetic forces will be important, if not prevalent, so that the version of fluid dynamics that must be used for many applications must be that which includes coupling to the magnetic and electric fields and the current. When plasma physics deals with particle interactions, fluid and MHD theories deal with the physics at large spatial and temporal scales.

Moreover, mainly because of large distances and small densities, Reynolds numbers in Astrophysics are large, and turbulent phenomena must be taken into account. Turbulence is often invoked in geophysics and astrophysics to help resolve some paradoxes brought forth by detailed observations, both spatially and in some cases temporally as well. In particular, it is found helpful in broadening line wings in stars, in preventing gravitational collapse in molecular clouds, in general in getting rid of extra angular momentum (essential in accretion disks) or energy, in provoking irregular and energetic events (the whole range of solar flares, from nanoflare to major eruption), in rendering the flows inhomogeneous and thus providing mixing of chemical elements as well as local conditions untypical of the average physical parameters and allowing for unexplained–otherwise abundances, and as a source of irregularity, for example in variable stars, and in understanding the variations of the magnetic fields in the solar and terrestrial dynamos.

In the absence of a general theoretical framework, one can choose to look at some sub–problems. In this Lecture, we consider the coupling of weakly interacting waves and derive in an abstract way the "kinetic equation" describing the interaction of Fourier modes to lowest order. We then consider an extension of such kinetic equations to the strong coupling case, but justify it only in an *ad hoc* way from a phenomenological point of view. In the second Lecture, we introduce some exact results and solutions in idealized cases, and attempt to follow structures by their curvature and torsion. The final Lecture is devoted to a variety of applications, including a consideration of the limitations on numerical experiments and the coupling to a magnetic field in the simple incompressible case, ending with a brief account of the role of turbulence in the interstellar medium at the scale of the kiloparsec.

An excellent illustration of the variety, complexity and (to some of us at least) beauty of structures that develop in turbulent flows is to be found in Van Dyke (1982).

1.2 Weak non–linearities

Except for incompressible neutral fluids, waves are present in a flow the interactions of which, at low amplitude, may dominate the dynamical evolution provided the characteristic time associated with such waves τ_w is substantially smaller than the time of interaction between waves τ_{int}. Many physical problems fall in this category $\epsilon = \tau_w/\tau_{int} << 1$, and a formalism can be developed somewhat independently of the physics but taking into account the structure of the problem. Typical cases for plasmas are that of *e.g.* electrostatic waves, drif waves, ion–acoustic waves, Langmuir waves or whistler waves (see Krall & Trivelpiece, 1973), with applications both in the cosmos and in the laboratory (for example in fusion devices such as tokamaks).

The following derivation follows closely the approach of Benney and Newell (1969) which can be consulted for more detail. Here, an attempt is made at stressing the structure of the problem and its solution. Let us write the time evolution of our one-dimensional problem as

$$\frac{\partial u}{\partial t} = \mathcal{L}_x u + \varepsilon \mathcal{N}_x(u; u) \; ; \tag{1.1}$$

$\mathcal{L}$ (*resp.* $\mathcal{N}$) is a linear (resp. quadratically non linear) operator and $\varepsilon \ll 1$. With $A(k) = \int e^{ik \cdot x} u(x) dx$ the corresponding Fourier amplitudes, we now define the frequency $\omega(k)$ associated with the linear operator (the dispersion relation) and the Fourier representation of the non-linear operator $H(m,n)$:

$$[\partial_t + i\omega(k)]A(k,t) = \varepsilon \int_{-\infty}^{\infty} H(m,n)A(m,t)A(n,t)\delta(k - m - n)dmdn \tag{1.2}$$

where ω is taken to be real. Using now the interaction representation $a(k,t) \equiv a_k$ with $A(k,t) = a_k e^{-i\omega(k)t}$ yields:

$$\frac{\partial a_k}{\partial t} = \varepsilon \int H_{mn} a_m a_n e^{i(\omega_k - \omega_m - \omega_n)t} \delta_{kmn} dmn \tag{1.3}$$

where new more concise notations have been used the definition of which is self-explanatory: $H_{mn} = H(m,n)$, $\omega_k = \omega(k)$, $\delta_{kmn} = \delta(k - m - n)$ and $d_{mn} = dmdn$.

We see that in this representation, one is "riding" on the wave and that the dependency of the Fourier amplitudes comes solely from the non–linear operator H_{mn}; for $\varepsilon \ll 1$, this variation is small and can be approximated through an expansion in ε, and exploiting the fact that except when $\omega_k = \omega_m + \omega_n$, *i.e.* at resonance, the phase factor oscillates rapidly and will allow for cancellations. For simple non–dispersive waves for which $\mathcal{L} = \bar{\alpha} \, u_x$ and thus $\omega_k = \bar{\alpha}.k$, the oscillating term disappears. Note that this is not true, however, in incompressible MHD (A. Newell, 1994, private communication) because of the form of the coupling terms in the MHD equations.

1.3 The ϵ–expansion

The second step of the procedure is to perform an ε –expansion

$$a_\ell = a_{0,\ell} + \varepsilon a_{1,\ell} + \varepsilon^2 a_{2,\ell} + \dots \tag{1.4}$$

and solve order by order. At lowest order, of course, $a_{0,\ell}$ is constant. At order one, solving for $a_{1,\ell}$:

$$a_{1,\ell} = \int H_{mn} a_{0,mn} \Delta(\omega_{\ell,mn}) \delta_{\ell,mn} dmn \; . \tag{1.5}$$

Here

$$\omega_{\ell,mn} = \omega(\ell) - \omega(m) - \omega(n) \tag{1.6}$$

and $\delta_{\ell,mn} = \delta(\omega_{\ell,mn})$; we define as well $a_{0,mn} = a_{0,m}a_{0,n}$. Finally, the time – dependent function Δ is given by:

$$\Delta(\omega_{\ell,mn}) = \int_0^t \exp[it\omega_{\ell,mn}]dt = \frac{\exp[i\omega_{\ell,mn}] - 1}{i\omega_{\ell,mn}} \quad . \tag{1.7}$$

Resonance occurs for $\omega_{\ell,mn} = 0$. Note that if we restrict the analysis to three–wave interactions (instead of a superposition of waves) with $\ell = \ell_1 + \ell_2$, we obtain:

$$a_1(\ell, t) = \frac{-i}{\omega(\ell_1 + \ell_2) - \omega(\ell_1) - \omega(\ell_2)} \quad ; \tag{1.8}$$

on the other hand, in the case of a random superposition of waves, resonances are the main contributors to the integral (1.5) but the singularity as in (1.8) at resonance may disappear through mixing.

Pursuing now at order ε^2 and keeping the notation concise (but imprecise), we have

$$\partial_t a_{2,\ell} = \int H a_0 a_1$$

with $a_1 = \int H a_0 \Delta$. Hence, $a_2 = \int H a_0 \int_0^t H a_0 \Delta$, recalling that a_0 is time–independent. With the new notations $a_{0,mpq} = a_{0,m}a_{0,p}a_{0,q}$ and similarly $H_{mnpq} = H_{mn}H_{pq}$ and defining now the function E as

$$E(x; y) = \int_0^t \Delta(x - y)e^{iyt}dt \quad , \tag{1.9}$$

we have:

$$a_{2,\ell} = \int H_{mnpq}a_{0,mpq}E(\omega_{\ell,mpq}; \omega_{\ell,mn})\delta_{\ell,mn}\delta_{n,pq}d_{mnpq} \tag{1.10}$$

with $d_{mnpq} = d_{mn}d_{pq}$.

1.4 Moments and cumulants

At this stage, we have to introduce the statistics of the random fields with a few definitions. A nth - order physical space cumulant is:

$$R_u^{(n)}(r, r^{(1)}...r^{(n-2)}, t) = < u(x,t)u(x+r,t)...u(x+r^{(n-2)},t) >$$

$$- \sum_\alpha < u(x,t)...u(x+r^{(\alpha)},t) >< u(x+r^{(\alpha+1)},t)...u(x+r^{(n-2)},t) > \tag{1.11}$$

In other words, $R_u^{(n)} \equiv 0$ $\forall n$ for a Gaussian field u. Taking now the Fourier transform and going directly to the interaction representation, one writes, with

$$k_\ell + k_{\ell'} + ... + k_{\ell(n-2)} = 0 \tag{1.12}$$

and with $dk_{\ell\ell'...\ell(n-2)} = dk_\ell...dk_{\ell(n-2)}$

$$R_u^{(n)}(r, r^{(1)}, ..., r^{(n-2)}, t) = \tag{1.13}$$

$$\int_{-\infty}^{\infty} q_{\ell\ell'\ldots\ell(n-2)}^{(n)} \exp\left[-it(\omega_{\ell\ell'\ldots\ell(n-2)} + k_\ell r + \ldots + k_{\ell(n-2)} r^{(n-2)})\right] dk_{\ell\ell'\ldots\ell(n-2)}$$

An ε-expansion can again be assumed for the reduced Fourier cumulants $q_{\ell\ell'\ldots\ell(n-2)}^{(n)}$ namely

$$q^{(n)} = q_0^{(n)} + \varepsilon q_1^{(n)} + \varepsilon^2 q_2^{(n)} + \ldots \tag{1.14}$$

At lowest order, they are time-independent:

$$q_{0\ell}^{(2)} = < a_0(\ell)a_0(\ell') > \delta(\ell + \ell') \tag{1.15}$$

$$q_{0\ell\ell'}^{(3)} = < a_0(\ell)a_0(\ell')a_0(\ell'') > \delta(\ell + \ell' + \ell'')$$

since $\partial_t a_0 \equiv 0$. At order ε, we obtain:

$$\begin{aligned}\delta(\ell + \ell')q_{1\ell}^{(2)} &= < a_{1,\ell}a_{0,\ell'} > + < a_{1,\ell'}a_{0,\ell} > \\ &= P_{\ell\ell'} < a_{1,\ell}a_{0,\ell'} >\end{aligned} \tag{1.16}$$

where $P_{\ell\ell'\ell''\ldots}$ indicates that a cyclic summation is performed on indices. Similarly

$$\delta(\ell + \ell' + \ell'')q_{1\ell\ell'}^{(3)} = P_{\ell\ell'\ell''} < a_1(\ell)a_0(\ell')a_0(\ell'') > \quad . \tag{1.17}$$

The next step is to solve for cumulants, using the fact that a_0 is constant. Thus one can write:

$$q_{1,\ell}^{(2)} = P_{\ell\ell'} \int H_{mn} q_{0.mn}^{(3)} \Delta(\omega_{\ell,mn}) \delta_{\ell,mn} dmn \tag{1.18}$$

and similarly for higher-order cumulants.

Until now, nothing – really – has been done apart from writing equations and defining terms in ϵ-expansions. The new assumption occurring now consists in introducing a new slow time scale $T_2 = \varepsilon^2 t$ in the cumulants; thus $q_0^{(N)}$ is replaced by:

$$q_0^{(N)} + \varepsilon q_{0;1}^{(N)} + \varepsilon^2 \left(q_{0;2}^{(N)} - t\frac{\delta q_0^{(N)}}{\delta T_2} \right) + \ldots \tag{1.19}$$

1.5 The closure at second order

As is standard in similar computations for ODE's (see for example, Bender & Orszag, 1978, Chapter II) the terms $\delta q_0^{(N)}/\delta T_2$ are chosen to remove secularities. The last step is to realize that since

$$\varepsilon = \tau_\omega/\tau_{NL} \ll 1 \tag{1.20}$$

and since we wish to average over many wave periods, we have to evaluate integrals of the form

$$\lim_{t\to\infty} \int f(k)\Delta(k,t)dk \tag{1.21}$$

where

$$\Delta(k) = \int_0^t e^{i\omega(k)t} dt \qquad (1.22)$$

contains the time–dependence, and $\omega(k)$ (*i.e.* the dispersion relation) is the link with the (linearized) physical problem. We now use the Lemma:

$$lim_{t\to\infty} \int f(k)\Delta(k,t)dk = \pi f(0) + iP_C \int \frac{f(k)}{k} dk \qquad (1.23)$$

where P_C stands for the Cauchy Principal Value integral. In other words, what we are really doing here is to replace

$$lim_{t\to\infty} \int \frac{sin\omega t}{\omega}$$

by $\pi\delta(\omega)$.

The end result of this procedure is to obtain a set of kinetic equations for the temporal evolution of cumulants $q^{(n)}$ of order n (Benney & Newell, 1969). For example, for $n = 2$, the kinetic equation for the energy spectrum takes the form:

$$\frac{\partial q^{(2)}(k,t)}{\partial t} = F(k,t) - D(k,t)q^{(2)}(k,t)dt \quad . \qquad (1.24)$$

For a steady solution, $q^{(2)} = F(k)/D(k)$ where $F(k)$ can in fact be regarded as a resultant forcing term for the modal energy distribution (with $F(k) >> 0$), and $D(k)$ a dissipation. Many applications of this formalism (or its variants) have been developed, in particular in plasma physics (see *e.g.* the book by Zakharov *et al.*, 1992).

As a final point, let us note that an alternative approach inspired from Celestial Mechanics – and relying on the technique of normal forms whereby the equations for amplitude and phases of the waves decouple – has been recently proposed (Frisch *et al.*, 1995) in the context of β–plane turbulence. The difference from what is proposed here stems from the functional form of the dispersion relation (namely $\omega(k) \sim 1/k$), resulting in special resonances; this leads to the appearance of an extra term in the kinetic equation.

1.6 From weak to strong coupling

Whereas in the beginning of this Lecture, we have dealt with a theory of weak turbulence dominated by waves, we shall here deal with a model of strong turbulence when it is the non linear interactions which shape the flow. There are several such models that have been derived over the last forty years, starting mostly from the theories developed by Kraichnan as for example the DIA (Kraichnan, 1959) or the TFM following the diagrammatic techniques of theoretical physics. A good introduction to the DIA (or rather to its russian version) can be found in Monin & Yaglom (1975) vol. II. A thorough description of such models can be found *e.g.* in Orszag (1974) and in Lesieur (1990).

1.7 Closure models for strong turbulence

1.7.1 The Navier–Stokes equations

What follows can be found in many places, but it may be useful to remind the reader of how simple the idea of such models is. The price to pay, of course, is that we do not **control** anymore the error made. Let us write the Navier–Stokes equations in the incompressible case, with $\nabla \cdot \mathbf{u} = 0$ where $\mathbf{u}$ is the velocity field:

$$\frac{D\mathbf{u}}{Dt} \equiv \frac{\partial \mathbf{u}}{\partial t} + \mathbf{u} \cdot \nabla \mathbf{u} = -\frac{1}{\rho_0}\nabla \mathcal{P} + \nu\nabla^2\mathbf{u} \quad ; \tag{1.25}$$

$\partial_t + \mathbf{u} \cdot \nabla$ is the Lagrangian derivative; $\mathcal{P}$ is the pressure, ρ_0 the constant density and ν the kinematic viscosity. The minus sign in front of the pressure stems from an analogy with elasticity (Lamb, 1932). An equivalent form of the Navier–Stokes equations with a modified pressure $\mathcal{P}^* = \mathcal{P} + u^2/2$, using the vector identities:

$$\varepsilon_{ijl}\varepsilon_{imn} = \delta_{jm}\delta_{ln} - \delta_{jn}\delta_{lm} \quad ,$$
$$\mathbf{a} \times (\mathbf{b} \times \mathbf{c}) = \mathbf{b}(\mathbf{a} \cdot \mathbf{c}) - \mathbf{c}(\mathbf{a} \cdot \mathbf{b}) \quad , \tag{1.26}$$
$$\nabla \times (\mathbf{a} \times \mathbf{b}) = \mathbf{a}(\nabla \cdot \mathbf{b}) - \mathbf{b}(\nabla \cdot \mathbf{a}) - \mathbf{a} \cdot \nabla\mathbf{b} + \mathbf{b} \cdot \nabla\mathbf{a} \quad ,$$

is

$$\frac{\partial \mathbf{u}}{\partial t} = -\frac{1}{\rho_0}\nabla \mathcal{P}^* + \mathbf{u} \times \omega + \nu\nabla^2\mathbf{u} \quad . \tag{1.27}$$

In order to obtain the time evolution of the total kinetic energy, one simply takes the dot product of the preceding equation with the velocity. When integrating over the whole volume of the fluid, one sees that (a) the nonlinear terms cancel out, *i.e.* they represent an exchange of energy among modes but do not lead *per se* to any dissipation; and (b) dissipation of the total kinetic energy is due to the square vorticity (or enstrophy). Indeed:

$$\nu \int \mathbf{u} \cdot \Delta\mathbf{u} \, d\mathbf{x} = -\nu \int \mathbf{u} \cdot (\nabla \times \omega) \, d\mathbf{x}$$
$$= -\nu \int \nabla \times \mathbf{u} \cdot \omega \, d\mathbf{x} = -\nu \int \omega^2 \, d\mathbf{x} \quad . \tag{1.28}$$

Similarly for a conducting fluid (see below the MHD equations, §1.8.2), dissipation of total magnetic energy is due to the square current (or magnetic enstrophy). However, this result does not hold pointwise (see also §2.2).

When adimensionalized, the Navier–Stokes equations depend only on one parameter, the Reynolds number R_V:

$$R_V = \frac{U_0 L_0}{\nu} \tag{1.29}$$

where U_0 and L_0 are the characteristic velocity and scale of the flow (for example the *r.m.s.* velocity and the integral scale, see §3.2.2 for the definition of the latter).

Taking the curl of the Navier–Stokes equation, we obtain the equation for the temporal evolution of the vorticity $\omega = \nabla \times \mathbf{u}$:

$$\frac{\partial \omega}{\partial t} = \nabla \times (\mathbf{u} \times \omega) + \nu\Delta\omega \quad . \tag{1.30}$$

or equivalently

$$\frac{D\omega}{Dt} = \frac{\partial \omega}{\partial t} + \boldsymbol{u} \cdot \nabla \omega = \omega \cdot \nabla \boldsymbol{u} + \nu \Delta \omega \ . \tag{1.31}$$

The latter form of the vorticity equation shows clearly the stretching of vorticity through velocity gradients.

1.7.2 The closure problem

Forgetting both indices and gradients as well as terms involving dissipation, the fluid equations in symbolic form are:

$$\frac{\partial u}{\partial t} = uu, \tag{1.32}$$

a form that stresses the nonlinearity (taken quadratic here). Because the field is assumed turbulent and thus random, it seems hopeless to try to solve for every details of its space–time variation; we shall rather deal with moments and write an evolution equation for the second and third–order ones:

$$\frac{\partial <uu>}{\partial t} = <uuu>$$
$$\frac{\partial <uuu>}{\partial t} = <uuuu> = \Sigma <uu><uu> + <uuuu>_c \qquad ; \tag{1.33}$$

where we have assumed a zero–mean velocity; also note that here, we use the notation $<uuuu>_c$ for the fourth–order cumulant (denoted $R_u^{(n)}$, with $n = 4$, in §1.4). The problem we face now is immediately apparent : it is a problem of closure, since evidently the time evolution of the moment of order n will involve the next–order moment. In weak turbulence, the closure arises naturally from the fast phase factor with contributions only at resonances. Here, we have to do something else. One possibility is to assume that $<uuuu>_c = 0$, a relationship which holds for a Gaussian field but leads to a model which was shown to yield negative energy spectra, a fact attributed to the unhindered build–up of triple correlations.

The next degree of sophistication in such models thus leads to postulate a simple form of damping of the third–order correlation $<uuu>$, in the following manner :

$$<uuuu>_c = -\mu_m <uuu> \tag{1.34}$$

where the rate μ_m at which triple moments decorrelate is going to be model–dependent. When $\mu_m = constant$ (hence a constant characteristic time for the nonlinearities to build–up), we arrive (after some steps) at the Markovian Random Coupling Model or MRCM; the energy spectrum that results $E(k) \sim k^{-2}$ is not consistent with Kolmogorov (1941a) phenomenology. However, it is an adequate spectrum for Burgers' equation

$$\partial_t u + u \partial_x u = \nu \partial_{xx}^2 u \tag{1.35}$$

which is a one–dimensional model of (compressible since $\partial_x u(x) \neq 0$) turbulent flows; it leads to the formation of shocks; indeed, the characteristic time of shock formation $L_0/\delta u$ (where δu is the velocity jump) is **independent** of the local scale ℓ (in that sense, the shock can be viewed as a coherent structure).

1.7.3 The EDQNM model

The next best assumption, then, leading to a successful and widely used model – the Eddy Damped Quasi Normal Approximation or in short EDQNM – is to assume a scale–dependence of the damping rate of triple correlations; hence we postulate a relationship $\mu_m = \mu(k, t)$ in which all physical characteristic rates will be included (dissipation, non–linear interactions and waves). For an incompressible neutral fluid

$$\mu_m = \mu(k, t) \sim \sqrt{k^3 E(k, t)} + \nu k^2 \quad ; \tag{1.36}$$

the nonlinear damping rate τ_{NL}^{-1} can be found by stating that the eddy turn–over time τ_{NL} is $\sim u(\ell)/\ell$ and $u(\ell) \sim \sqrt{kE(k)}$ with, omitting factors of 2π, $\ell \sim k^{-1}$. In the inertial range, this evaluation of $\mu(k)$ is compatible with a Kolmogorov (1941a) or in short K41 spectrum. Finally, one can also determine μ_m through an auxiliary equation, as done in the "test field" model or TFM of Kraichnan (1971).

Going back to the model, we now have :

$$\frac{\partial < u(t)u(t) >}{\partial t} \sim \int_0^t \exp\left[-\mu(k, t - \tau)\right] < u(\tau)u(\tau) >< u(\tau)u(\tau) > d\tau \quad . \tag{1.37}$$

There is in fact a dependence on wave vectors omitted here, as well as a dependence on the precise form of the nonlinear interactions. The final step in obtaining the EDQNM closure consists in markovianizing the preceding equation and solving by writing

$$\theta(k, p, q, t) = \int_0^t \exp\left[-\int_\tau^t \mu(k, p, q, t_1) dt_1\right] d\tau \quad ,$$

together with the evaluation of the correlations $< u(\tau)u(\tau) >$ at time $\tau = t$.

In order to be a little more explicit, an example, for three–dimensional Navier–Stokes incompressible turbulence of the EDQNM closure is given below. One first writes:

$$\mu(k, p, q, t) = \mu(k, t) + \mu(p, t) + \mu(q, t)$$
$$\mu(k, t) = \sqrt{k^3 E(k, t)} + \nu k^2 \tag{1.38}$$

where both the linear (viscous) and nonlinear (turnover) characteristic rates have been included and where, in the isotropic and homogeneous case, the energy spectrum depends only on wavenumber. Defining

$$\theta(k, p, q, t) = \mu^{-1}(k, p, q, t) \quad , \tag{1.39}$$

we obtain

$$\left(\frac{\partial}{\partial t} + 2\nu k^2\right) E(k, t) = \int\int_{\Delta k} \theta(k, p, q, t) \, kp^{-1}q^{-1} b(k, p, q) \, E(q, t) \times \tag{1.40}$$
$$[E(p, t) - E(k, t)] dp dq$$

with the Euler physics concentrated in the geometrical coefficient $b(k, p, q)$ given by

$$b(k, p, q) = pk^{-1}(\tilde{x}\tilde{y} + \tilde{z}^3) \tag{1.41}$$

(Orszag, 1974; see also Leith (1971) for the two–dimensional version); the three interacting wave vectors are such that $\boldsymbol{k} = \boldsymbol{p} + \boldsymbol{q}$ (similar to the resonance condition for waves in linear theory); $\tilde{x} = \cos(\boldsymbol{p}, \boldsymbol{q})$ (and $\tilde{y}$ and $\tilde{z}$ defined similarly by circular permutation of the three wavevectors); Δk is the area of the (p, q) plane such that the three wave vectors $\boldsymbol{k}, \boldsymbol{p}$ and $\boldsymbol{q}$ form a triangle (hence the name triad interactions). The first term on the *rhs* is a creation term, whereby energy at mode k is being transferred to modes p and q; the second term is an annihilation term linear in $E(k, t)$, and leading to a turbulent eddy viscosity (its expression at lowest order can be obtained by setting in equations (1.40) and (1.41) the ratio k/q as a small parameter, *i.e.* looking at the effect of small scales $\sim q^{-1}, p \sim q$ on large scales $\sim k^{-1}$). It can be checked that when $\nu \equiv 0$

$$\partial_t \int \mathrm{E}(k, t) dk = 0 \quad ;$$

this implies conservation of the kinetic energy by the nonlinear terms, a structural property preserved by the closure scheme.

1.8 Phenomenology of the inertial range

1.8.1 The Kolmogorov (1941a) law

At large scales, the flow is dominated by the instability mechanism that gives rise to it in the first place: this is the energy–containing range. On the other hand, at small scales dissipative processes will prevail and the energy spectrum $\mathrm{E}(k)$ will decay exponentially. In between, there may exist an "inertial" range through which energy goes at a fixed rate Π and in which $\mathrm{E}(k)$ only depends on both Π and k, but not on the number of steps needed to go from the energy–range down to k :

$$\mathrm{E}(k) \sim \Pi^a k^b \quad . \tag{1.42}$$

These hypotheses, due to Kolmogorov, immediately give

$$\mathrm{E}(k) \sim \Pi^{2/3} k^{-5/3} \tag{1.43}$$

which is the celebrated K41 law. From the expression for the energy spectrum above, we can immediately deduce the scale–dependence of the eddy turnover time

$$\tau_{NL}(k) = 1/\sqrt{k^3 \mathrm{E}(k)} \sim k^{-2/3} \quad . \tag{1.44}$$

Note that $\tau_{NL}(k) \to 0$ as $k \to \infty$.

If we now want to evaluate at what wavenumber k_D dissipative and nonlinear processes are comparable, we can say that $\tau_{diss}(k_D) = \tau_{NL}(k_D)$ which yields

$$k_D/k_0 \sim (\Pi/\nu^3)^{1/4} \quad . \tag{1.45}$$

This estimate of the dissipative length $\ell_D = 2\pi/k_D \sim R_V^{-3/4}$ (in units of the energy–containing scale $\ell_0 = 2\pi/k_0$) is called the Kolmogorov length; it is useful, for example in dealing with the numerization of the Navier–Stokes equations (and of turbulent flows in general, see §3.2). Also note that this scale might be thought of as the characteristic scale of vorticity (and in MHD, with the modification introduced in §1.8.3, of magnetic currents). Indeed, with an energy spectrum $E(k) \sim k^{-5/3}$, the vorticity follows a $k^2 E(k) \sim k^{1/3}$ law until the exponential decay associated with dissipation takes over; thus most of the vorticity is at a peak wavenumber determined by the balance between this 1/3 law and dissipation, a balance which occurs at small scale.

Several modifications of the preceding phenomenological argument have been presented; for example, one may include the effect of compressibility, or the fact that the number of steps down the cascade might count after all, in the latter case in an attempt to model the phenomenon of intermittency (see §3.8).

1.8.2 The MHD equations

The MHD equations derive from Maxwell's equations, with the assumption that velocities are sub–relativistic, hence the displacement current can be neglected (see *e.g.* Braginskii, 1965; Roberts, 1967). In the incompressible case they are given by:

$$\frac{\partial \boldsymbol{u}}{\partial t} + \boldsymbol{u} \cdot \nabla \boldsymbol{u} = -\frac{1}{\rho_0} \nabla \mathcal{P} + \nu \nabla^2 \boldsymbol{u} + \boldsymbol{j} \times \boldsymbol{b}$$

$$\frac{\partial \boldsymbol{b}}{\partial t} = \nabla \times (\boldsymbol{u} \times \boldsymbol{b}) + \eta \nabla^2 \boldsymbol{b} \tag{1.46}$$

$$\nabla \cdot \boldsymbol{u} = 0 \ ; \ \nabla \cdot \mathbf{b} = 0 \ ,$$

where $\boldsymbol{b} = \boldsymbol{B}/\sqrt{\mu_0 \rho_0}$ is the Alfvén velocity, μ_0 the permeability and η is the magnetic diffusivity. Note the analogy between the induction equation for $\boldsymbol{b}$ and that for the vorticity. The ratio $P^M = \nu/\eta$ is the magnetic Prandtl number; the current density is $\boldsymbol{j} = \nabla \times \boldsymbol{b}$.

Besides the kinetic Reynolds number R_V, we can also define the magnetic Reynolds number R_M:

$$R_M = \frac{U_0 L_0}{\eta} \ . \tag{1.47}$$

Unlike a uniform velocity field $\mathcal{V}_0$ which can be eliminated using Galilean invariance, a uniform magnetic field $\mathcal{B}_0$ will give rise to Alfvén waves (and in the compressible case, to fast and slow magneto–acoustic waves). But, moreover, the non–linear dynamics of the flow will *also* be affected by a large–scale quasi-uniform random magnetic field B_0 because nonlinear transfer will be hampered : indeed, nonlinear interactions **only** occur between one $z^+ = v + b$ eddy and one $z^- = v - b$ eddy which travel in opposite directions along B_0. The fact that no $z^+ z^+$ nor $z^- z^-$ interactions are allowed can be seen by rewriting the MHD equations above in terms of the Elsässer variables $\boldsymbol{z}^{\pm}$:

$$\frac{\partial \mathbf{z}^+}{\partial t} + \mathbf{z}^- . \nabla \mathbf{z}^+ = -\nabla p^* + \nu_1 \nabla^2 \mathbf{z}^+ + \nu_2 \nabla^2 \mathbf{z}^-, \tag{1.48a}$$

$$\frac{\partial \mathbf{z}^-}{\partial t} + \mathbf{z}^+ . \nabla \mathbf{z}^- = -\nabla p^* + \nu_1 \nabla^2 \mathbf{z}^- + \nu_2 \nabla^2 \mathbf{z}^+ \tag{1.48b}$$

$$\nabla . \mathbf{z}^+ = 0 \tag{1.48c}$$

$$\nabla . \mathbf{z}^- = 0 \tag{1.48d}$$

with $\nu_1 = (\nu + \eta)/2$ and $\nu_2 = (\nu - \eta)/2$; as usual, $p^* = p/\rho + b^2/2$ is the total pressure. The corresponding equations for the generalized vorticities $\omega^{\pm}$ are given in §3.7.1.

A one–dimensional model similar to that of Burgers (see equation (1.35)) but for MHD flows can be found in Thomas (1967), although in this case $\partial_x b(x) \neq 0$ which is not satisfactory if we stand by the non existence of magnetic monopoles. In the dispersive context of the Hall effect with equal magnetic and thermal pressures, one can also consider the Hada model (1993), with now $b(x) = b_y(x) + ib_z(x)$ a complex variable combination of the transverse components of the magnetic perturbation to a uniform $\mathcal{B}_0\hat{x}$ field.

1.8.3 Energy transfer in the presence of a uniform magnetic field

In MHD, a modification of the Kolmogorov phenomenology has been introduced independently by Iroshnikov (1963) and Kraichnan (1965) to take into account the fact that a strong quasi–uniform magnetic field has a non–trivial effect on the dynamics. We shall call it, in short, the IK theory for MHD.

Applying the ansatz that in such a case the characteristic time of energy transfer τ_{tr} is reduced in a $\tau_{NL}/\tau_A \gg 1$ ratio, with

$$\tau_A = \ell/B_0 \tag{1.49}$$

the Alfvén time as before, we have :

$$\tau_{tr}(k) = \tau_{NL}(k)^2/\tau_A(k) \ . \tag{1.50}$$

Dimensional analysis thus leads to the IK energy spectrum for MHD;

$$E(k) \sim (\Pi B_0)^{1/2} k^{-3/2} \ . \tag{1.51}$$

This spectral law can in fact be modified when taking into account the amount of correlations between velocity and magnetic field, the above case corresponding to the uncorrelated case (Grappin et al., 1983). But such modifications are felt only when the correlation coefficient, properly normalized, is close to unity.

This $-3/2$ spectrum gives for the transfer time in MHD a $k^{-3/4}$ dependence, as opposed to the $-2/3$ law for K41. Note that the simplest way to take into account this effect in the framework of closures such as the EDQNM is to simply add to the eddy–damping rate $\mu(k)$ the Alfvén rate $\tau_A^{-1}(k)$ (see Pouquet (1993) and references therein).

The wavenumber at which dissipation becomes significant in MHD is obtained in a similar manner as for the K41 case by equating characteristic times:

$$k_{D_{MHD}} \sim (\Pi/B_0\nu^2)^{1/3} \tag{1.52}$$

where a unit magnetic Prandtl number (equal viscosity and magnetic diffusivity) has been assumed. Note again the different scaling with Reynolds number ($k_{D_{MHD}} \sim R^{-2/3}$) from the K41 case.

Let us take an example. For a bipolar flow emerging from a new–born star through the interstellar medium, the transverse scale of the jet is $\sim 10^{16}$ cm, and the turbulent fluid velocity is ~ 10 kms^{-1}, or roughly 10% of the jet velocity; plugging in the magnetic diffusivity (with a $T^{3/2}$ dependence on temperature), we arrive at a magnetic Reynolds number of roughly 10^{12}. If Kolmogorov phenomenology applies to such a flow, this means that the dissipative length is 10^9 smaller than the integral scale, or 10^7 cm. On the other hand, if incompressible MHD effects prevail, the thickness of current layers is now 10^8 cm. Finally, in the case of shocks, the time of formation of which is independent of the scale of turbulent eddies, the scaling of the dissipative length is $R^{-1/2}$ since $\ell_D^2/\nu = constant$ (this estimate follows as well from inspection of equation (2.20) in §2.4.2 for a Burgers' vortex where advection is balanced by dissipation, as also in the flux expulsion mechanism where the $\mathbf{b} \cdot \nabla \mathbf{v}$ term is balanced by diffusion, with $\nabla \mathbf{v} = \mathcal{O}(1)$). For our bipolar jet, the shocks would thus be something like 10^{10} cm thick. It is noteworthy that, in such examples, at fixed viscosity, shocks are fat structures, and nonlinear mode coupling can produce smaller scales.

1.9 The self–similar temporal decay of energy

In the spirit of the K41 spectral law of energy distribution among Fourier modes, Kolmogorov (1941b) also derived the law of temporal decay of kinetic energy, assuming it is self–similar in time:

$$E(t) \sim t^{-\alpha_s} \ . \tag{1.53a}$$

This means that one does not consider here the final period of decay, which is exponential. Assume that the integral scale λ_I (see for its definition equation (3.1) in §3) follows a power law as well

$$\lambda_I \sim t^\beta \ , \tag{1.53b}$$

(where β will turn out to be positive: as energy decays, less weight in the integral (3.2) comes from the small scales and λ_I grows); assume further that the energy spectrum in the large scales is

$$E(k) \sim k^s \tag{1.53c}$$

with $1 < s \leq 4$ (an hypothesis linked to the regularity of the velocity (or vorticity) field as $k \to 0$); then one can show that

$$\alpha_s = \frac{2(s+1)}{s+3} \tag{1.54a}$$

and

$$\beta_s = \frac{2}{s+3} \tag{1.54b}$$

by using the constancy of $E\lambda_I$ and the fact that $\dot{E} = -\Pi$, and the Kolmogorov expression for the energy dissipation rate $\Pi = u^3/\lambda_I$ (see also Monin & Yaglom, 1975; Lesieur, 1990).

This law was derived by Kolmogorov for $s = 4$ corresponding to the production of a large–scale eddy through the beating of two small–scale eddies and by Saffman (1967) for $s = 2$, corresponding now to equipartition of modes in the large scales. Experiments are in agreement with these predictions with $E(t) \sim t^{-1.4}$ (see Chasnov (1995) for the axisymmetric case). When the integral scale is fixed ($\beta = 0$), experiments indicate a $E(t) \sim t^{-2}$ law (Smith *et al.*, 1993), in agreement with the above analysis with $s \to \infty$ (see also Lohse, 1994). In fact the derivation of these laws dates back in the latter case to Taylor (1935) and in the more general case to von Karman and Howarth (1938). Following Hossain *et al.* (1995), the procedure consists in writing the two evolution equations $\dot{u^2} = -u^3/\lambda_I$ and $\dot{\lambda_I} = \beta' u$. Self–similar evolution laws identical to those given above follow with $\beta' = 1/(s+1)$.

These laws can be modified to take into account the observed slowing–down of nonlinear transfer (and hence, of dissipation) in stratified flows (Kimura & Herring, 1995), or in incompressible MHD because of Alfvén waves. For example, in MHD one can use the IK phenomenology with now $\dot{E^T} = -\Pi \sim z^4 \lambda_I / B_0$ and with $E^T \sim z^2$ the total (kinetic plus magnetic) energy; with the same hypothesis of self–similar decay as before, one finds (Politano & Pouquet, 1995b):

$$E^T(t) \sim t^{-\alpha_{B,s}} \quad ; \quad \alpha_{B,s} = \frac{s+1}{s+2} \tag{1.55a}$$

and

$$\lambda_I(t) \sim t^{\beta_{B,s}} \quad ; \quad \beta_{B,s} = \frac{1}{s+2} \ . \tag{1.55b}$$

The role that a uniform magnetic field B_0 can play in the slowing–down of transfer to small scales has been considered in Hossain *et al.* (1995) leading to a different evaluation of the power–law temporal decay of energy than that presented here; the numerical simulations performed by these authors in three dimensions at low resolution do not as yet allow to differentiate the possible regimes (although their result does agree with a t^{-1} law); this point clearly requires further investigation.

Furthermore, it is to be noted that both for MHD flows and for stratified flows, the formation of vorticity filaments which occurs in three–dimensional turbulent flows (see §2.4) is impeded; is the more rapid energy decay in isotropic neutral fluids linked to the instability (bursting) of these filaments ? Then, if filaments disappear at high Reynolds numbers as sometimes conjectured, will the decay law of energy lessen as well ? These points might be checked by experiments.

2. From statistics to structures

2.1 Introduction

Up to now, the description we have given of a turbulent field has dealt with statistical moments of the velocity field, and in particular the energy spectrum. But in fact, to an observer or an experimentalist (and this includes numerical experiments), the flow is best described with structures in physical space and, in the incompressible case of a neutral fluid, this means dealing with vortices (see Van Dyke, 1982), in the compressible case with shocks as well, and in MHD also with flux tubes. The von Karman vortex street that develops behind a cylinder, and the way it is destroyed as the Reynolds number increases is well known. Convective cells are another common occurrence both in the laboratory and for natural flows, in the atmosphere and in the solar photosphere. As we move from the convective threshold, defects develop (see Newell, Passot & Souli, 1990) with patterns that become progressively more complex.

In order to describe complex turbulent flows, one may look at small scales by a visualization of vorticity for example. The striking phenomenon becomes the intermittency (both in time and in space) of such flows. Long quiescent plages are interrupted at random interval by sharp peaks. Such a feature was not included in the K41 phenomenology and is sometimes thought to lead to steeper spectra (see for example the β model of Frisch et al, 1978). But this point in fact is presently debated. Measurements in wind tunnels as well as in natural flows tell us that (a) K41 is a good approximation and (b) there is a flattening of the spectrum prior to entering the exponential drop-off, flattening newly discovered and attributed to a "bottleneck" phenomenon (Falkovich, 1994; Lohse & Müller–Groeling, 1995). In astrophysics, the $-5/3$ law seems to be fulfilled on an inordinate range of scales (Amstrong et al., 1981). Satellite observations of the solar wind in situ also lead to such a spectrum, although several other regimes are present (depending on whether one deals with fast and slow winds, far or close to the sun, and whether compressibility is important or not).

2.2 Betchov's relation

2.2.1 A kinematic relationship between strain and vorticity

What kind of structure can emerge from the dynamics of an incompressible neutral fluid ? In 1956, Betchov wrote a simple relation between averaged square vorticity and the eigenvalues of the strain tensor S_{ij} noted a, b and c, which gave a first indication. We follow here his derivation.

Write the rate of deformation tensor $\partial_j u_i$ in the basis where the strain $S_{ij} = \frac{1}{2}(\partial_j u_i + \partial_i u_j)$ is diagonal:

$$u_{j,i} = \partial_i u_j = \begin{pmatrix} a & \mathcal{C}_- & \mathcal{B}_- \\ \mathcal{C} & b & \mathcal{A} \\ \mathcal{B} & \mathcal{A}_- & c \end{pmatrix} \tag{2.1}$$

with $a + b + c = 0$ for an incompressible fluid, and with $\mathcal{A}_- = -\mathcal{A}$ (and similarly for $\mathcal{B}_-$ and $\mathcal{C}_-$). We shall also assume homogeneity and isotropy. It can be shown that the set of these three scalars $(\mathcal{A}, \mathcal{B}, \mathcal{C})$ is in fact equal to half the vorticity $\omega = \nabla \times \boldsymbol{u}$ (hence vorticity is a "pseudo" vector, i.e. a vector that transforms like a tensor). A first result derived by Betchov is that, on average, strain and vorticity are equal. This can be done by evaluating both

$$< u_{i,j} u_{i,j} > = a^2 + b^2 + c^2 + 2(\mathcal{A}^2 + \mathcal{B}^2 + \mathcal{C}^2) \tag{2.2}$$

and

$$< u_{i,j} u_{j,i} > = < \partial_j (u_i \partial_i u_j) > - \left\langle u_i \frac{\partial^2 u_j}{\partial x_i \partial x_j} \right\rangle = 0 \tag{2.3}$$

$$= a^2 + b^2 + c^2 - 2(\mathcal{A}^2 + \mathcal{B}^2 + \mathcal{C}^2) \ .$$

Defining $\sigma^2 = a^2 + b^2 + c^2$, and since $2(\mathcal{A}^2 + \mathcal{B}^2 + \mathcal{C}^2) = \omega^2/2$, we obtain that on average

$$< \sigma^2 > = < \omega^2/2 > \ . \tag{2.4}$$

Similar relationships – which do not involve the dynamical equations at this point but are of a purely kinematic character – can also be derived in incompressible MHD, on the magnetic field, and of course on the Elsässer variables as well (Pouquet, 1994).

This relationship, however, does **not** hold **pointwise**. As we saw in §1.7, the volume–averaged square vorticity is proportional to the dissipation of kinetic energy; this result is slightly paradoxical since vorticity, really, is an anti–symmetric (energy–conserving) tensor: the antisymmetric part of the velocity gradient tensor (the vorticity) simply rotates locally the velocity vector. What the above relation tells us is that, because of the divergenceless incompressibility assumption on the velocity (an assumption which of course carries over to the magnetic field as well), on average the norms of the symmetric and antisymmetric parts of the velocity gradient tensor are proportional. On the other hand, pointwise dissipation is proportional to $S_{ij}S_{ij}$ (see *e.g.* Batchelor, 1967, §3.4).

2.2.2 A dynamical relationship between strain and vorticity

We now use explicitly the dynamical equations, but omit dissipation terms. In the fluid case, the temporal evolution of mean–square vorticity is:

$$
\begin{aligned}
\partial_T < u_{i,j} u_{i,j} > &= \partial_T < \omega^2 > \\
&= 4\partial_T < \mathcal{A}^2 + \mathcal{B}^2 + \mathcal{C}^2 > \\
&= 2 < u_{i,j}\partial_j\partial_T u_i > \\
&= 2 < u_{i,j}\partial_j(-u_k\partial_k u_i) > \\
&= -2 < u_{i,j} u_{k,j} u_{i,k} > -2 < u_{i,j} u_k u_{i,k,j} >
\end{aligned}
\tag{2.5}
$$

+ pressure (and viscosity) terms.

There are a few steps in this computation: using isotropy, you can show that

$$
< u_k u_{i,kj} u_{i,j} > = 0 \quad ;
$$

on the other hand, one has

$$
< u_{i,j} u_{k,j} u_{i,k} > = < u_{i,j} u_{j,k} u_{i,k} > = < a^3 + b^3 + c^3 > -3 < a\mathcal{A}^2 + b\mathcal{B}^2 + c\mathcal{C}^2 > \quad .
\tag{2.6a}
$$

Use now the fact that

$$
< u_{i,j} u_{j,k} u_{k,i} > = 0 = < a^3 + b^3 + c^3 > +3 < a\mathcal{A}^2 + b\mathcal{B}^2 + c\mathcal{C}^2 > \quad ;
\tag{2.6b}
$$

furthermore, since $a + b + c = 0$, it follows that $< a^3 + b^3 + c^3 > = 3 < abc >$. After a few lines, the interesting Betchov relation obtains:

$$
\partial_T < \omega^2 > = -3 < abc > \quad .
\tag{2.7}
$$

We now assume (with no loss of generality) that $a \geq 0$ and $c \leq 0$ since the flow is incompressible; the intermediate eigenvalue b can be of any sign. But we recall that turbulence is associated with vortex stretching and thus on average growth of enstrophy $< \omega^2 >$. This in turn implies that on average the second eigenvalue of the strain tensor be positive:

$$
< b > > 0 \quad ,
\tag{2.8}
$$

a prediction made by Betchov and a fact verified by present–day numerical simulations (with however, $< b > < < a >$, see also §2.4.4).

From result (2.7) we deduce that, with two positive eigenvalues, a blob of vorticity at $t = 0$ will be stretched into a sheet. But is that sheet stable ? Numerical investigations clearly indicate that the first inviscid phase of evolution of a flow is an exponential decrease in scales with indeed the formation of vorticity sheets (see for example Brachet et al. (1992); see also §2.4). But such sheets are unstable, for example through Kelvin–Helmoltz or through a self–focusing mechanism (Lin & Corcos, 1984; Neu, 1984; Passot et al., 1995) of a vortex sheet embedded in a large-scale strain. At later times, vortex filaments or vortex tubes form (see e.g. Vincent & Meneguzzi, 1991; 1994) that are long–lived and whose statistics are presently under study (Vincent & Meneguzzi, op. cit.; Jimenez et al., 1993; see also Porter, Pouquet & Woodward (1994) for the compressible case). One open question is whether vorticity filaments persist as the Reynolds number is increased. From experiments (Douady et al., 1991) and from the observation of atmospheric flows, intense vortices are known to break down and even burst (Escudier, 1982; Granger, 1993; and Keller (1994) in the compressible case). Note that one can reconcile the a priori contradictory facts that, on the one hand the Betchov's relation holds implying that on average the second eigenvalue of strain is positive and thus turbulence stretches vorticity into sheets, and on the other hand that the most striking features of turbulent flows are the vortex filaments; the solution of the paradox is that the filaments, although at high vorticity, contain only a small percentage of the total (volume averaged) enstrophy, as shown e.g. in Jimenez et al. (1993).

Another striking feature emerging from numerical simulations (Kerr, 1985) is the strong alignment of vorticity with e_2, the eigenvector of S_{ij} associated with the intermediate eigenvalue (see also the model of §2.4.4). When vortex filaments are formed, this can be understood as bi-dimensionalization of the swirling flow around that filament; other models can be found such as in Vieillefosse (1982; 1984).

2.3 Pressure as a diagnostic tool

In an incompressible fluid, the pressure field is, in a sense, what maintains the velocity divergenceless. In particular, writing the Navier–Stokes equations in Fourier space for velocity modes $\hat{v}(k)$, the pressure acts as a projection operator

$$\mathcal{P}_{ij}(k) = \delta_{ij} - \frac{k_i k_j}{k^2}$$

onto the perpendicular to the wave vector k. Indeed, for a vector v one has $\mathcal{P}_{ij}(k)v_j = v_i - [(k \cdot v)/k]\hat{k}_i$ where $\hat{k}$ denotes the unit vector along k. For compressible flows, this projection is useful as well when decomposing a velocity field into its solenoidal (divergence–less) v^S and longitudinal (curl–less) v^C components:

$$v = v^S + v^C$$

with $\nabla \cdot v^S = 0$ and $\nabla \times v^C = 0$. In the latter case, one can introduce a potential ϕ such that $v^C = \nabla\phi$ hence $\nabla \cdot v = \nabla^2\phi$; solving in Fourier space, one recovers $v_i^S(k) = \mathcal{P}_{ij}(k)v_j(k)$.

In configuration space, pressure can be useful too, as we see now.

2.3.1 The fluid case

Recently, it was proposed (Brachet, 1990) to use pressure depletion to visualize vortex filaments in the laboratory. Indeed taking the divergence of the Euler equations, we obtain:

$$\frac{\partial^2 \mathcal{P}}{\partial x_i^2} + (\partial_i u_j)(\partial_j u_i) = 0 \ . \tag{2.9}$$

Writing as usual the velocity gradient matrix $\partial_j u_i$ as the sum of a symmetric part S_{ij} (strain) and an antisymmetric part V_{ij} namely:

$$\partial_j u_i = \frac{1}{2}(\partial_j u_i + \partial_i u_j) + \frac{1}{2}(\partial_j u_i - \partial_i u_j) = S_{ij} + V_{ij} \tag{2.10}$$

we then have $4S^2 = (\partial_i u_j + \partial_j u_i)^2$ and $4V^2 = (\partial_i u_j - \partial_j u_i)^2$ leading to (setting $\sigma^2 = 2S^2$)

$$2\Delta \mathcal{P} = \omega^2 - \sigma^2 \ . \tag{2.11}$$

Thus, low pressures are regions of strong vorticity (intuitively, think in terms of the centrifugal force) and high pressures are regions of strong strain which are also regions of strong dissipation. In fact, an electrostatic analogy can be drawn here. Vorticity filaments can now be visualized and studied using such a technique (Douady et al., 1991).

2.3.2 The MHD case

We can extend this result to the MHD case. Defining

$$\sigma^{\pm} = \sigma^V \pm \sigma^M \tag{2.12}$$

the generalized strain $\partial_j z_i^{\pm}$ (associated with the $z^{\pm}$ Elsässer variables) and

$$\omega^{\pm} = \nabla \times z^{\pm} = \omega \pm j \tag{2.13}$$

the generalized vorticities, we obtain:

$$2\Delta \mathcal{P} = \omega^+ \omega^- - \sigma^+ \sigma^- \ . \tag{2.14}$$

In the original variables (velocity and magnetic field), this reads:

$$2\Delta \mathcal{P} = (\omega^2 - \sigma_V^2) - (j^2 - \sigma_M^2) \ .$$

Note that when z^+ and z^- are comparable, hence

$$E^R = z^{+2} - z^{-2} \equiv 2E^C \sim 0 \tag{2.15}$$

i.e. when velocity–magnetic field correlations are weak, then $\omega^+ \sim \omega^- \sim \bar{\omega}$ and $\sigma^+ \sim \sigma^- \sim \bar{\sigma}$. In that case, we recover the pure fluid formulation, where low pressures are associated with high generalized $\omega^{\pm}$ vorticity.

We know from closure models of turbulence (Grappin et al., 1983) as well as from numerical simulations (see e.g. Passot et al., 1990) that the correlation coefficient

$$\rho_C = \frac{2 < \boldsymbol{u} \cdot \boldsymbol{b} >}{u^2 + b^2} \tag{2.16}$$

may be strong in portions of the flow corresponding to the large energy–containing eddies (as well as in very small scales), but is weak in the vicinity of the dissipation wavenumber. Hence, at scales of the order of the width of current and vorticity sheets, $2\Delta\mathcal{P} = \bar{\omega}^2 - \bar{\sigma}^2$, and thus the same type of pressure diagnostic as in the fluid case will allow to visualize dissipative structures in MHD as well.

2.4 Examples of vortex filaments

We give here a few simple examples of vortex structures.

2.4.1 The viscous core of a vortex

Take at $t = 0$

$$\omega_0(\boldsymbol{x}) = \Gamma_0\delta(x)\delta(y) \ ; \tag{2.17}$$

hence $\boldsymbol{u}(\boldsymbol{x} = 0) = 0$ and $\boldsymbol{u} \times \boldsymbol{\omega} = 0 \ \forall\boldsymbol{x}$. We thus simply have to solve $\partial_T\omega = \nu\nabla^2\omega$ leading to

$$\omega(\boldsymbol{x}, T) = (\Gamma_0/4\pi\nu T)\exp(-r^2/4\nu T) \ . \tag{2.18}$$

At fixed time, the radius of the filament is thinner the higher the Reynolds number R_V ; at fixed viscosity, the filament spreads with time. More general cases can be done as well (Saffman, 1992, §13.3). For a turbulent vortex, there is an overshoot of circulation leading to negative vorticity at large distances, leading in turn to an axisymmetric instability (Saffman, *op. cit.* p. 257).

Similarly, a magnetic flux tube will have a radius with a $R_M^{-1/2}$ scaling (see also §3.7.2).

2.4.2 Burgers' vortex with viscosity

Take a simple straining velocity field

$$\boldsymbol{u} = (0, \gamma_s y, -\gamma_s z) \tag{2.19}$$

and a z–dependent vorticity with only a y-component. A steady–state solution obtains readily:

$$\omega(z) = \omega_0 \exp\left[-\gamma_s z^2/2\nu\right]\hat{\boldsymbol{e}}_y \tag{2.20}$$

resulting from a balance between stretching and spreading at all times. The thickness of such vortex layers scales like $R_V^{-1/2}$, as in the flux expulsion problem in MHD.

This steady solution can be easily extended to MHD using the Elsässer variables $\omega^{\pm} = \nabla \times z^{\pm}$, with $\omega^{+} \sim \hat{\boldsymbol{e}}_y \exp[(-\gamma_s/2\nu)z^2]$, or dealing with the magnetic field itself, using the analogy with vorticity (see §3.7.2).

2.4.3 Self–similar solution of the inviscid problem

Let us take a uniform two–dimensional strain $\boldsymbol{u} = (0, \gamma_s y, -\gamma_s z)$ as before, so of course $\nabla \cdot \boldsymbol{u} = 0$ and $\nabla \times \boldsymbol{u} = 0$. It can be checked easily that the vorticity

$$\omega(\boldsymbol{x}, T) = e^{\gamma_s T}g(ze^{\gamma_s T})\hat{\boldsymbol{e}}_y \tag{2.21}$$

is a solution of Euler equation for that given straining field. This shows that such an external strain leads to exponential growth of the vorticity in layers.

Let us now put $\gamma_s \equiv 1$ to simplify notations, and add to the velocity the Biot–Savart induced velocity due to the vorticity.

2.4.4 An inviscid model of alignment

Following Brachet et al. (1992), we take a velocity field $v = v_s + v_\omega$ with $v_s = (0, y, -z)$ a strain and $v_\omega = (f(ze^T), 0, 0)$ the velocity induced by the vorticity $\omega = (0, 2\bar{\alpha}, 0)$ with

$$\bar{\alpha} = e^T f'(ze^T) = \partial_z v_\omega \; . \tag{2.22}$$

Computing now the strain matrix eigenvalues, one finds

$$\begin{aligned} \lambda_1(T) &= 0.5 + \sqrt{\bar{\alpha}^2 + 1/4} \\ \lambda_2(T) &= 1 \\ \lambda_3(T) &= -0.5 - \sqrt{\bar{\alpha}^2 + 1/4} \end{aligned} \tag{2.23}$$

and similarly for the eigenvectors $v_1 = (\bar{\alpha}, 0, -0.5 + \sqrt{\bar{\alpha}^2 + 1/4})$; $v_2 = (0, 1, 0)$ and $v_3 = (\bar{\alpha}, 0, -0.5 - \sqrt{\bar{\alpha}^2 + 1/4})$. Recalling that $\bar{\alpha} \sim e^T$, we see that $\lambda_1 \sim e^T$.

Two remarks are in order: (a) the vorticity is parallel to the second eigenvector v_2 for all times ; and (b) the eigenvalue associated with v_2 is the largest one for early times (up to $\bar{\alpha} = \sqrt{2}$) ; but when the induced velocity field becomes large, vorticity becomes aligned with the intermediate eigenvector, as indicated by numerical simulations (Kerr, 1985). When a slight variation of vorticity in the x direction is taken into account inducing a non–zero z–component of the Biot–Savart velocity field, interesting instabilities may develop (Lin & Corcos, 1984; Neu, 1984; Passot et al., 1995) that result in a filamentation of the vorticity sheet.

The decomposition of the velocity into strain and vorticity–induced fields is done in Pullin & Saffman (1992) in cylindrical coordinates. They also compute directly the skewness and show that it is proportional to the strain: without strain, S_k is zero (a gaussian field has no skewness).

2.5 The motion of a line element in an inviscid fluid

2.5.1 The Frenet–Serret frame

A line in space can be parametrized by time T and by a local abcissa λ so that we have $r(\lambda, T)$ of coordinates x_i. The vector l along that curve is

$$l = \partial r / \partial \lambda = \ell \hat{t} \tag{2.24}$$

with ℓ the length and $\hat{t}$ the tangent unit vector; the velocity field is $u = dr/dT$. We shall follow here the derivation given in Drummond I. & Münch W. (1991) for the time evolution of the curvature of the line. The case of surfaces is treated in Pope (1988) in the context of combustion, where the deformation of surfaces is linked to the rate of mixing and reaction in the turbulent fluid. The length ℓ of the element is given by:

$$\ell = \frac{\partial s}{\partial \lambda} = \sqrt{\sum_i \ell_i^2} \tag{2.25}$$

where s is the curvilinear distance measured along the curve. One introduces the Frenet–Serret local frame defined by the set of orthonormal vectors $(\hat{t}, \hat{n}, \hat{b})$ where $\hat{t} = \partial r / \partial s$ is the tangent vector, $\hat{n}$ is the normal vector pointing towards the center of curvature of the line and $\hat{b} = \hat{t} \times \hat{n}$ completes the set. When moving along the curve, one has :

$$\frac{\partial}{\partial s}\begin{pmatrix} \hat{t} \\ \hat{n} \\ \hat{b} \end{pmatrix} = \begin{pmatrix} 0 & \kappa & 0 \\ \kappa_- & 0 & \tau \\ 0 & \tau_- & 0 \end{pmatrix}\begin{pmatrix} \hat{t} \\ \hat{n} \\ \hat{b} \end{pmatrix} = \hat{t} \cdot \nabla \begin{pmatrix} \hat{t} \\ \hat{n} \\ \hat{b} \end{pmatrix} \qquad (2.26)$$

with $\kappa_- = -\kappa$ and $\tau_- = -\tau$ (the Frenet–Serret relationships).

The matrix above is anti–symmetric (in order to conserve the norm) ; κ and τ are respectively the curvature and torsion of the line measured locally in space (with $R = 1/\kappa$ the radius of curvature and $T = 1/\tau$ the radius of torsion). Start at a point O on the curve. To first order, this curve can be seen as a straight line – its tangent. To second order, while moving along the curve from the point O, the curve bends around with curvature κ ; in the plane $(\hat{t}, \hat{n})$, the curve is assimilated to a circle of radius of curvature $R = 1/\kappa$ (a positive number). To third order (or equivalently, going further away from the point O) the curve may in fact warp out of the $(\hat{t}, \hat{n})$ plane into the $\pm\hat{b}$ direction; locally around point O it can then be considered as an helix of pitch angle δ with $\tan\delta = \tau/\kappa$. Note that since the curve may warp in either direction along $\hat{b}$, τ can be of either sign (it is a pseudo-scalar). In fact, it can be shown by using a Taylor expansion around point O (see e.g. Klingenberg, 1978), assuming that the curve is differentiable, that

$$\mathbf{l}(s) - \mathbf{l}(s_0) = \hat{t}(s_0)[\Delta s - \frac{(\Delta s)^3}{6}\kappa^2(s_0)] + \hat{n}(s_0)[\frac{(\Delta s)^2}{2}\kappa(s_0) + \frac{(\Delta s)^3}{6}\dot{\kappa}(s_0)]$$
$$+ \hat{b}(s_0)[\frac{(\Delta s)^3}{6}\kappa(s_0)\tau(s_0)] + o((\Delta s)^3) \qquad (2.27)$$

where $\Delta s = s - s_0$.

2.5.2 Temporal evolution of curvature

Since the line vector $\mathbf{l}$ of length ℓ and its tangent $\hat{t}$ are of course related by $\mathbf{l} = \ell\hat{t}$, the time variation of its components are given by:

$$\frac{dl_i}{dT} \equiv \dot{l}_i = \dot{\ell}\hat{t} + \ell\dot{\hat{t}} = W_{ij}l_j \qquad (2.28)$$

with

$$W_{ij} = \frac{\partial u_i}{\partial x_j} \qquad (2.29)$$

the velocity gradient matrix. From here on, the temporal derivative of a unit Frenet vector $\hat{v}$ will be noted $\dot{v}$ which stands for $\dot{\hat{v}}$, where $\hat{v}$ stands for $(\hat{t}, \hat{n}, \hat{b})$.

Second and third order variations of the line element are defined as:

$$\mathbf{a}^{(2)} = \frac{\partial \mathbf{l}}{\partial \lambda} \; ; \; \mathbf{a}^{(3)} = \frac{\partial \mathbf{a}^{(2)}}{\partial \lambda} \; . \qquad (2.30)$$

The time derivative of $\mathbf{a}^{(2)}$ is readily obtained using the chain rule; defining

$$W_{ijk} = \frac{\partial}{\partial x_k}W_{ij}$$

we have:

$$\dot{a}_i^{(2)} = W_{ij}a_j^{(2)} + W_{ijk}l_jl_k \quad , \tag{2.31}$$

and similarly:

$$\dot{a}_i^{(3)} = W_{ij}a_j^{(3)} + 3W_{ijk}l_ja_k^{(2)} + W_{ijkm}l_jl_kl_m \quad , \tag{2.32}$$

where $W_{ijkm} = \frac{\partial}{\partial x_m}W_{ijk}$ (Drummond & Münch, 1991).

From the Frenet–Serret transformation matrix, using $\hat{t} = \ell^{-1}l$ and differentiating with respect to s or λ with:

$$\frac{\partial}{\partial s} = \ell^{-1}\frac{\partial}{\partial\lambda} \quad , \tag{2.33}$$

one obtains:

$$\frac{\partial\hat{t}_i}{\partial s} = \kappa\hat{n}_i = \ell^{-2}(a_i^{(2)} - \ell^{-2}l_il_ja_j^{(2)}) \quad . \tag{2.34}$$

When taking now the time derivative of $\kappa\hat{n}_i$ and the dot product with $\hat{n}_i$, a first expression for the temporal derivative of the curvature of a material line follows:

$$\hat{n}_i(\dot{\kappa}\hat{n}_i + \kappa\dot{n}_i) = \dot{\kappa} \quad . \tag{2.35}$$

This equality stems from the fact that $\mid\hat{n}\mid = 1$ and $\hat{n}_i\dot{n}_i = \frac{1}{2}\frac{d}{dT}(\hat{n}^2) = 0$. When differentiating the expression for $\partial\hat{t}_i/\partial s$ given above and taking the dot product with $\hat{n}_i$ we have:

$$\dot{\kappa} = \hat{n}_i\ell^{-2}(-2\ell^{-1}\dot{\ell}a_i^{(2)} + \dot{a}_i^{(2)} - \ell^{-2}l_ja_j^{(2)}l_i) \quad , \tag{2.36}$$

using the orthogonality of the basis system ($\hat{n}_il_i = \hat{n}_i\ell\hat{t}_i = 0$). Noting that $\dot{\ell} = \ell W_{ij}\hat{t}_i\hat{t}_j$ and replacing, one obtains the time evolution for the curvature of a line element in an inviscid fluid:

$$\dot{\kappa} = W_{ij}(-2\hat{t}_i\hat{t}_j + \hat{n}_i\hat{n}_j)\kappa + W_{ijk}\hat{n}_i\hat{t}_j\hat{t}_k \quad . \tag{2.37}$$

When the material line is a vorticity line, only the symmetric part S_{ij} of the velocity gradient tensor W_{ij} enters the dynamical equation. Indeed, we write again:

$$\partial_ju_i = W_{ij} = \frac{1}{2}(\partial_iu_j + \partial_ju_i) + \frac{1}{2}(\partial_ju_i - \partial_iu_j) = S_{ij} + V_{ij} \quad .$$

The antisymmetric part $2V_{ij}$ can be written as $-\epsilon_{ijk}\omega_k$; hence:

$$D_T\omega = \omega\cdot\nabla u = \omega_j(S_{ij} - \frac{1}{2}\epsilon_{ijk}\omega_k) = \omega_jS_{ij}. \tag{2.38}$$

The temporal evolution of the curvature in the case of a vortex line is thus

$$\dot{\kappa} = -\alpha\kappa - \tau S_b + \hat{t}\cdot\nabla S_n \tag{2.39}$$

(Constantin, Procaccia & Segel, 1994); this equation is the analog of the general case for a magnetic flux line given in equation (2.36), but with the full velocity gradient operator replaced by its symmetric part S_{ij} with:

$$\alpha = \hat{t}\cdot S\cdot\hat{t} \quad , \tag{2.40}$$

and

$$S_n = \hat{n}\cdot S\cdot\hat{t} \quad , \quad S_b = \hat{b}\cdot S\cdot\hat{t} \quad . \tag{2.41}$$

2.5.3 Temporal evolution of torsion and of the Frenet frame

The temporal derivative of the torsion of a material line can also be obtained following the same type of arguments (see also Betchov, 1965). In the case of a vortex line (Constantin *et al.*, 1994) it reads:

$$\dot{\tau} = -\alpha\tau + S_b\kappa + \hat{t}\cdot\nabla N \ , \tag{2.42}$$

with

$$\kappa N = \hat{t}\cdot\nabla S_b + \tau S_n \ . \tag{2.43}$$

This result can be obtained in several ways. We first note that curvature and torsion are defined in the local Frenet frame associated with a point on the material line, a line which varies with time as given by the dynamical equations (*e.g.* Navier-Stokes, or MHD ...). Equivalently, we can derive the time evolution of the Frenet vectors. Indeed, starting from the Frenet matrix and differentiating with time, we obtain:

$$D_T\frac{\partial\hat{t}_i}{\partial s} = \dot{\kappa}\hat{n}_i + \kappa\dot{n}_i$$

$$D_T\frac{\partial\hat{n}_i}{\partial s} = -\dot{\kappa}\hat{t}_i - \kappa\dot{t}_i + \dot{\tau}\hat{b}_i + \tau\dot{b}_i \tag{2.44}$$

$$D_T\frac{\partial\hat{b}_i}{\partial s} = -\dot{\tau}\hat{n}_i - \tau\dot{n}_i \ .$$

Using the following definitions for $\dot{n}_i$

$$\dot{n} \equiv N_t\hat{t} + N_b\hat{b} \tag{2.45}$$

we rewrite $D_T\frac{\partial\hat{t}_i}{\partial s}$ as

$$D_T\frac{\partial\hat{t}_i}{\partial s} = \dot{\kappa}\hat{n}_i + \kappa N_t\hat{t}_i + \kappa N_b\hat{b}_i \ . \tag{2.46}$$

Note that the Lagrangian derivative $D_T = \partial_T + u\cdot\nabla$ is nothing but the chain-rule of temporal derivative d_T, a fact implicitly used here:

$$\frac{d}{dT} = \frac{\partial}{\partial T} + \sum_i\frac{dx_i}{dT}\frac{\partial}{\partial x_i} \ . \tag{2.47}$$

We see that the time derivatives of curvature obtains by taking the dot product of equation (2.46) with $\hat{n}$ (and similarly for torsion, using (2.44c)). On the other hand, the time derivative of the Frenet normal unit vector obtains by projecting on $\hat{t}$ and $\hat{b}$ (and projecting on $\hat{t}$ and $\hat{n}$ for $\hat{b}$).

An important word of caution, however: the ∂_s and $D_T = \partial_T + u\cdot\nabla$ derivatives **do not commute**; instead, one can derive the following relationship (see Constantin *et al.* (1994) for the symmetric case of a vortex line where γ in the equation below is replaced by the expression for α given in (2.40)):

$$D_T\frac{\partial f}{\partial s} = \frac{\partial}{\partial s}D_Tf - \gamma\frac{\partial f}{\partial s} \ , \tag{2.48}$$

the ∂_s differentiation is taken along the material line with tangent $\hat{t}$ (for example, a vortex line or a magnetic flux line) *i.e.* where $D_TX = X\cdot\nabla u$ (with e.g. $X = \omega =$

$w\hat{\omega} = w\hat{t}$ or in MHD $\boldsymbol{X} = \boldsymbol{B} = B\hat{t}$ with w and B respectively the amplitudes of the vorticity and magnetic field).

Using (2.48) above with $f = \hat{t}_i$ yields:

$$D_T \frac{\partial \hat{t}_i}{\partial s} = -\gamma \frac{\partial \hat{t}_i}{\partial s} + \frac{\partial}{\partial s} D_T \hat{t}_i \ . \tag{2.49}$$

To evaluate $D_T \hat{t}_i$ we come back to the definition of the material line vector $\boldsymbol{l} = \ell\hat{t}$; as we saw before, we have $\dot{l}_i = W_{ij}l_j = \ell\dot{t}_i + \hat{t}_i\dot{\ell}$. We now define the components of $\dot{t}$ on the $\hat{n}$ and $\hat{b}$ vectors respectively as W_n and W_b, namely:

$$\dot{t} \equiv W_n\hat{n} + W_b\hat{b} \ . \tag{2.50}$$

Thus:

$$\dot{t}_i = W_{ij}\hat{t}_j - \hat{t}_i W_{jm}\hat{t}_j\hat{t}_m, \tag{2.51}$$

and by projecting on the unit axes:

$$\begin{aligned} W_n &= W_{ij}\hat{n}_i\hat{t}_j \\ W_b &= W_{ij}\hat{b}_i\hat{t}_j \end{aligned} \tag{2.52}$$

which are the analogous relationships to (2.40) and (2.41) derived by Constantin *et al.* (1994), but in the general (non symmetric) case.

Coming back now to equation (2.46), and identifying the components of the unit vectors leads to:

$$\begin{aligned} \dot{\kappa} &= -\gamma\kappa - W_b\tau + \hat{t}\cdot\nabla W_n \\ N_t &= -W_n \\ \kappa N_b &= \tau W_n + \hat{t}\cdot\nabla W_b \end{aligned} \tag{2.53}$$

using (2.49), (2.50) and the definition of the Frenet matrix.

The time derivative of the bi-normal vector readily obtains from the anti-symmetry property of the transformation matrix (because as we know the Frenet vectors must remain of unit length):

$$D_T \begin{pmatrix} \hat{t} \\ \hat{n} \\ \hat{b} \end{pmatrix} = \begin{pmatrix} 0 & W_n & W_b \\ W_{-n} & 0 & N_b \\ W_{-b} & N_{-b} & 0 \end{pmatrix} \begin{pmatrix} \hat{t} \\ \hat{n} \\ \hat{b} \end{pmatrix} \tag{2.54}$$

with $W_{-n} = -W_n$, $W_{-b} = -W_b$ and $N_{-b} = -N_b$. Similarly to the case of curvature, in order to obtain the temporal evolution of the torsion of a field line, we start with $\partial\hat{b}/\partial s = -\tau\hat{n}$, differentiate again with time and take the dot product with $\hat{n}$, using equations (2.26) and (2.48), together with (2.54):

$$\dot{\tau} = -\gamma\tau + W_b\kappa + \hat{t}\cdot\nabla N_b \tag{2.55}$$

using

$$\gamma = W_{ij}\hat{t}_i\hat{t}_j \ . \tag{2.56}$$

Together with

$$\dot{\kappa} = -\gamma\kappa + W_{ijk}\hat{n}_i\hat{t}_j\hat{t}_k + W_b\kappa \ , \tag{2.57}$$

this set of equations (2.55) and (2.57) is similar to that obtained in Constantin *et al.* (1994) except that the full velocity gradient matrix appears, as opposed to simply the (symmetric) strain in the special case of a vortex line.

Finally we can remark that, on the basis that torsion is a pseudo–scalar, one could expect its temporal development to be correlated with that of helicity $H = <\boldsymbol{u}\cdot\boldsymbol{\omega}>$ (or equivalently in MHD $H^M = <\boldsymbol{a}\cdot\boldsymbol{b}>$ where $\boldsymbol{a}$ is the magnetic potential, $\boldsymbol{b} = \nabla\times\boldsymbol{a}$). However, the numerical experiments performed by Drummond & Münch (1991) on a model of a turbulent flow indicate that torsion tends to increase irrespective of the amount of helicity, which simply relates to the amount of left– and right–handed torsion. This point clearly needs further investigation with high-resolution numerical simulations. On the other hand, torsion can be considered like curvature in the $(\hat{n}, \hat{b})$ plane, in which case the link with helicity is not so clear.

2.5.4 The role of viscosity on the evolution of vortex lines

We follow here the analysis done in Constantin *et al.* (1994). Defining

$$\boldsymbol{\omega} = w\hat{\omega}\ ,$$

it is straightforward to show that the linear dissipative part of the Navier–Stokes equations written for the vorticity $D_T\boldsymbol{\omega} = \nu\Delta\boldsymbol{\omega}$ transforms into:

$$D_T w = \nu P_R$$
$$D_T\hat{\omega} = \nu w^{-1}\boldsymbol{R} \tag{2.58}$$

where

$$P_R = \hat{\omega}\cdot\Delta\boldsymbol{\omega} \tag{2.59a}$$

and

$$\boldsymbol{R} = \Delta\boldsymbol{\omega} - \hat{\omega}P_R\ . \tag{2.59b}$$

Obviously, P_R is the projection of the Laplacian $\Delta\boldsymbol{\omega}$ on the tangent vector $\hat{\omega}$. When $P_R > 0$ *i.e.* when the angle between vorticity and $\Delta\boldsymbol{\omega}$ is between $\pm\pi/2$, enstrophy is **produced** by viscosity (the more so the more $\Delta\boldsymbol{\omega}$ is aligned with $\hat{\omega}$, and furthermore in that case $D_t\hat{\omega}\sim 0$). Using

$$\Delta(\lambda\boldsymbol{v}) = \lambda\Delta\boldsymbol{v} + \boldsymbol{v}\Delta\lambda + 2(\nabla\lambda\cdot\nabla)\boldsymbol{v}$$

with $\lambda = 1/w$ and $v = \boldsymbol{\omega}$, yields:

$$\Delta\boldsymbol{\omega} = w\Delta\hat{\omega} + \hat{\omega}\Delta w + 2(\nabla w\cdot\nabla)\hat{\omega}\ .$$

On the other hand, it can be shown that

$$\hat{\omega}\cdot\Delta\hat{\omega} = -\mid\nabla\hat{\omega}\mid^2\ .$$

One then arrives at:

$$P_R = \Delta w - w\mid\nabla\hat{\omega}\mid^2 \tag{2.60}$$

from which one can also deduce that dissipation of enstrophy is **minimized** when vortex lines are parallel (hence $\nabla\hat{\omega}$ is negligible). In fact, in a vortex tube with constant vorticity w and with parallel vortex lines, there is no dissipation of vorticity as is well known (dissipation occurs at its boundaries when the tube blends with the background vorticity field). One should recall here that steady field configurations can be found using the principle of minimization of dissipation of energy (Montgomery & Philipps, 1989) or a simpler form of minimization of energy, as in the Taylor principle in MHD (Taylor, 1974; 1993).

2.5.5 Discussion

As we have seen, from the inviscid equations for the vorticity vector in the pure fluid case one derives two equations for the vorticity magnitude w and for the vorticity tangent unit vector $\hat{\omega}$:

$$
\begin{aligned}
D_T w &= \alpha w \\
D_T \hat{\omega} &= -\alpha \hat{\omega} + S_{ij} \hat{\omega}_j \ .
\end{aligned}
\tag{2.61}
$$

The viscous parts of the temporal evolution of $\hat{\omega}$ and w are given in (2.58). Hence, growth in amplitude means a lesser variation in angle and vice–versa. Equivalently, for the curvature and torsion of a vortex line written in matrix form we have:

$$
D_T \begin{pmatrix} \kappa \\ \tau \end{pmatrix} = - \begin{pmatrix} \alpha & S_b \\ -S_b & \alpha \end{pmatrix} \begin{pmatrix} \kappa \\ \tau \end{pmatrix} + Y
\tag{2.62}
$$

where in Y are bundled up the non–homogeneous terms. In this latter equation, the matrix clearly decomposes into a symmetric part which, assuming $\alpha > 0$ (*i.e.* where vortex stretching occurs) leads to an exponential decay of the (κ, τ) geometrical coefficients of the line, and an antisymmetric part which exchanges curvature and torsion. As noted in Constantin *et al.* (1994), neglecting the non linear term Y, one has $D_T(\kappa^2 + \tau^2) = D_T E^G = -2\alpha E^G$ whereas $D_T E^w = +2\alpha E^w$ where $E^w = w^2$: in this linear approximation, geometry and magnitude are antithetic but nonlinearities could alter this conclusion. A numerical study of this problem would be of interest.

Finally, the full nonlinear dynamics of curvature and torsion given in the preceding section is bound to lead to quite complex a behavior which would be interesting to study numerically. Several simplifications have been derived in the past, on the assumption that vortex filaments form, and that they are strong and straight; this leads to a variety of approximations of the Biot–Savart law. One of the most successful (or at least widely used) is the so–called "*Local Induction Approximation*" or LIA (Hama, 1962; see also Ricca, 1991 for a history of the rediscovery of the equation). The LIA equation leads to solitonic behavior since it is equivalent to the Non–Linear Schrödinger (NLS) equation (Hasimoto, 1972). One can in fact introduce a whole spectrum of approximations (see Kida, 1981; Fukumoto and Miyazaki, 1991; Ricca, 1992; 1994; 1995; Klein *et al.*, 1992), the dynamics of which are not fully understood, but which can be related to the propagation of waves along vortex filaments as observed in the laboratory (Hopfinger, Browand & Gagne, 1982) or in numerical experiments of turbulent flows (Vincent, private communication).

2.6 The question of singularity

Whether 3D flows develop or not a singularity in a finite time is an open question. Betchov was able to show that $D_t < \omega^2 > \sim S_k < \omega^2 >^{3/2}$ where S_k is the skewness $< (\partial_x u)^3 > / < (\partial_x u)^2 >^{3/2}$ (normalized third–order moment). If S_k is constant, blow–up obtains in a finite time. Similarly, one can write the equation for the time–evolution of the rate of deformation tensor:

$$\frac{\partial}{\partial t} W_{ij} = W_{ij} W_{ij} + (\partial_i \partial_j - \frac{1}{3} \delta_{ij} \partial^2) \mathcal{P} \qquad (2.63)$$

where $\mathcal{P}$ is the pressure and where viscosity is assumed equal to zero. Neglecting the last term, one obtains again a finite–time singularity (Léorat, 1975 ; Vieillefosse, 1982, 1984 ; Cantwell, 1993). In Pelz and Boratav (1995), the symmetry imposed at all time to the flow locally enforces the isotropy of the pressure Hessian $\partial_{ij}^2 \mathcal{P}$ which in turn may lead to a singularity; it seems indeed to be approached by the flow (although not necessarily in the vorticity at first, but in the strain). It was shown recently that with flows with special symmetries (namely $u_x = f(y)$, $u_y = f(z)$ and $u_z = f(x)$ where f may be taken as a Burgers vortex for example), singularities occur but the diagonal terms of the pressure Hessian are in fact zero in this case (Bhattacharjee & Wang, 1992; Bhattacharjee, Ng & Wang, 1995); furthermore, this flow leads naturally to the development of spiral structures and to a Kolmogorov spectrum, as in the model of Lundgren (1982).

Note that if the dynamics is dominated by the self–induction of an ensemble of isolated vortex filaments, and if the LIA is a sufficient modelization of the dynamics, then the ensuing behavior is governed by the NLS (integrable) equation. However, the interaction of a vortex filament with other filaments and/or the background vorticity may be essential in accelerating the dynamical evolution.

A detailed analysis of the invariants $Q = \Delta \mathcal{P}/2$ and $R = - \det W$ of the velocity gradient matrix (note that the other invariant $P = \partial_i u_i \equiv 0$ for an incompressible flow) also yields information on the possible development of singularities (Soria *et al.*, 1994). However, isotropy of the pressure Hessian in the generic case is certainly not ensured and this question remains open (see also Kerr, 1993; 1995; Beale *et al.*, 1985).

3. Mostly MHD, and mostly 3D

3.1 Introduction

There are numerous observations of magnetic fields in the Universe. They are dynamically important (Parker, 1979), and are often observed in equipartition with the velocity: $V_A \sim V_{rms}$ (and the thermal pressure as well), where $V_A = B/\sqrt{4\pi\rho_0}$ (with B the magnetic induction and ρ_0 the density) is the Alfvén velocity, and V_{rms} is based on the turbulent velocity field. Earth and planets, the sun and stars (Priest, 1985), the Milky Way and galaxies, most of them have sizable fields. There is presently some controversy on the efficiency of the dynamo mechanism of generation of large–scale magnetic fields for high magnetic Reynolds numbers R_M, and some authors think that such fields are primordial.

One of the best example of magnetic turbulence can be found in the visualization of the magnetic field in the radio lobes of Cygnus A: observed at high resolution with VLBI, eddies appear with a size which is a portion of the lobe itself. In the solar wind, Matthaeus & Goldstein (1982) have shown that the concepts developed in MHD turbulence seem to apply (in particular, the quasi–equipartition between kinetic and magnetic energy, an energy spectrum following a $-5/3$ law, and an inverse magnetic helicity cascade). More recently, intermittency has been shown to be present in the solar wind (Burlaga, 1991; Marsch, 1993), a point that will be discussed in §3.8 (see also Grappin *et al.*, 1991), but a conclusion that crucially depends on the temporal resolution (*i.e.* on the length of the data set in order to lessen the error bars on high–order structure functions).

The magnetic field is often structured in flux tubes, as observed at the border of solar granulations, in the solar corona, and possibly in the interstellar medium as well (Heiles, 1988). The galactic central lobe (Sofue & Handa, 1984) in our galaxy is another example of a magnetic filament at large–scale. Such filaments are observed as well in numerical simulations of kinematic dynamos (Galloway & Frisch, 1986). The full non–linear dynamo regime is less clear. We already know that the inverse cascade of magnetic helicity (Pouquet *et al.*, 1976; Meneguzzi *et al.*, 1981; see also Pouquet (1993) and references therein) leads to large–scale magnetic filaments. These results have been extended to the compresible regime as well (Horiuchi & Sato, 1988; Zhu *et al.*, 1995).

In the laboratory, however, experiments (in mercury, or with sodium or potassium) are limited to low magnetic Reynolds number (although the kinetic Reynolds number is large), with the exception of the liquid coolant of breeder reactors where $R_M \sim 30$. Thus numerical experiments, together with modeling, are going to play an important role in our path to understanding the complex dynamics of a conducting fluid even though such experiments – by which we mean integrating the primitive equations (Navier–Stokes, MHD) – are limited to moderate values of the Reynolds numbers. However, in dimension two in MHD, the small–scale dynamics may be similar on average to its three–dimensional counterpart (in contrast to the fluid case), so that it can be taken as a model of MHD turbulence. It can be studied numerically at higher Reynolds numbers than in three dimensions. Let us see why.

3.2 Limitations of numerical experiments

Let us take a regular cartesian grid of 2,000 points *per* linear dimension. On these eight billion points are attached the various fields (velocity, density, magnetic field ...) so that this problem is already too large for presently available computers with a few Giga words of memory. What Reynolds number can be simulated on such a grid ?

The ratio of wavenumbers k_{max}/k_{min} is equal to $1,000$. Given that both the energy containing range and the dissipation range must be described on some "reasonable" span of wavenumbers (that is, the relationships $k_0/k_{min} \sim c_R$ for the former, and $k_{max}/k_D \sim c'_R$ for the latter, with $c_R > 1$ and $c'_R > 1$ must be fulfilled), and recalling that $k_D/k_0 \sim R^{3/4}$ within the framework of the K41 phenomenology (see §1.8.1), we deduce that we can reasonably achieve Reynolds numbers of $\sim 1,000$ on such a grid, to be compared with values between 10^6 and 10^9 (with the exception of the geo-dynamo for which $R_M \sim 10^3$). Conversely, a Reynolds number of 10^8 would require a grid in excess of 10^{20} points, unless the flow can be reasonably assumed to be independent of one coordinate (as might be the case *e.g.* when a strong rotation or a strong magnetic field is present).

The situation is not as desperate as it would appear. A flow cannot be turbulent when no inertial range is present. But, as soon as there exists a reasonable range of wavenumbers unaffected by both the energy–containing eddies and the dissipative eddies, the inertial dynamics may be assumed to be proceeding unimpeded, and self–similarity can be applied. So that the minimum requirement (the presence of an inertial range) is less stringent than actually modeling high Reynolds number flows. Unfortunately, transitions to different regimes may occur at high Reynolds numbers (such as in convection, at $R \sim 10^5$).

A possibility to enhance the Reynolds numbers reached in a computation is to consider that in very turbulent flows, no dissipation occurs except in well defined (but not necessarily well understood) structures, such as shocks or vortex sheets. Working on the non-dissipative equations (Euler, ideal MHD) thus may give the false impression that one is computing at a Reynolds number close to infinity. Apart from trivial considerations of dissipative temporal schemes such as the Lax–Wendroff, another issue already mentioned arises, namely the necessity of an inertial range. The extent of such a range may be determined – somewhat empirically – by computing the ratio of the integral scale λ_I to the Taylor scale λ_T defined as:

$$\lambda_I = \frac{C_1}{E} \int_0^\infty \frac{\mathrm{E}(k)}{k} dk$$
$$\lambda_T^{-2} = \frac{C_2}{E} \int_0^\infty k^2 \mathrm{E}(k) dk = C_2 \frac{\Omega}{E}$$

$$(3.1)$$

where E and Ω are respectively the kinetic energy and enstrophy with

$$E = \int_0^\infty \mathrm{E}(k) dk$$
$$C_1 = \frac{3\pi}{4}$$
$$C_2 = 5 \ .$$

$$(3.2)$$

Here we follow standard definitions (see for example Jimenez *et al.*, 1993). For fluid flows, $(\lambda_I/\lambda_T)^2 \sim \mathrm{R}_\lambda$ where $\mathrm{R}_\lambda = U_0 \lambda_T/\nu$ is the Reynolds number based on the Taylor scale. The advantage of evaluating the effective resolution of a numerical simulation using

$$R_T = (\lambda_I/\lambda_T)^2 \qquad (3.3)$$

is that it is based solely on intrinsic properties of the flow due to its internal dynamics; moreover, it does not require a knowledge of the viscosity coefficient (real or numerical). It should however be noted that the Taylor scale which is based on the enstrophy Ω weighs on scales peaked around k_D^{-1} (for $\mathrm{E}(k) \sim k^{-5/3}$, $\Omega(k) \sim k^{1/3}$ up to $k \sim k_D$) so that in fact dissipative processes themselves do enter in the definition of R_T. To give an example, $R_T \sim 60$ (with a Taylor Reynolds number for that flow of $\mathrm{R}_\lambda \sim 168$) for a computation on a grid of 512^3 points using a pseudo–spectral code integrating the Navier–Stokes equations in the incompressible case (Jimenez *et al.*, 1993). On the other hand, for a computation on a regular grid of 512^3 points using the PPM code (Piecewise Parabolic Method, see Woodward, 1986) for homogeneous decaying compressible flows, one obtains $R_T = 55$. That these numbers, at the same resolution, are very close although the methods are vastly different, is not surprising. Indeed, similar numbers should obtain irrespective of the numerical method (with regular grids; adaptive grids are a different matter), because R_T measures how well the flow is handled independently of both large scale and small scale non–inertial effects. What changes in these methods is the amount of dissipation between two wavenumbers in a given ratio in the inertial range (apart from eddy viscosity), since they use different algorithms of dissipation, as in the methods of hyperviscosity (see below).

3.3 The identification of structures

Vortex sheets and filaments have been identified visually (for the case of the laboratory, see §2.3 for a brief account). In three dimensions, obvious difficulties arise in order to be able to view the flow in perspective (ray–tracing volume–rendering is one solution). But another problem arises as well, that of their *automatic* identification in order to study their statistical properties in a systematic way. Thus, algorithms must be devised to identify them blindly. For example, one can construct progressively the "skeleton" of elongated tubes (Villaseñor & Vincent, 1992). Working from the perturbation of exact solutions such as Fourier modes in the linear case of waves, and solitons in the nonlinear integrable case is another possibility (Osborne, 1993). Indeed, Fourier series appear ill–adapted to sharp structures such as fronts, but wavelets may be one answer since they order structures both by their location in configuration space and by their scale. The definition of a vortex embedded in a turbulent flow itself poses problem (see Jeong & Hussain (1995) for a review of various criteria used to define the vortex). Thus, the radius of a vortex, and the scaling of this radius with Reynolds number is not known, although there exist several theoretical predictions; should one take the radius at half peak velocity ? vorticity ? pressure ? or a definition based more on geometry such as the one proposed by Constantin *et al.* (1994) dealing with the gradient of the direction of vorticity $|\nabla\hat{\omega}|^{-1}$ where $\hat{\omega}$ is the unit tangent vector to vorticity (see §2.5) ? One can also think in terms of the spectrum of eigenvalues of the velocity gradient matrix (Chong *et al.*, 1990; Porter *et al.*, 1992). The need to be quantitative about the statistical properties of coherent structures embedded within three–dimensional turbulent flows cannot be fulfilled without first establishing the robustness of the detection algorithms.

3.4 Numerical modelizations of flows

Numerous and clever methods have been devised to circumvent the problem due to the limitations of present–day computers. Sophisticated codes have been written in the fluid community using the Euler equations and treating the flow as a superposition of shock tubes (the incompressible case does not fall into that category); such codes are presently being implemented in MHD, with inclusion of more detailed physics such as ambipolar drift, self–gravity, radiative transfer and chemistry. Vortex methods are used for incompressible flows: in two dimensions using for example contour surgery (Dritschel, 1993), and in three dimensions, the attempts at simulating flows using vorticity as the fundamental variable and reconstructing the fluid velocity from the Biot–Savart law

$$\boldsymbol{u}(x,y,z,t) = \frac{1}{4\pi} \int \int \int_{\mathcal{V}} \frac{(\boldsymbol{x} - \boldsymbol{x}')}{\mid x - x' \mid^3} \ \omega(x')d^3\boldsymbol{x}' \tag{3.4}$$

with some degree of approximation (Chorin, 1994 and references therein). This approach may be powerful, when combined with Euler–type compressible codes. Analytical approximations have been derived which might be useful in this context, for example the LIA or Localized Induction Approximation (Kida, 1981; Ricca, 1991, 1992, 1994), as already mentioned.

Other possibilities to circumvent the small memory of computers today is to use refined grids, or finite element methods. One can also attempt to reduce the number of modes explicitly treated in the computation by resorting to sparse methods (Grossman & Lohse, 1991; Vazquez & Scalo, 1992; Meneguzzi et al., 1995). Another example of the reduce–mode methodology is to take only a few number of modes (the Lorenz model being the extremum along these lines) and study the effect of truncation on the results, a method used in convection and in more general bifurcation problems close to criticality.

One can also use higher power of the Laplacian operator or a "hyperviscosity" model in which the dissipative Laplacian operator written in Fourier space νk^2 is replaced by $\nu(k)k^{2\alpha}$ with $\alpha \in \mathcal{N}$ and $\alpha > 1$; the modified viscosity $\nu(k)$ can be either a constant or a nonlinear viscosity coefficient dependent on the velocity field (Passot et al, 1990). However such methods lead to a strong accumulation of energy in the early dissipation range (Jimenez, 1994; Porter et al, 1994; see also Yao et al. (1995) which includes a comparison of contour surgery and hyperviscosity methods in two dimensions). Finally, transport coefficients may also prove very useful in modeling flows incorporating eddy viscosity coefficients – which can be nonlinear – derived for example from closure models (see for example Yoshizawa, 1990; Lesieur, 1990).

3.5 Two–dimensional flows and reconnection

3.5.1 The two–dimensional case

For non–conducting flows, the two–dimensional geometry is vastly different from three dimensions because of the lack of vortex stretching, leading to a direct cascade of enstrophy towards small scales, and an inverse cascade of energy towards large scales (Onsager, 1949; see for a review Kraichnan & Montgomery, 1980; see also Robert (1991); Zeitlin, 1991). However in MHD vorticity is not conserved by the equations, and energy cascades to the small scales as in dimension three. High resolution runs of two–dimensional MHD turbulence (see the lectures of Biskamp, this volume) indicate the formation of vortex and current sheets highly localized in space and leading for long times to isolated structures (Kinney & McWilliams, 1995; see also Isichenko, 1992; Isichenko & Gruzinov, 1994).

In the forced case, there is a release of energy intermittent in space, and in time as well (Einaudi et al., 1995), possibly leading to a workable mechanism of heating of the solar corona that encompasses the whole spectrum of flares and eruptions as one phenomenon. Indeed, a large–scale quasi–uniform magnetic field tends to suppress the turbulence along it, and renders the flow quasi two–dimensional. MHD dissipative phenomena have been thought to be playing a role in the heating of stellar corona (Einaudi & Velli, 1993; Vlahos, 1993), but one difficulty comes from the fact that the Reynolds number is very large.

Stated simply, the question is: how can we heat a medium (through dissipation of kinetic and magnetic energy) when the viscosity is so low ? In other words, is the limit of energy dissipation $\mathcal{D} = \nu\Omega$ non–zero as $\nu \to 0$ (assuming for simplicity that $\nu = \eta$) ? In the inviscid case, numerical simulations at moderate resolutions, and thus for times that are not large (Frisch *et al.*, 1983), indicate that $\mathcal{D} \to 0$, and this result holds for early times in viscous flows. But several studies (Biskamp & Welter, 1989; Politano *et al.*, 1989; Passot *et al.*, 1990) looking at this scaling for later times (both with the primitive MHD equations, and also using a hyperviscosity method) indicate that $\mathcal{D} \to \mathcal{D}_0$ independent of Reynolds number (see also Mikic *et al.*, 1989; DeLuca & Craig, 1992). In other words, the early development of current sheets and vorticity quadrupoles in two dimensions appears to be non–singular, whereas their disruption through tearing might be. This problem is delicate but should be pursued using the resolution that computers offer nowadays.

In incompressible MHD, it has been shown that the presence of a sufficiently strong uniform magnetic field B_0 (strong enough to suppress magnetic null–points where small scales are likely to develop) impedes the formation of finite–time singularities in two–dimensional geometry with periodic boundary conditions (Frisch *et al.*, 1983). The three–dimensional periodic case remains for the most part largely unexplored. But of course, even when B_0 is strong enough to suppress null–points, nevertheless there can be in the flow numerous locally two–dimensional X–points embedded in the three–dimensional fluid. Furthermore, in the presence of B_0, anisotropy develops (Shebalin *et al.*, 1983; Oughton *et al.*, 1994; see also Strauss, 1993; and Matthaeus *et al.* (1994) for the occurence of anisotropy in the solar wind). No quantitative evaluation of how much two–dimensionality must be taken into account in the temporal evolution of a 3D MHD turbulent flow has been performed yet. In this context, the question remains open of the relevance to reconnection of the existence of zeros of the magnetic field (Greene, 1993; Lau & Finn, 1990) or whether three–dimensional reconnection happens mostly at two–dimensional X–points configurations (Hesse & Schindler, 1988). Recent numerical simulations indicate that both configurations with comparable current sheets occur (Politano *et al.*, 1995).

3.5.2 Reconnection as an inviscid topological event

Reconnection is a cut–and–paste operation on field lines, which can be the magnetic field lines, or current, or vorticity. There is presently in the fluid community a lot of interest in reconnection of vortex filaments. The preferred configuration seems to be that of two anti–parallel filaments, say horizontal to start with. If of infinite extent, the fluid cannot, in that configuration, moves from top to bottom. After reconnection has taken place, the filaments locally are still antiparallel, but vertical, and a drastic change of topology has taken place, with the fluid moving from top to bottom but not from left to right anymore (all arrows can be reversed). Paradoxically this phenomenon, at least in its first phase, may be thought of as inviscid: it does not need viscosity, but corresponds to a change of topology in the solution of the problem (although the development of small scales in the vicinity of the reconnection event leads, in the presence of viscosity, to local heating). This idea has been recently put forward (Shinoda & Fujiwara, 1993; Nore *et al.*, 1995) by using an analogy with the nonlinear Schrödinger equation (or NLS):

$$i\psi_t + \Omega_{NLS}\psi + \alpha_{NLS}\nabla^2\psi - \beta_{NLS}|\psi|^2\psi = 0 \quad ;$$

in Shinoda & Fujiwara (1993), the derivative is taken with respect to the curvilinear coordinate s along the filament, and $\Omega, \alpha, \beta_{NLS}$ take the value $[0, 1, -1/2]$, whereas in Nore *et al.* (1995), the triplet of parameters is $[1, 1/2, 1]$; the change of sign of β_{NLS} between the two formulations is important, since when $\beta_{NLS} > 0$, the equation is defocusing whereas it is focusing in the opposite case. This equation is integrable (leading to solitonic solutions), and furthermore it is hamiltonian: the higher derivative term is dispersive, not dissipative. Such a dispersive term provides, nevertheless, a cut–off at small scales (in the same spirit that a uniform magnetic field does, as in Frisch *et al.*, 1983). This equation has been known to be related to barotropic ($\mathcal{P} \sim \rho^2$) fluid dynamics for $\beta_{NLS} > 0$ (except for a modified pressure term, and except for the fact that, because of its strong dispersion, no shocks form). Note that in the compressible fluid analogy $\psi = \sqrt{\rho}e^{i\phi}$ where ρ is the density of the fluid and $v = \nabla\phi$. The NLS equation is also known (Hasimoto, 1972) to describe the evolution of a vortex filament under the LIA hypothesis (see §2.5 and §3.4). In this latter case, $\psi(s) = \kappa(s)\exp\left[i\int^s \tau(s')ds'\right]$, with as before κ the curvature of the line and τ its torsion; so the amplitude of the soliton is the curvature of the filament. Reconnection occurs, in the Hamiltonian non dissipative system of NLS, where $\psi = 0$; this corresponds to $\kappa = 0$, thus to two strong and straight filaments. Note that, with this configuration, this is where the strongest (1D) gradients occur. But at $\psi = 0$, there is an ambiguity on the torsion which is defined within a factor of 2π; this allows for the change of topology. A quantum analogy in the case of vortex rings can also be drawn (Shinoda & Fujiwara, 1993): here, the two initial vortex rings correspond to a configuration with two atoms, and the final state to a diatomic molecule. Although the system is conservative, there is nevertheless a loss of energy of the vortex (but not of the total system constituted of filaments plus background vorticity), through emission of acoustic waves (and in the quantum analogy, radiation), corresponding to the analogous loss of energy in the dissipative case.

3.5.3 Reconnection as a diagnostic for instability

Are sheets and filaments the two ingredients of a dynamical cascade of energy ? Vortex filaments are known to be unstable in the experimental context, through *e.g.* bursting. This bursting event may be present as well in numerical simulations, corresponding to the large region created by the reconnection process (Nore *et al.* (1995) private communication). In fact, the term (2.59b) in the decomposition of the equation for the viscous evolution of vorticity was proposed by Kida & Takaoka (1991) as a diagnostic for reconnection. In this light, it should be recalled that in the numerical experiments of Boratav *et al.* (1992), during reconnection of two filaments, sheets are formed at an intermediate stage (an observation also made by Lundgren, 1995). Since the filaments stem from the instability of vortex sheets, the first two steps of a self–similar process of cascading of energy à la Kolmogorov (sheets to filaments to sheets, ...) are now in place. Since furthermore the rolling up of vortex sheets into spiral structures is known analytically to lead to a K41 spectrum (Lundgren, 1982), with a bottleneck at the onset of the dissipation (leading possibly to a k^{-1} spectrum (Townsend, 1951) due to the presence of intense and straight vortex filaments – but with little of the total enstrophy associated with them), many ingredients of a more detailed vision of the Kolmogorov cascade are in place. However we shall be led in §3.8 to take into account the departure from K41 stemming from intermittent events – a departure, though, that is too weak at the level of the energy spectrum to account for much, but a departure that is clearly observed for higher orders of the structure functions. One could, of course, also consider more complex objects in–between sheets and filaments, of intermediate dimension, and view the corrections to the K41 spectrum as a signature of these fractal objects.

It is then not necessarily surprising that in the case of decaying two–dimensional MHD flows the level of dissipation – in the latter temporal phase clearly associated with reconnection events and the disruption of strong current sheets – is found to be independent of the Reynolds number, although locally the number of magnetic islands may depend on the Reynolds number because the aspect ratio of the initial layer depends on it (Meneguzzi et al., 1995). These issues need to be clarified further.

3.6 A short–cut on the dynamo problem

Many books and review articles are devoted to the dynamo problem (see for example Moffatt, 1978; Parker, 1979; Zeldovich et al., 1983; Childress, 1992; Roberts, 1993), i.e. to the generation of a magnetic field by turbulence, and we shall just give here a brief overview.

The induction equation given in §1.8.2 can be rewritten in terms of the magnetic potential a (with $B = \nabla \times a$) which, in the framework of the $[B - \omega]$ analogy, plays in a sense the role of the velocity; it obeys the equation:

$$\frac{\partial a}{\partial t} = -u \cdot \nabla a + u_j \nabla a_j - \nabla \phi + \eta \nabla^2 a \tag{3.5}$$

where ϕ is a gauge scalar field which can be viewed as a pressure. The first two terms on the right hand side can also be written as $u \times B$. In the Coulomb gauge ($\nabla \cdot a = 0$), this scalar ϕ obtains from a Poisson equation by taking the divergence of $\partial_t a$, leading in the non dissipative case to $\nabla^2 \phi = \nabla \cdot (u \times B)$. Other choices of gauge, however, are possible and may be useful in the numerical context (see e.g. Brandenburg et al., 1995a).

3.6.1 The two–dimensional case

In a strictly two–dimensional problem $a = a\hat{e}_z$ and $\partial_z \equiv 0$ so that the magnetic potential is simply advected (except for diffusion), an equation which gives rise to the anti–dynamo theorem in two dimensions since, integrating over the 2D volume, one obtains:

$$\partial_t \int a^2 d^2 x = -2\eta \int b^2 d^2 x \tag{3.6}$$

with the gradient terms giving no contributions to the integrals assuming zero field at infinity; hence:

$$\int_0^T < B^2(t) > \, dt \leq \frac{< a_0^2 >}{2\eta} \tag{3.7}$$

where $< a_0^2 > \equiv < a^2(t = 0) >$; hence, $lim_{t \to \infty} < B^2(t) > = 0$ (see Zeldovich et al., 1983). Note however that this does **not** mean that for intermediate times, the magnetic energy does not grow. In fact it does because of line stretching by velocity gradients. Indeed, growth of magnetic energy is observed for a duration which increases with Reynolds number, as shown *e.g.* in the framework of two–point closure calculations (Pouquet, 1978).

3.6.2 The mean–field dynamo

The equation for the magnetic field evolution when the velocity field is given (kinematic problem) is identical to that for the vorticity equation (except that of course for the vorticity, the problem is never kinematic in the sense that $\omega = \nabla \times u$): stretching of vorticity and magnetic field lines occur through velocity gradients, provided the Reynolds number is sufficiently large. Since, for the velocity, a $k^{-5/3}$ energy spectrum results, it may be supposed that in this phase the magnetic energy spectrum also follows a $k^{1/3}$ law (see Moffatt, 1961), a fact consolidated by numerical experiments (Brandenburg *et al.* 1995b). On the other hand, the growing magnetic field could develop singular (small–scale) structures, as in the context of the fast dynamo. This growth of a small–scale field which may reach equipartition with the kinetic turbulence first in the small scales is probably to be associated with a fast dynamo (see below).

In the context of multiple–scale expansion, it can be shown that in the kinematic context a mean field $\bar{B}$ obeys the following dynamo equation (see *e.g.* Roberts, 1993):

$$\frac{\partial \bar{B}}{\partial t} = \nabla \times (\alpha_{turb} \bar{B}) + \beta_{turb} \Delta \bar{B} \tag{3.8}$$

where α_{turb} and β_{turb} are given independently as transport coefficients evaluated as averages over the small–scale turbulent velocity field; in particular, α_{turb} is proportional to its kinetic helicity $< u \cdot \omega >$ (Steenbeck, Krause & Rädler, 1966). In that approach, the small–scale magnetic field itself is supposed to be non–growing, an hypothesis which may prove false if the mechanism of small–scale fast dynamo is at work. The mean–field approach gives rise to transport coefficients which can be, and are, used directly in numerical computations, as a model of the more complex underlying physics, for example in the context of a modelization of the solar convection zone. In the nonlinear regime, saturation effects can be modeled in which quenching of the α coefficient occurs through a dependence of α on the magnetic field; in the context of second–order closures, the instability of the large–scale magnetic field is governed by a combination of kinetic and magnetic helicity (Pouquet *et al.*, 1976; see also Gruzinov & Diamond, 1995).

3.6.3 The fast dynamo

Mean–field dynamo does not require any chaos but a topologically non–trivial velocity: it is the kinetic helicity of the small scales which is the motor of the growth of the large–scale field in the kinematic regime (and the inverse cascade of magnetic helicity in the non–linear regime). On the other hand, a dynamo is called "fast" when the limit, when $R_M \to \infty$, of the growth rate of the magnetic field is non–zero, with no reference to helicity. This case is of obvious interest in astrophysics. How do the two dynamo mechanisms (α and fast) compete or combine is an open challenging problem with many aspects.

Fast dynamos are associated with chaotic properties of the flow, here the analogy first drawn by Arnold in the '60s being between the magnetic induction equation and the equation for a small displacement in the parameter space of the fluid (*e.g.* the butterfly effect in weather prediction). In fact, theorems exist to ensure that if no chaos is present in the velocity field (as measured by topological entropy, which is related to the spectrum of Lyapunov exponents), then no fast dynamo can occur (see Vishik, 1989; Childress, 1992).

There is now good numerical evidence for such dynamos (Arnold & Korkina, 1983; Galloway & Frisch, 1986; Galanti *et al.*, 1993), more convincingly so in the case of a "2.5 D" (*i.e.* three components but one ignorable coordinate) time–dependent velocity field (Galloway & Proctor, 1993) because very high Reynolds numbers can then be achieved (up to 10^5 in spherical geometry). However, in dimension three, there is no evidence that the asymptotic high Reynolds number regime is achieved; for example, in Galanti *et al.* (1993), one sees that the large discrepancy between the growth rate of a dynamo embedded in either a large–scale flow ($k_0 = 1$), or a smaller–scale flow ($k_0 = 2$) diminishes, albeit slowly, with R^M. The large discrepancy at low R^M may be characteristic of a transitory regime only. This could be attributed to the fact that the asymptotic fast dynamo is one with many folding and reversals in the magnetic structures that become in the limit $R^M \to \infty$ infinitely thin; these small–scale lobes in turn give the predominant contribution to the magnetic energy (with a $E^M(k) \sim k^0$ spectrum) (although most of the magnetic flux cancels).

3.6.4 Discussion

As already stated, one open problem is to combine these two approaches for the generation of magnetic fields: chaos on the one hand (and Beltrami flows – *i.e.* maximally helical $\mathbf{v} = \pm \nabla \times \mathbf{v}$ flows are chaotic), and helical flows on the other hand. In particular, the growth of a small–scale magnetic field prevents one from writing the usual kinematic mean–field equation (as in eq. (3.8)), but the growth of the mean–field itself could presumably prevent further growth of the small–scale field so that the two fields cohabit.

It is sometimes advocated that the mean field obtained through the dynamo process kills the turbulence that gave rise to it, at levels much below equipartition. In this case, the observed magnetic fields should be of primordial origin (see *e.g.* Kulsrud & Anderson (1992) in the case of the galactic dynamo, and references therein). These arguments fail to convince. First of all, care should be taken before concluding: (a) on the one hand, the level of saturation may depend on several factors, one of which is scale separation (see Galanti *et al.* (1992) for numerical evidence at moderate Reynolds numbers; such computations should be pursued at higher R); (b) the processes by which large–scale magnetic excitation grows **DO NOT** lead to saturation nor to a steady–state, because they are linked to an inverse cascade, so that mixing temporal and spatial averages is dangerous. They DO NOT take into account the fact that the *non–linear* growth of magnetic field is, in the last analysis, governed by the non–dissipative mechanism of inverse cascade to large scales: magnetic helicity $H^M = <\mathbf{a} \cdot \mathbf{B}>$ in three dimensions (and square magnetic potential in two dimensions where a dynamo mechanism is also at work, for intermediate times only, of course). Thus, NO steady–state can be reached until the largest scale of the flow is excited, at which point boundary conditions play a determinant role. There

but see stribling —
claim of $<\vec{A} \cdot \vec{B}>$

are models based on two–point closures of MHD turbulence which indicate that the underlying cause of the growth of a large scale magnetic field is indeed linked to the inverse cascade of magnetic helicity (Pouquet et al., 1976), a phenomenon likely unimpeded by the growth of a small–scale field. Phenomenology (Kraichnan, 1979) and numerical simulations (Meneguzzi et al., 1981) at moderate resolutions indicate that indeed the magnetic helicity grows linearly with time, so that an ergodic hypothesis linking time and space averaging cannot apply. This problem can and should be tackled on present day computers granted **sufficient** resolution be used (a 512^3 grid to start with).

3.7 Fluid analysis of MHD

3.7.1 Vorticity equation

In addition to vortex stretching, the Lorentz force $j \times B$ provides a further mechanism of vortex production (or more precisely its curl). Due to the symmetry of the MHD equations, it may be useful to discuss this phenomenon in terms of the Elsässer variables $z^{\pm} = v \pm b$; after some algebra, the equations for the generalized vorticities $\omega^{\pm} = \omega \pm j$ read:

$$\frac{\partial \omega^+}{\partial t} + z^-.\nabla \omega^+ = \omega^+.\nabla z^- + \Sigma_m \nabla z_m^+ \times \nabla z_m^- \qquad (3.9)$$

where dissipative terms have been omitted, and where the equation for ω^- obtains by exchange of $\pm$ (Pouquet, 1994; see also references therein). To derive equation (3.9), the following identities may be of help:

$$\nabla \times (\lambda b) = \lambda \nabla \times b - b \times \nabla \lambda \ ,$$
$$\nabla \cdot (\lambda b) = \lambda \nabla \cdot b + b \cdot \nabla \lambda \ ,$$
$$\nabla \cdot (a \times b) = b \cdot (\nabla \times a) - a \cdot (\nabla \times b) \ .$$

Let us extend the terminology of vortex stretching to the production of $\pm$ vorticity. We see that the l.h.s. of (3.9) is a Lagrangian derivative, following the z^- field. On the r.h.s. of (3.9), two terms appear. The first one is similar to vortex stretching by z^-–gradients, and the second one is a source term. It might in fact be instructive to write the equation for the evolution of the $\partial_i z_j^{\pm}$ gradient matrices themselves. Numerical simulations of MHD flows in three dimensions indicate that the total enstrophy $< \omega^2 + j^2 >$ grows faster in an MHD flow than in the equivalent random flow with B $\equiv 0$ (Politano et al, 1995), a growth that can be presumably attributed, at early times, to this production term. On the other hand, for long times, the question remains open of the role of this extra term (Kinney et al., 1993) which, if non–zero, breaks the flux freezing of the generalized vorticities.

Similarly, the kinematics part of Betchov's relation derived in the fluid case in §2.2 follows immediately when one supposes homogeneity and incompressibility; thus one has $< \omega^2/2 >=< \sigma_V^2 >$ and $< j^2/2 >=< \sigma_M^2 >$ where the σ_{ij} are proportional to the symmetric parts of the gradient matrices $\partial_i u_j$ and $\partial_i b_j$. Similar relationships obtain in the $z^{\pm}$ variables (Pouquet, 1994).

3.7.2 The Burgers' vortex in MHD

As already mentioned in §2.4.2, we can take two $z^{\pm}$ large–scale strains set off–center; namely one assumes for z^{+} a strain of the form $(0, \gamma_s^+ y, -\gamma_s^+ z)$ and for z^{-} one can take $(0, \gamma_s^-(y + y_0), -\gamma_s^-(z + z_0))$ with $\gamma_s^+ \neq \pm\gamma_s^-$. Note that the coupling term in the equation for the evolution of the $\omega^{\pm}$ vorticities (see preceding Section) is zero with these assumptions. We find (with again $\nu = \eta$)

$$\omega^{\pm}(z) = \omega_0^{\pm} \hat{e}_y \exp[-\gamma_s^{\mp}(z + z_0)^2/2\nu] \ . \tag{3.10}$$

When $y_0 \equiv 0$, we have $z^- = qz^+$ with $q = \gamma_s^-/\gamma_s^+$, hence $v = (q + 1)b/(q - 1)$ which reduces the problem to a rather trivial case since $u//b$. Similarly, a self–similar solution of the form

$$\omega^{\pm} = (0, \exp[\gamma_s^{\pm}t]g(z\exp[\gamma_s^{\pm}t]), 0) \tag{3.11}$$

arises in the absence of dissipation, as also described in §2.4.3 for the pure fluid case. Note that when the large-scale straining fields $z^{\pm}$ are taken colinear with

$$z^+ = f^+(z)\hat{e}_x \tag{3.12}$$

and

$$z^- = -f^-(y)\hat{e}_x \ ,$$

the resulting vorticity fields $\omega^+ = \hat{e}_y \partial f^+/\partial z$ and $\omega^- = -\hat{e}_z \partial f^-/\partial y$ are not colinear; this again represents a steady solution of the inviscid problem for which the cross–terms in the vorticity equations cancel out.

Another possibility in MHD is of course to use the analogy between the vorticity equation (1.30) and the induction equation (1.46) and write the Burgers' solution for B (see *e.g.* Moffatt, 1978, §3.4). It has been shown (Bajer, 1995) that for a magnetic Prandtl number P^M different from unity, the solution is no more $B = \omega$, and furthermore that the dissipation of the magnetized vortex in the equilibrium solution depends in a significant way on P^M.

3.7.3 The Serret–Frenet analysis in the MHD case

Equations (2.54) apply to a Frenet frame associated with a vortex line (with the symmetric operator S_{ij}) or a magnetic field line with the full operator W_{ij}. We can in fact also derive equations for curvature and torsion of a vortex line when it is no more material but is submitted to a force F. The resulting evolution equation for the vorticity is written as:

$$D_T\omega = \omega \cdot \nabla u + F = \hat{\omega}D_T w + wD_T\hat{\omega} \tag{3.13}$$

where $\omega = w\hat{\omega}$, as usual. The evolution of its amplitude readily obtains:

$$wD_T w = \omega_i(\omega_j\partial_j u_i + F_i). \tag{3.14}$$

Thus the direction of the vorticity vector changes with time as:

$$\begin{aligned} D_T\hat{\omega} &= \hat{\omega} \cdot \nabla u + w^{-1}F_{\perp}\hat{\omega} - \alpha\hat{\omega} \\ &= (S_n + F_n w^{-1})\hat{n} + (S_b + F_b w^{-1})\hat{b} \end{aligned} \tag{3.15}$$

where $\boldsymbol{F}_\perp = F_n \hat{n} + F_b \hat{b}$ is the projection perpendicular to the direction of the vorticity of the force acting upon it.

It is easy to show that in the case of a non–material line, instead of equation (2.48), we have:

$$D_T \frac{\partial f}{\partial s} = \frac{\partial}{\partial s} D_T f - \gamma \frac{\partial f}{\partial s} + w^{-1}[\boldsymbol{F} \cdot \nabla f - (\hat{\omega} \cdot \boldsymbol{F})\hat{\omega} \cdot \nabla f] \; . \tag{3.16}$$

Associating now, as before, a Frenet frame with a vortex line of curvature κ_ω and torsion τ_ω and following the steps of §2.5.2, one obtains:

$$\begin{aligned} \dot{\kappa}_\omega &= -\alpha \kappa_\omega - \tau_\omega S_b^F + \hat{\omega} \cdot \nabla S_n^F \\ S_h^F &= S_h + F_h w^{-1} \quad , \quad h = n, b \end{aligned} \tag{3.17}$$

(see also Brandenburg, Procaccia & Segel, 1995c) and

$$\dot{\tau}_\omega = -\alpha \tau_\omega + S_b^F \kappa_\omega + \hat{\omega} \cdot \nabla N^F \tag{3.18a}$$

with N^F defined by

$$\kappa_\omega N^F = \hat{\omega} \cdot \nabla S_b^F + \tau_\omega S_n^F. \tag{3.18b}$$

The force $\boldsymbol{F}$ in the case of MHD is the curl of the Lorentz force or $\boldsymbol{F}_B \equiv \nabla \times (\boldsymbol{j} \times \boldsymbol{B})$ where $\boldsymbol{j} = \nabla \times \boldsymbol{B}$ is the current. Note that equations (3.18a,b) apply as well to the compressible case with now:

$$\boldsymbol{F} = \boldsymbol{F}_C = -\omega \nabla \cdot \boldsymbol{u} + \nabla \mathcal{P} \times \nabla(1/\rho) \tag{3.19}$$

with $\mathcal{P}$ and ρ respectively the pressure and density, the last term on the r.h.s. being the baroclinic term arising at the intersection of shocks, or in the vicinity of curved shocks. Indeed, the incompressibility condition has nowhere been used in the preceding analysis except in writing the flux–freezing condition as $\partial_T \boldsymbol{X} = \nabla \times (\boldsymbol{u} \times \boldsymbol{X})$ leading to $D_T \boldsymbol{X} = \boldsymbol{X} \cdot \nabla \boldsymbol{u}$.

In the presence of dissipation, conclusions similar to those of §2.5.4 can also be drawn in MHD for magnetic energy and the geometry (curvature and torsion) of a magnetic field line, with simply α replaced by γ i.e. with the full $\partial_i u_j$ matrix acting upon the magnetic field. Hence, growth of magnetic enstrophy – or magnetic energy – is accompanied by decay at the same rate (at least as long as the inhomogeneous term can be neglected) of curvature and torsion: the stronger the tubes, the more rigid, the straighter. This can be readily understood if one recalls the analogy of a magnetic field line as a string: as you stretch the string by pulling on it, you store more (potential) energy in it, you also straighten it.

3.8 Intermittency

3.8.1 Examples of intermittent flows

The intermittency of a turbulent flow is linked to the observation that intense small–scale structures are sparse, both in time and space (see for example Meneveau & Sreenivasan, 1991); this phenomenon is well documented in fluid turbulence, in the laboratory and in the atmospheric context, such as for Clear Air Turbulence (CAT) in aviation, and in meteorology with the occurrence of tornados and cyclonic storms for which vorticity and its outbursts clearly play a role. Other examples come from the solar wind, observed in detail with a series of satellites), and from solar physics. In the latter case, it concerns the observation that the spectrum of solar eruptions and flares, when looking for example at the number of events according to either their duration or their energy, follow a power–law distribution that can be interpreted in terms of self–organized criticality as for avalanches or the falling of sand piles (Lu & Hamilton, 1991; Lu et al., 1993; Lu, 1995abc). Intermittency (in space, of the dissipative layers of current and vorticity, and in time of the burst of reconnection occurring at different levels) may very well be the agent of criticality. Further away, in the interstellar medium, observations are less detailed; nevertheless, several signs of intermittency have been identified, such as the detection of exponential wings in the probability distribution functions of the velocity of clouds; these wings indicate that extreme events of velocity differences of large amplitude occur substantially more often than in the Gaussian case. This departure from Gaussianity for fluid turbulence in the laboratory or in numerical simulations is measured in terms of the skewness $\mathcal{S}_k$ and flatness $\mathcal{F}_l$ factors, namely $< (\partial_x u)^n > / < (\partial_x u)^2 >^{n/2}$ (normalized third–order and fourth–order moments with respectively $n = 3$ and $n = 4$). A predictive theory of turbulence should allow to recover the values of such coefficients (for example, $\mathcal{S}_k \sim -0.4$ whereas for a Gaussian field, $\mathcal{S}_k = 0$), a point not yet reached today. The intermittency of the velocity field in the interstellar medium may be essential to understand some puzzling observations, like the occurrence of molecules (like the radical OH) not expected at the temperature of the cold clouds ($\sim 10K$). But intermittency provides a source of strong localized heating which may excite these molecules (see Falgarone, 1995).

3.8.2 Scaling laws of structure functions

Intermittency is often thought to yield corrections to the K41 spectrum; one such example can be found in the β model of Frisch et al. (1978). Similar heuristic analysis have been carried out in MHD as well (Carbone, 1993; 1994; see also Biskamp, 1994). Departures from the classical K41 (or in MHD, IK – see §1.8.3) theories are presumably small but it may be the case that when looking at higher moments of the turbulent fields such corrections may be more easily measurable. One deals with structure functions of the turbulent fields, and to simplify notations, we shall be concerned only with the one–dimensional (or longitudinal) case; assuming now that velocity differences at scale ℓ follow

$$< \delta v_\ell > \sim \ell^\varsigma \ , \tag{3.20}$$

one writes as well for the p–order structure function

$$< \delta v_\ell^p > \sim \ell^{\varsigma_p} \tag{3.21}$$

and similarly for the transfer function

$$< \epsilon_\ell^p > \sim \ell^{\tau_p} \ . \tag{3.22}$$

Together with

$$\bar{\epsilon} = \bar{v}^3 / \ell \tag{3.23}$$

which relates the mean flux and the r.m.s. velocity, we also assume Kolmogorov (1962) refined similarity hypothesis

$$\varepsilon_\ell = v_\ell^3/\ell \ . \tag{3.24}$$

This leads immediately to the following relationship:

$$\zeta_p = p/3 + \tau_{p/3} \ . \tag{3.25}$$

When no intermittency is taken into account, $\tau_p \equiv 0 \ \forall p$.

Note that exact results for scaling exponents (Eyink, 1995) can be derived, as well as bounds for high Reynolds number flows (Procaccia & Constantin, 1993). We also have $\zeta_3 \equiv 1$, hence $\delta v_\ell^3 \sim \ell$, since the mean flux by definition is assumed constant in the inertial range; in fact, $\zeta_3 \equiv 1$ can also be shown from first principles, assuming isotropy, homogeneity, incompressibility and stationarity (Monin & Yaglom, 1975, Volume II, Chapter 7). A similar relationship obtains for a scalar (say the temperature θ) passively advected by the velocity field, namely that $< \delta\theta_\ell^2 \delta v_\ell > \sim \ell$.

In MHD, the equivalent non–intermittent law for the direct cascade of energy assuming Iroshnikov–Kraichnan phenomenology reads $\zeta_p^{(B)} = p/4$ (Biskamp, 1994; Carbone, 1994). The refined similarity hypothesis similar to that of Kolmogorov (1962) in MHD in the framework of the IK theory thus reads:

$$\zeta_p^{(B)} = p/4 + \tau_{p/4}^{(B)} \ . \tag{3.26}$$

The relationship $\zeta_4^{(B)} = 1$, however, does not seem to arise in a simple way from the primitive equations.

3.8.3 The She–Lévêque (SL) model of intermittency

Of the models derived in the Navier–Stokes case, none fit well at high p (Vincent & Meneguzzi, 1991). However, such is not the case for the new model derived in She & Lévêque (1994) (or SL model) which relies on three hypothesis. Defining

$$\epsilon_l^{(p+1)} = < \epsilon_l^{p+2} > / < \epsilon_l^{p+1} > \ , \tag{3.27}$$

the assumed scale–dependency reads:

$$\epsilon_l^{(p+1)} = A_p \epsilon_l^{(p)\beta} \epsilon_l^{(\infty)1-\beta} \tag{3.28}$$

where $0 < \beta < 1$ (see also equation 3.32); in fact in SL, $\beta \equiv 2/3$ and

$$\epsilon_l^{(\infty)} \sim v_0^2/t_l \tag{3.29}$$

is an estimate of the maximum amount of energy that can be dissipated in the most intermittent structures in a time $t_l \sim l^{2/3}$, in accordance with standard K41 phenomenology. This is equivalent to assuming that the divergence of the energy flux as $l \to 0$ follows a 2/3 anomalous scaling law. This 2/3 law in turn leads to

$$\tau_p = -2p/3 + C_0 + f(p) \ , \tag{3.30}$$

$C_0 \equiv 2$ being interpreted as the co–dimension of dissipative structures, taken to be filaments (or tubes) in dimension three for incompressible Navier–Stokes fluids, and

$$f(p) = -C_0 \beta^p \ . \tag{3.31}$$

Once C_0 and β are determined on physical grounds, the resulting SL model is parameter–free. It reads:

$$\zeta_p = p/9 + 2[1 - \left(\frac{2}{3}\right)^{p/3}] \ .$$

In particular, the correction to the K41 spectrum is small: $E(k) \sim k^{-(5/3+C_{SL})}$ with $C_{SL} \sim 0.03$. In other words, $\zeta_2^{(K41)} = 2/3$ whereas $\zeta_2^{(SL)} \sim 0.696$. But discrepancies at higher orders arise quite rapidly; for example, $\zeta_{10}^{(K41)} = 10/3$ whereas $\zeta_{10}^{(SL)} \sim 2.59$, and the experiments of Benzi et al. (1993) give $\zeta_{10} \sim 2.60$ using the Extended Self–Similarity (ESS) hypothesis allowing for a better estimate of exponents, an hypothesis which may be working as well in MHD (Grauer & Marliani, 1995).

It should be noted that grey–scale images of the vorticity at early times in numerical simulations of turbulent flows reveal clearly spiral structures at the border of vortex sheets, which will later be collapsing into filaments. These structures are central to the Lundgren (1982) model which is the only model based on an object that yields a $-5/3$ energy spectrum (see also Pullin & Saffman, 1992). At later times, because of the intensification of vorticity into filaments, these less intense spiral structures are smeared out within the background vorticity. However, it is known that the filaments of vorticity, although containing most of the most intense vorticity (and thus being prevalent in a visualization based on thresholding at high amplitudes), on the other hand contain little of the total enstrophy, as shown by probability distribution functions. Most of the vorticity is in a background field, which may be structured preferentially into sheets. So, in a sense, in the $\mathcal{L}_{max}$ norm, the vorticity field is filamentary, but in the $\mathcal{L}_2$ norm, it may be organized mostly in spirals. It is possible that the same can be said for dissipation or energy transfer, which are the building blocks of the SL theory. Of course, when going to higher powers of the structure functions, the most intense features will come to dominate, and departures from a pure Kolmogorov law gets stronger.

In the equation (3.28), β measures the degree of efficiency of energy transfer from scale to scale. Defining $\pi_l = \epsilon_l / \epsilon_l^{(\infty)}$ as a normalized transfer rate, this equation can also be written

$$< \pi_l^p > = B_p < \pi_l >^{\frac{1-\beta^p}{1-\beta}} \ , \tag{3.32}$$

leading to a $\log_\beta$ Poisson statistics for $B_p = 1 \ \forall p$ (see Dubrulle, 1994). For $\beta = 1$, a hierarchical relationship holds, namely $< \pi_l^p > = < \pi_l >^p$ corresponding to K41, whereas for $\beta = 0$ one finds $< \pi_l^p > = < \pi_l > \ \forall p$, an extreme case of intermittency since it implies that all the dissipation is concentrated in one single structure of characteristic transfer rate $\epsilon_l = \epsilon_l^{(\infty)}$.

3.8.4 The generalized SL model and its application to MHD

The SL model leads to an excellent agreement with experimental data. It can be easily extended to MHD in the framework of the IK theory described in §1.8.2, in the case when the correlations between velocity and magnetic field can be considered weak (Grauer & Marliani, 1994; Politano & Pouquet, 1995). The corresponding model for the exponents of the structure functions now read, assuming a scaling of time compatible with the IK theory, and further assuming that dissipative structures in MHD are sheets (see below):

$$\zeta_p^{(B,S)} = p/8 + 1 - \frac{1}{2^{p/4}} \quad . \tag{3.33}$$

The original SL model is parameter–free, but has two main assumptions underlying it: (a) the characteristic scaling of dissipative structures follow classical K41 phenomenology; and (b) the most intense dissipative structures are filaments. In fact, the SL model can be obviously generalized, by relaxing those two constraints, and in that case depends on two parameters, β and the typical temporal scaling $t_\ell \sim \ell^x$, a generalization valid both in the fluid case (Dubrulle, 1994; She & Waymire, 1994) as in the MHD case (Politano & Pouquet, 1995), with $t_\ell \sim \ell^{x_B}$.

The two–parameter SL model in the fluid case reads:

$$\zeta_p = \frac{p}{3}(1 - x) + C_0[1 - \beta^{p/3}]$$
$$C_0 = \frac{x}{1 - \beta} \tag{3.34}$$

and in MHD:

$$\zeta_p^B = \frac{p}{4}(1 - x_B) + C_0^B(1 - \beta_B^{p/4})$$
$$C_0^B = \frac{x_B}{1 - \beta_B} \tag{3.35}$$

with both $\beta \neq 1$ and $\beta_B \neq 1$.

This generalization of the SL model in the case of MHD, when compared with solar wind data, leads to the conclusion that $\zeta_p^B \sim \zeta_p^{(B,S)}$, and thus that dissipative structures – the codimension of which is determined by $C_0^B = f(\beta_B, x_B)$ – correspond to sheets, as observed in three–dimensional numerical simulations at moderate resolutions in a regime where kinetic and magnetic energies are comparable (Politano et al., 1995). If such is the case, it is remarkable that in the fluid problem and in the MHD–IK version, the two parameters of the SL model (x and β) are equal to each other, but differ from one problem to the next. In fact, when applying the SL phenomenology to shell models of turbulence, $x = \beta$ when the shell model has both the kinetic energy and a simplified form of kinetic helicity as invariants (Frick et al., 1995).

Comparison with the turbulence structure functions observed in the solar wind and computed from the Voyager experiment at $\sim 8.5 AU$ (Burlaga, 1991) is encouraging but error bars are still too large. At smaller distance (Marsch, 1993), the results are more puzzling. It is to be hoped that the data coming from the Ulysses spacecraft out of the ecliptic plane and integrating for a long time within the same sector of the solar wind, over a coronal hole will allow for a better test of MHD models of intermittency (see Ruzmaikin et al., 1995). The magnetosphere may be another area where models of intermittent MHD flows can be tested, as well as cometary physics.

3.9 Turbulence in the interstellar medium

As a conclusion, this Section is devoted to a phenomenological description of the interstellar medium (ISM), viewed as an example of a complex astrophysical fluid (Heiles *et al.*, 1993; McKee *et al.*, 1993) where some of the ideas on fully developed homogeneous turbulence can prove to be useful, as stressed for example in Scalo (1984; 1985) and Falgarone (1995); other vast areas of application of fluid dynamics to astrophysics are, of course, that dealing with convective flows, with jets and accretion disks, as well as cosmology; in planetology where the concepts issued from chaos are essential, and in magnetospheric physics, where plasmas instabilities have to be taken into account, ... An account of gas dynamics in astrophysics can be found in Shu (1992).

When looking at images of the ISM obtained at high resolution, the resemblance with our own sky is striking: a multitude of sharp filamentary structures, and local condensations – the cloud complexes, with a self–similarity of such a description on a large range of scales, in excess of $1,000$. The physics (and chemistry) within the ISM is highly complex, and differ at different scales, densities and temperatures. Rotation is present, as well as gravity, compressibility and magnetic fields. The flow is supersonic (around Mach four) and sub–Alfvenic (by roughly a factor two). Apart from its gravitational potential, energy is being put into the system at large scale because of galactic shear, and at small scale because of ionization winds emanating from OB stars, as well as from the more energetic events linked with supernovae. At the galactic scale of the kiloparsec, it is structured in at least three phases: a hot $(10^4 K)$ tenuous (.1 particle per cm^{-3}) phase, a cold $(10K)$ dense $(100cm^{-3})$ phase of Giant Molecular Clouds of a scale of ~ 40 parsec (themselves structured into smaller (1 parsec) molecular clouds, themselves structured in cold dense cores holding a few solar masses, ...), together with expanding HII regions; a super–hot $(10^6 K)$ coronal phase is also present, although there is disagreement between authors as to its extent, both from the point of view of volume and of mass, as well as to its dynamical importance, being very tenuous.

Reynolds numbers are immense, and in fact dissipation is thought to intervene only through the mechanism of ambipolar drift (or ambipolar diffusion), although this mechanism, highly nonlinear, leads as well to the formation of sharp fronts (Brandenburg & Zweibel, 1994; also see for the one–dimensional case Hénon (1981; 1984), in the framework of the formation of the very thin rings of Saturn).

All ingredients (ionization winds to feed the turbulence at small scale, propagating fronts to stir–up the medium at large scale and allow for the formation of density fluctuations, gravitational potential to allow for the collapse of such clumps which in turn yield more stars, more winds, ...) are necessary in this energetic cycle of the ISM, including rotation and magnetic fields (Vazquez–Semadeni *et al.*, 1995ab). It should be stressed that cooling and heating of the gas are acting on a substantially faster time scale than a typical hydrodynamical time, but in fact tend to equilibrate so that the global dynamics is in fact shaped by the turbulence.

If linear analysis and a quasi–static vision of the ISM is useful theoretically to understand the ISM (Mouschovias, 1976ab; Elmegreen, 1991, 1994), nevertheless the action of nonlinear terms does arise as well, including the complex and somewhat contradictory effects of the magnetic field which acts both as a brake to collapse (through an enhanced pressure for example), and as an acceleration to collapse because of the confinement of magnetic field lines, a "pressure–cooker" effect.

Turbulence may be sufficient to prevent the collapse of the whole flow (the age of clouds is roughly ten times their free–fall time based on the linear Jeans analysis), either through a turbulent pressure, or because the formation of shocks is more rapid than the collapsing of clumps (Léorat *et al.*, 1990). But it is also known that the magnetic field, among other things, may prevent turbulent motions to develop thoroughly. In the incompressible case (see §1.9), the temporal decay of energy is slowed down compared to the pure fluid case, in part because nonlinear interactions are hampered by the presence of a uniform magnetic field (see §1.8.3). On the other hand, in a compressible flow the magnetic field has a tendency to weaken shocks. And when the medium is dispersive – as for example with ambipolar drift or the Hall term in either a weakly or a partially ionized medium – the steepening of shocks can be balanced by the dispersive effects. Soliton equations are an extreme case where analytical solutions are known that represent such a balance between nonlinearities and dispersion, and thus are sometimes called nonlinear waves (see for the development of such a point of view Adams *et al.* (1994), and also Adams & Fatuzzo (1993)).

So, is the interstellar medium to be regarded as fully turbulent, or is the observed turbulence really made up of Alfvén and magnetosonic waves, or solitons ? But then, what is the difference between waves and turbulence ?

Back to Chapter one ...

Acknowledgements

I am thankful to J.-D. Fournier, A. Newell and H. Politano for useful discussions. I also wish to thank J.S. Davis for helping greatly with the bibliography, and C. Caseneuve and V. Chéron for typing part of these Notes. The people with whom I have collaborated over the years, as well as the organizers of this School have all provided me with a stimulating environment. May they also be thanked.

These Lectures owe much to a variety of papers which have been plagiarised with various degrees of thoroughness. The first part of Lecture 1 follows the Benney and Newell (1969) original approach on weak turbulence. Section 2.5 is inspired from Drummond and Münch (1991) as well as Constantin, Procaccia and Segel (1994). In all cases, all mistakes are mine.

This work has received partial financial support from EEC contract ERBCHRXCT930410.

References

Adams F. & Fatuzzo M. 1993 *Astrophys. J.* **403**, 142.

Adams F., Fatuzzo M. & R. Watkins 1994 *Astrophys. J.* **426**, 629.

Armstrong, J., Cordes, J. & Rickett B. 1981 *Nature* **291**, 561.

Arnol'd, V.I., Zeldovich, Ya. B., Ruzmaikin, A.A. & Sokolov, D.D. 1982 *Sov. Phys. JETP* **54** 1083.

Arnold V.I. & Korkina E.I. 1983 *Vestn. Mosk. Univ. Mat. Meckh.* **38**, 43.

Bajer K. 1995 in "Small–scale structures in fluids and MHD", M. Meneguzzi, A. Pouquet & P.L. Sulem Editors, Notes in Physics **462**, Springer–Verlag.

Batchelor G. 1967 **An Introduction to Fluid Dynamics**, Cambridge University Press.

Beale J., Kato T. & Majda A. 1985 *Commun. Math Phys.*. **94**, 61.

Benney J. & Newell A. 1969 *Stud. Appl. Math.* **48**, 29.

Bender C. & Orszag S. 1978 **Advanced Mathematical Methods for Scientists and Engineers**, McGraw-Hill.

Benzi R., S. Ciliberto, R. Tripicciona, C. Baudet, F. Massaioli & S. Succi 1993 *Phys. Rev. E* **48**, R29.

Betchov R. 1956 *J. Fluid Mech.* **1**, 467.

Betchov R. 1965 *J. Fluid Mech.* **22**, 471.

Bhattacharjee A. & Wang X. 1992 *Phys. Rev. Lett.* **69**, 2196.

Bhattacharjee A., Ng C.S. & Wang X. 1995 *Phys. Rev. E*, to appear.

Biskamp D. 1994 **Nonlinear Magnetohydrodynamics**, Cambridge University Press.

Biskamp, D. & Welter H. 1989 *Phys. Fluids B* **1**, 1964.

Brachet, M.E. 1990 *Compte Rendus Acad. Sci. Paris* **311**, 375.

Brachet, M.E., Meneguzzi, M., Vincent, A., Politano, H. & Sulem, P.L. 1992 *Phys. Fluids A* **4**, 2845.

Braginskii S. in *Reviews of Plasma Physics* 1965 Vol. 1, p. 205, M. Leontovich Ed., Consultants Bureau, New–York.

Brandenburg A. & E. Zweibel 1994, *Astrophys. Lett.* **427** L91.

Brandenburg A., Nordlund, A., Stein, R. & Torkelsson U. 1995a in "Small–scale structures in fluids and MHD", M. Meneguzzi, A. Pouquet & P.L. Sulem Editors, Notes in Physics **462**, Springer–Verlag.

Brandenburg, A., Jennings, R., Nordlund, A., Rieutord, M., Stein, R. & Tuominen I. 1995b "Magnetic Structures in a dynamo simulation", *J. Fluid Mech.*, to appear.

Brandenburg A., Procaccia I. & Segel D. 1995c *Phys. of Plasmas*, **2**, 1148.

Burlaga L. 1991 *J. Geophys. Res.* **96**, 5847.

Cantwell B. 1993 *Phys. Fluids A* **5**, 2008.

Carbone V. 1993 *Phys. Rev. Lett.* **71**, 1546.

Carbone V. 1994 *Phys. Rev. E* **50**, R671.

Chasnov J. 1995 *Phys. Fluids* **7**, 600.

Childress, S. 1992 *Fast Dynamo Theory* in **Topological Aspects of the Dynamics of Fluids and Plasmas** p. 111 H.K. Moffatt *et al.* Eds, Kluwer Academic Press.

Chong, M., Perry A. & Cantwell B. 1990 in **Topological Fluid Dynamics**, p. 408, K.K. Moffatt & A. Tsinober Eds., Cambridge University Press.

Chorin A. **Vorticity and turbulence**, *Applied Mathematical Sciences*, **103**, Springer Verlag, 1994.

Constantin P., Procaccia I. & Segel D. 1995 *Phys. Rev. E* **51**, 3207.

DeLuca E. & Craig I. 1992 *Astrophys. J.* **390**, 679.

Douady, S., Couderc, Y. & Brachet M.-E. 1991 *Phys. Rev. Lett.* **67**, 982.

Dritschel D. 1993 Phys. Fluids **A5**, 984.

Drummond I. & Münch W. 1991 *J. Fluid Mech.* **225**, 529.

Dubrulle B. 1994 *Phys. Rev. Lett.* **73**, 959.

Einaudi G. & Velli M., "Coronal heating mechanisms", European Solar Physics meeting, Catania 1993, Springer-Verlag, in press.

Einaudi, G., Velli, M., Politano, H. & A. Pouquet, 1995: "Current sheet formation in solar active regions", *Astrophys. J. Letters*, to appear.

Elmegreen, B. G. 1991 *Astrophys. J.* **378**, 139.

Elmegreen, B. G. 1994 *Astrophys. J.* **433**, 39.

Escudier M. 1982 in **Intense Atmospheric Vortices** p. 247, L. Bengtsson & J. Lighthill Eds, Springer–Verlag.

Eyink G. 1995 *Phys. Rev. Lett.* **74**, 3800.

Falgarone E. 1995 in "Small–scale structures in fluids and MHD", M. Meneguzzi, A. Pouquet & P.L. Sulem Editors, Notes in Physics **462**, Springer–Verlag.

Falgarone E., D.C. Lis, T.G. Philips, D. Porter, Pouquet, A. & P. Woodward 1994 *Astrophys. J.*, **436**, 728.

Falkovich G. 1994 Phys. Fluids **6**, 1411.

Frick, P., Dubrulle, B. & Babiano A. 1995 *Phys. Rev. E*, to appear.

Frisch, U., P.L. Sulem & M. Nelkin 1978 *J. Fluid Mech.*, **87**, 719.

Frisch, U., A. Pouquet, P.L. Sulem & M. Meneguzzi 1983 *J. Mécanique Théor. Appl.* **20**, 191.

Frisch, U., B. Legras & B. Villone 1995 *Physica D*, to appear.

Fukumoto Y. & Miyazaki T. 1991 *J. Fluid Mech.* **222**, 369.

Galanti, B., Sulem, P.L. & Pouquet, A. 1992 *Geophys. Astrophys. Fluid Dyn.* **66** 183.

Galanti, B., Sulem, P.L. & Pouquet, A. 1993 in *Theory of Solar and Planetary Dynamos*, p. 99, M. Proctor, P. Mattheus & A. Rucklidge Eds., Cambridge University Press (1993).

Galloway, D.J. & Frisch U. 1986 *Geophys. Astrophys. Fluid Dyn.* **36**, 53.

Galloway, D.J. & Proctor, M.R.E. 1993, *Nature*, **356**, 691.

Granger R. 1993 *J. Fluid Mech.* **246**, 653.

Grappin R., A. Pouquet, & J. Léorat 1983 *Astron. Astrophys.* **126**, 51.

Grappin R., M. Velli & A. Mangeney 1991 *Ann. Geophys.* **9**, 416.

Grauer R., J. Krug & C. Marliani 1994 *Phys. Lett. A*, **195**, 335.

Grauer R. & C. Marliani 1995 *Phys. of Plasmas*, **2**, 41.

Greene J. 1993 *Phys. Fluids B* **5**, 2355.

Grossman S. & D. Lohse 1991 *Phys. Rev. Lett.* **67**, 445.

Gruzinov A. & Diamond Γ. 1995 *Phys. Plasmas* **2**, 1941.

Hada T. 1993 *Geophys. Res. Lett.* **20**, 2415.

Hama F. 1962 *Phys. Fluids* **5**, 1156.

Hasimoto H. 1972 *J. Fluid Mech.* **51**, 477.

Heiles C. 1988 *Astrophys. J.* **324**, 321.

Heiles C., Goodman, A. A., McKee, C. F., & Zweibel, E. G. 1993 in **Protostars and Planets** bf III, ed. E. H. Levy & J. I. Lunine (Univ. of Arizona Press), 279.

Hénon, M. 1981 *Nature* **293**, 33.

Hénon, M. 1984 IAU Colloquium **75**, 363; A. Brahic & Cepadeus Eds.

Hesse M. & K. Schindler 1988 *J. Geophys. Res.* **93**, 5559.

Hopfinger, E., Browand F. & Gagne, Y. 1982 *J. Fluid Mech.* **125**, 505.

Horiuchi R. & Sato T. 1988 *Phys. Fluids* **33**, 1142.

Hossain M., Gray P., Pontius D., Matthaeus W. & Oughton S. 1995 *Phys. Fluids*, to appear.

Iroshnikov P. 1963 *Sov. Astron.* **7**, 566.

Isichenko M. 1992 *Rev. Mod. Phys.* **64**, 961.

Isichenko M. & Gruzinov M. 1994 *Phys. Plasmas* **1**, 1802.

Jeong J. & Hussain F. 1995 *J. Fluid Mech.* **285**, 69.

Jimenez J. 1994 *J. Fluid Mech.* **279**, 169.

Jimenez J., Wray A., Saffman P.G. & Rogallo R. 1993 *J. Fluid Mech.* **255**, 65.

Keller J. 1994 *Phys. Fluids* **6**, 1515.

Kerr R. 1985 *J. Fluid Mech.* **153**, 31.

Kerr R. 1993 *Phys. Fluids* **A5**, 1725.

Kerr R. 1995 in "Small–scale structures in fluids and MHD", M. Meneguzzi, A. Pouquet & P.L. Sulem Editors, Notes in Physics **462**, Springer–Verlag.

Kida S. 1981 *J. Fluid Mech.* **112**, 397.

Kida S. & Takaoka 1991 *J. Phys. Soc. Japan* **60**, 2184.

Kimura Y. & Herring J. 1995 in "Small–scale structures in fluids and MHD", M. Meneguzzi, A. Pouquet & P.L. Sulem Editors, Notes in Physics **462**, Springer–Verlag.

Kinney R., T. Tajima, J.C. McWilliams & N. Petviashvili 1993 *Phys. Plasmas*, **1**, 260.

Kinney R. & McWilliams J.C. 1995 in "Small–scale structures in fluids and MHD", M. Meneguzzi, A. Pouquet & P.L. Sulem Editors, Notes in Physics **462**, Springer–Verlag.

Klein, R., Majda, A. & McLaughlin R. 1992 *Phys. Fluids* **A4**, 2271.

Klingenberg, W. 1978 **A Course in Differential Geometry**, GTM **51**, Springer–Verlag.

Kolmogorov A. 1941a *Dokl. Akad. Nauk SSSR* **30**, 299.

Kolmogorov A. 1941b *Dokl. Akad. Nauk SSSR* **31**, 538.

Kolmogorov A. 1962 *J. Fluid Mech.* **13**, 82.

Kraichnan R.H. 1959 *J. Fluid Mech.* **5**, 497.

Kraichnan R.H. 1965 *Phys. Fluids* **8**, 1385.

Kraichnan R.H. 1971 *J. Fluid Mech.* **47**, 525.

Kraichnan, R.H. 1973 *J. Fluid Mech.*, **59**, 745.

Kraichnan, R.H. 1979 *Phys. Rev. Lett.*, **42**, 1677.

Kraichnan, R.H. & Montgomery D. 1980 *Rep. Prog. Phys.* **43** 547.

Krall N. & Trivelpiece A. 1973 **Principles of Plasma Physics**, McGraw–Hill.

Kulsrud R. & Anderson S. 1992 Astrophys J. **396**, 606.

Lamb 1932 **Hydrodynamics**, sixth edition, Cambridge University Press.

Lau Y–T & Finn J. 1990 *Astrophys. J.* **350**, 672.

Leith C. 1971 *J. Atmos. Sci.* **28**, 145.

Léorat, J. 1975 Thèse, Université Paris VII.

Léorat, J., T. Passot, & A. Pouquet 1990 *Monthly Not. R.A.S.* **243**, 293.

Lesieur M. 1990 **Turbulence in Fluids**, Second Edition, Kluwer.

Lohse D. 1994 *Phys. Rev. Lett.* **73**, 3223.

Lohse D. & Müller–Groeling 1995 *Phys. Fluids* in press.

Lin S. & G. Corcos 1984 *J. Fluid Mech.* **141**, 139.

Lu, E. & Hamilton R. 1991 *Astrophys. J.* **380**, L89.

Lu, E., Hamilton R., McTierman M. & Bromund K. 1993 *Astrophys. J.* **412**, 841.

Lu, E. 1995a *Phys. Rev. Lett.* **74**, 2511.

Lu, E. 1995b *Astrophys. J. Lett.* **446**, L109.

Lu, E. 1995c *Astrophys. J.* **447**, 416.

Lundgren T. 1982 *Phys. Fluids* **25**, 2193.

Lundgren T. & Mansour M. 1995 in "Small–scale structures in fluids and MHD", M. Meneguzzi, A. Pouquet & P.L. Sulem Editors, Notes in Physics **462**, Springer–Verlag.

McKee, C. F., Zweibel, E. G., Goodman, A. A., & Heiles, C. 1993 in **Protostars and Planets** bf III, ed. E. H. Levy & J. I. Lunine (Univ. of Arizona Press), 327.

Marsch E. 1993 *Ann. Geophys.* **11**, 227.

Matthaeus, W. & Goldstein M. 1982 *J. Geophys. Res.* **87**, 6011.

Matthaeus, W., Bieber J. & Zank G. 1994 Bartol Preprint BA–94–67, University of Delaware.

Meneguzzi M., U. Frisch & A. Pouquet 1981 *Phys. Rev. Lett.* **47**, 1060.

Meneguzzi M., Politano, H., Pouquet, A. & M. Zolver 1995 "A Sparse Mode Spectral Method for the Simulation of Turbulent Flows" *J. Comp. Phys.* to appear.

Meneveau C. & K. Sreenivasan 1991 *J. Fluid Mech.* **224**, 429.

Mikić Z., D. Schnack & G. Van Hoven 1989 *Astrophys. J.* **338**, 1148.

Moffatt H.K. 1961 *J. Fluid Mech.* **11**, 625.

Moffatt H.K. 1978 **Magnetic Field Generation in Electrically Conducting Fluids**, Cambridge University Press.

Monin A. & A. Yaglom 1975 **Statistical Fluid Mechanics**, Volume II. Edited by J. Lumley, The MIT Press. Originally published in 1965 in Russian by Nauka Press.

Montgomery D. & Philipps L. 1989 *Phys. Rev.* **A40**, 1515.

Mouschovias, T. 1976a *Astrophys. J.* **206**, 753.

Mouschovias, T. 1976b *Astrophys. J.* **207**, 141.

Neu, J.C. 1984 *J. Fluid Mech.*, **143**, 253.

Newell, A., Passot, T. & Souli, M. 1990 *J.Fluid Mech.* **220**, 197.

Nore C., Abid, M. & Brachet M. 1995 in "Small–scale structures in fluids and MHD", M. Meneguzzi, A. Pouquet & P.L. Sulem Editors, Notes in Physics **462**, Springer–Verlag.

Onsager M. 1949 *Nuovo Cimento Supp.* **6**, 279.

Orszag S. 1974 *Lectures on the Statistical Theory of Turbulence*, Flow Research Report **31** (also Les Houches Proceedings, 1973).

Oughton, S., Priest, E. & Matthaeus W. 1994 *J. Fluid Mech.* **280**, 95.

Osborne A. 1993 *Phys. Rev. E* **48**, 296.

Parker E. 1979 **Cosmical Magnetic Fields: their Origin and Activity**, Clarendon Press, Oxford.

Passot, T., Politano, H., Pouquet, A. & Sulem, P.L. 1990 *Theor. Comp. Fluid Dyn.*, **1**, 47.

Passot, T., H. Politano, P–L. Sulem, J–R. Angilella & M. Meneguzzi 1995 *J. Fluid Mech.* **282**, 313.

Pelz R. & Boratav O. 1995 in "Small–scale structures in fluids and MHD", M. Meneguzzi, A. Pouquet & P.L. Sulem Editors, Notes in Physics **462**, Springer–Verlag.

Politano, H., Pouquet, A. & Sulem, P.L. 1989 *Phys. Fluids B* **1**, 2230.

Politano H., Pouquet A. & Sulem P.L. 1995 "Emergent Structures of three–dimensional incompressible MHD Flows" *Phys. Plasmas* **2**, 2931.

Politano H. & Pouquet, A. 1995a *Phys. Rev. E* **52**, 636.

Politano H. & Pouquet, A. 1995b, *Phys. Rev. Lett.*, to appear.

Pope S. 1988 *J. Engng. Sci.* **26**, 445.

Porter D., Pouquet, A. & Woodward, P. 1992 *Theor. Comp. Fluid Dyn.* **4**, 13.

Porter D., Pouquet, A. & Woodward, P. 1994 *Phys. Fluids A* **6**, 2133.

Porter D., Pouquet, A. & Woodward, P. 1995 in "Small–scale structures in fluids and MHD", M. Meneguzzi, A. Pouquet & P.L. Sulem Editors, Notes in Physics **462**, Springer–Verlag.

Pouquet A. 1978 *J. Fluid Mech.* **88**, 1.

Pouquet A. 1993 in **Les Houches Summer School on Astrophysical Fluid Dynamics**, July 1987; also Preprint High Altitude Observatory, December 1987; Les Houches, Session **XLVII**, p. 139; Eds. J. P. Zahn & J. Zinn–Justin, Elsevier.

Pouquet A. 1994 in *Research Trends in Physics*, American Institute of Physics, V. Stefan Ed., Springer–Verlag.

Pouquet A., U. Frisch, & J. Léorat 1976 *J. Fluid Mech.*, **77**, 321.

Priest E. 1985 *Rep. Prog. Phys.* **48**, 955.

Procaccia I. & P. Constantin 1993 *Phys. Rev. Lett.* **70**, 3416.

Pullin D. & P. Saffman 1992 *Phys. Fluids A* **5**, 126.

Ricca R. 1991 Nature **352**, 561.

Ricca R. 1992 Phys. Fluids **A4**, 938.

Ricca R. 1994 J. Fluid Mech. **273**, 241.

Ricca R. 1995 in "Small–scale structures in fluids and MHD", M. Meneguzzi, A. Pouquet & P.L. Sulem Editors, Notes in Physics **462**, Springer–Verlag.

Robert J. 1991 *J. Stat. Phys.* **65**, 531.

Roberts P.H. 1967 **An Introduction to Magnetohydrodynamics**, Longmans.

Roberts P.H. 1993 in **Les Houches Summer School on Astrophysical Fluid Dynamics**, July 1987, Session **XLVII**; Eds. J. P. Zahn & J. Zinn–Justin, Elsevier.

Ruzmaikin A., Feynman, J., Goldstein, B. & Smith E. 1995 *J. Geophys. Res.* **100**, 3395.

Saffman P. 1967 *Phys. Fluids* **10**, 1349.

Saffman P. 1992 **Vortex Dynamics**, Cambridge University Press.

Scalo J. 1984 *Astrophys. J.* **277**, 566.

Scalo J. 1985 in **Protostars and Planets** Vol. II, p. 201, D. Black & M. Matthews Eds, Tucson, University of Arizona Press, Tucson.

She Z.S. & E. Lévêque 1994 *Phys. Rev. Lett.* **72**, 336.

Shebalin, J., Matthaeus, W. & Montgomery D. 1983 *J. Plasma Phys.* **29**, 525.

Shinoda, M. & Fujiwara T. 1993 in **Unstable and Turbulent Motion of Fluid**, p. 79, S. Kida Ed., World Scientific.

Shu F. 1992 **The Physics of Astrophysics: Gas Dynamics**, Vol. II, University Science Books, Mill Valley, California.

Smith, M., Donelly R., Goldenfeld N & Vinen W. 1993 *Phys. Rev. Lett.* **71**, 2583.

Sofue & Handa 1984 *Nature* **310**, 568.

Soria, J., Sondegaard, R., Cantwell, B., Chong, M. & Perry S. 1994 *Phys. Fluids* **6**, 871.

Steenbeck, M., Krause, F. & Rädler, K.H. 1966 "A calculation of the mean electromotive force in an electrically conducting fluid in turbulent motion under the influence of Coriolis forces", *Z. Naturforsch* **21a**, 369–376. See also the translation in P.H. Roberts & M. Stix, "The turbulent dynamo", Technical Notes NCAR–TN/IA–60, pp.29–47, Boulder, Colorado (1971).

Strauss H. 1993 *Geophys. Res. Lett.* **20**, 325.

Taylor G.I. 1935 *Proc. Roy. Soc. London A* **151**, 421.

Taylor J.B. 1974 *Phys. Rev. Lett.* **33**, 1139.

Taylor J.B. 1993 *Phys. Fluids B* **5**, 3893.

Thomas J. 1967 *Phys. Fluids* **11**, 1245.

Townsend A. 1951 *Proc. Roy. Soc. London A* **208**, 534.

Van Dyke M. 1982 **An Album of Fluid Motions**, Parabolic Press, Stanford.

Vazquez–Semadeni E. & J. Scalo 1992 *Phys. Rev. Lett.* **68** 2921.

Vazquez–Semadeni E., Passot, T. & Pouquet, A. 1995ab *Astrophys. J.*, **441**, 702; and "A Turbulent Model for the Interstellar medium. II: Magnetic Fields and Rotation" 1995 *Astrophys. J.*, December 20 issue.

Vieillefosse P. 1982 *J. Phys.* **43**, 837.

Vieillefosse P. 1984 *Physica A* **125**, 150.

Villaseñor J. & A. Vincent 1992 CVGIP. **Image Understanding 55**, 27.

Vincent, A. & Meneguzzi, M. 1991 *J. Fluid Mech.*, **225**, 1.

Vincent, A. & Meneguzzi, M. 1994 *J. Fluid Mech.*, **258**, 245.

Vishik, M.M. 1989 *Geophys. Astrophys. Fluid Dyn.* **29**, 151.

Vlahos, L. 1993 *Adv. Space Res.* **13**, 122.

von Karman T. & Howarth L. 1938 *Proc. Roy. Soc. London A* **164**, 192.

Woodward P. 1986 **Numerical Methods for Astrophysicists**," in *Astrophysical Radiation Hydrodynamics*, K.-H. Winkler & M. L. Norman eds., p. 245, Reidel.

Yao, H., Zabusky, N. & Dritschel D. 1995 *Phys. Fluids* **7**, 539.

Yoshizawa A. 1990 *Phys. Fluids A* **2**, 838.

Zakharov E., L'vov V. & Falkovich G. 1992 **Kolmogorov Spectra of Turbulence I: Wave Turbulence**, Springer–Verlag.

Zeitlin V. 1991 *Physica D* **49**, 353.

Zeldovich. Ya, Ruzmaikin, A. & Sokoloff D. 1983 **Magnetic Fields in Astrophysics**, Gordon Breach.

Zhu S–P, Horiuchi R., Sato T, and the Complexity Simulation Group (K. Watanabe, T. Hayashi, Y. Todo, T. Watanabe, A. Kageyama & H. Takamaru) 1995 *Phys. Rev. E* **51**, 6047.

Radio Astronomical Diagnostics

Arnold O. Benz

Institute of Astronomy, ETH CH-8092 Zurich, Switzerland

Abstract. A brief introduction into the diagnostic capabilities and results of radio waves from coronal plasmas of the Sun and other late-type stars is presented. These coronal emissions show that the plasma is in a dynamic state with time scale down to a few tens of microseconds. Gyrosynchrotron emission in flares reveal the presence of relativistic electrons, which, in active stars, seem to persist even during quasi-quiet (quiescent) intervals. Coherent emissions of solar electron beams by the two-stream instability has been discovered up to 8 GHz. Particularly efficient emitters are trapped electrons having a loss-cone distribution. This is probably the most frequent cause of highly polarized stellar radio flares. Of greatest interest are emissions by unstable currents and shocks, which have been identified in the solar corona. A general introduction into some basic theories, but not a review, is given, illustrated with recent observations. A more extended introduction can be found in Benz (1993).

Keywords. Radio radiation: continuum - Stars: radio radiation - Sun: radio radiation - Sun: radio bursts - Plasma physics: general - Plasma physics: radiation theory

1 Solar and Stellar Coronae

The solar corona has become a scientific problem of first importance when W. Grotrian and B. Edlén suggested in 1939 that it consists of a thin plasma of some million degrees temperature. This unexpectedly high temperature is still an enigma today. It is one of the reasons why the solar corona will be a primary target of solar research in the coming years.

Soft X-ray motion pictures show ceaseless variability in active regions. Loops brighten up and disappear within a few hours. About 1500 X-ray bright points light up per day in quiet regions and coronal holes. They consist of several parallel loops and have a mean lifetime of eight hours. The various soft X-ray brightenings represent enhanced thermal radiation from particular loops and suggest discrete and major energy inputs. The associated process may have different forms. Are they enough to heat the solar corona? The answer seems to become more difficult, the more we know about the corona. Additional forms of energy input, in particular by waves and stationary electric currents, must also be considered. The heating problem of the solar corona can only be solved by a better understanding of all the various processes (Pallavicini 1987; Narain & Ulmschneider 1991).

The discovery of a low-level microwave emission ($\gtrsim 1$ GHz) from apparently non-flaring, single dMe stars by Gary and Linsky (1981) has added another mystery to coronal physics. The radiation is 2–3 orders of magnitude more luminous than the quiet Sun caused by thermal bremsstrahlung of the coronal plasma. Spectral investigations and VLBI measurements of the source size have shown that the radio emissionhas a brightness temperature in excess of 10^9K. It must originate from more energetic electrons than the X-rays, which show temperatures of the order of 10^7K. The radio emission is probably non-thermal and most likely produced by the gyrosynchrotron (= mildly relativistic synchrotron) process. The different word 'quiescent' has been chosen for stars to indicate the difference from the solar non-flare radio emissions and to emphasize the possibility that the observed low-level variability (on time scales longer than ten minutes) may in fact be a superposition of many flare-like events. The observed brightness temperature of some 10^9 K is compatible with an interpretation by gyrosynchrotron radiation of a population of energetic ($\gtrsim 100$ keV) electrons spiraling in the coronal magnetic field. Such a particle population does not exist, at least not at this energy level, in the quiet Sun. However, the presence of such particles is ubiquitous during solar flares. They do not noticeably contribute to the quiet solar radio emission. The gyrosynchrotron emission, however, permanently dominates the microwave emission of dMe stars, surpassing the thermal solar radio luminosity by orders of magnitude.

Figure 1 is the gist of a decade of painstaking measurements by many observers and instruments. The quiescent soft X-ray and microwave emissions are compared not only for different stars and types of stars, but also with the temporary sources of solar flares. The correlation between thermal emission and radiation of non-thermal particles is obvious and stretches over many orders of magnitude. It tells us that the continuous presence of accelerated particles and coronal heating have something in common. Figure 1 supports the conjecture that a common process, the dynamo in the stellar interior, drives both the coronal heating and flares. Moreover, the similarity of flares and coronae suggests a more direct relation between the thermal constituent of coronal plasmas and powerful, possibly violent acceleration processes in the corona. The heating of the coronae of rapidly rotating stars and flares both accelerate electrons and may be similar processes!

Are coronal plasmas heated by flares? The answer is not simple, as there are different types of flare-like, dynamic processes even in a relatively inactive corona like the Sun's. There is substantial variability in the X-ray and radio emissions of most stellar coronae over a variety of time scales. The observed variability is in the form of individual flares and slow variations of the quiescent background. Solar flares show a wide range of amplitude (up 10 orders of magnitude), whereas the variations of the quiescent component on time scales of hours do not exceed amplitudes of 50%. Continuous low-amplitude, short-period variability (microflares) has been proposed for the heating of stellar coronae, but the observational evidence is still meager.

In conclusion, observations – both solar and stellar – strongly suggest that it is not possible to comprehend coronae without understanding their dynamic phenomena. Apart from the heating problem, coronal processes offer numerous and exciting challenges for plasma astrophysics in general. Some of them will be discussed in the following.

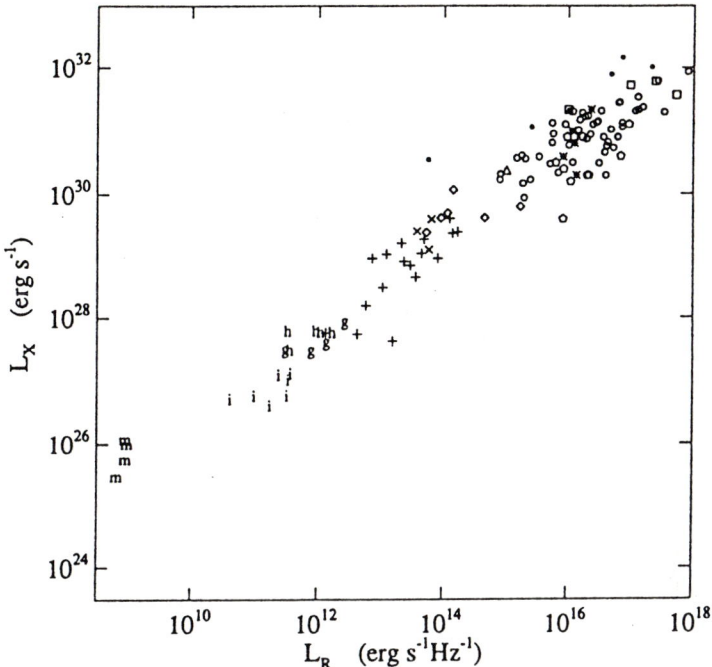

Fig. 1: The soft X-ray luminosity is plotted vs. the radio luminosity at a frequency of 5–10 GHz. Different types of stars are compared with the peak flux of solar flares. Key to the symbols: m solar microflare; i intermediate impulsive solar flare; h gradual solar flare with dominating large impulsive phase; g pure gradual solar flare; + dM(e) stars; × dK(e) stars; ◇ BY Dra binaries; ◯ RS CVn binaries; ○ RS CVn binaries with two giants; △ AB Dor; * Algols; Ū FK Com stars; (pentagon) post-T Tau stars (Benz & Güdel 1994).

2 Cold Plasma Modes and Instabilities

A plasma can sustain many different waves as it has many eigenmodes. Each mode is characterized by a frequency ω and a wave vector $\mathbf{k}$ and represents a possible linear wave, which can be excited by a disturbance. The waves may be grouped according to the basic approximations made to derive them.

Let's first study high-frequency waves, radiowaves being an example of them. They primarily involve oscillating electrons; thermal effects, like pressure, can be neglected. The plasma can be considered to be a fluid. For each species α there is an equation for particle and momentum conservation,

$$\frac{\partial n^\alpha}{\partial t} + \nabla \cdot (n^\alpha \mathbf{V}^\alpha) = 0$$

$$\frac{\partial \mathbf{V}^\alpha}{\partial t} + (\mathbf{V}^\alpha \cdot \nabla)\mathbf{V}^\alpha = \frac{q_\alpha}{m_\alpha}(\mathbf{E} + \frac{1}{c}\mathbf{V}^\alpha \times \mathbf{B}^\alpha) \quad .$$

In addition we are using the full set of Maxwell's equations in their classical form,

$$\nabla \cdot \mathbf{B} = 0 \qquad \nabla \times \mathbf{E} = -\frac{1}{c}\frac{\partial \mathbf{B}}{\partial t} \qquad \rho^* := \sum_\alpha q_\alpha n_\alpha$$

$$\nabla \cdot \mathbf{E} = 4\pi \rho^* \qquad \nabla \times \mathbf{B} = \frac{4\pi}{c}\mathbf{J} + \frac{1}{c}\frac{\partial \mathbf{E}}{\partial t} \qquad \mathbf{J} := \sum_\alpha q_\alpha n_\alpha \mathbf{V}^\alpha \quad .$$

This system of differential equations is non-linear. Only small disturbances are of interest, we thus linearize the system by assuming that all variables have the form $A = A_0 + A_1(\mathbf{x}, t)$. The zero-order variables describe a stationary, homogeneous background. We shall furthermore assume that $\mathbf{E}_0 = 0$, $\mathbf{J}_0 = 0$, and $\rho_0^* = 0$. Linear here means that the approximation is to first order, thus the products of two first-order terms are neglected. Hence

$$\frac{\partial n_1^\alpha}{\partial t} + \mathbf{V}_0^\alpha \cdot \nabla n_1^\alpha + n_0^\alpha \nabla \cdot \mathbf{V}_1^\alpha = 0$$

$$\frac{\partial \mathbf{V}_1^\alpha}{\partial t} + (\mathbf{V}_0^\alpha \cdot \nabla)\mathbf{V}_1^\alpha = \frac{q_\alpha}{m_\alpha}(\mathbf{E}_1 + \frac{1}{c}\mathbf{V}_1^\alpha \times \mathbf{B}_0 + \frac{1}{c}\mathbf{V}_0^\alpha \times \mathbf{B}_1)$$

$$\nabla \times \mathbf{E}_1 = -\frac{1}{c}\frac{\partial \mathbf{B}_1}{\partial t}$$

$$\nabla \cdot \mathbf{E}_1 = 4\pi \rho_1^*$$

$$\rho_1^* = \sum_\alpha q_\alpha n_1^\alpha$$

$$\nabla \times \mathbf{B}_1 = \frac{4\pi}{c}\mathbf{J}_1 + \frac{1}{c}\frac{\partial \mathbf{E}_1}{\partial t}$$

$$\nabla \cdot \mathbf{B}_1 = 0$$

$$\mathbf{J}_1 = \sum_\alpha q_\alpha(n_1^\alpha \mathbf{V}_0^\alpha + n_0^\alpha \mathbf{V}_1^\alpha) \quad .$$

We now assume that the disturbance has the form of a wave: $A_1 = \bar{A}_1 \exp[i(\mathbf{k}x - \omega t)]$. All derivatives then become factors,

$$\frac{\partial}{\partial t} \to -i\omega, \ \nabla \to i\mathbf{k} \quad .$$

The system of equations is ordinary and linear. We eliminate all variables except E. The result is a system of three linear, homogeneous equations,

$$(k^2 \hat{\mathbf{1}} - \mathbf{k} \circ \mathbf{k} - \frac{\omega^2}{c^2}\hat{\epsilon}) * \mathbf{E}_1 = 0 \quad .$$

This equation only has a non-trivial solution if the determinant of the system equals zero. We thus require

$$\det[(\frac{ck}{\omega})^2\hat{1} - (\frac{c}{\omega})^2\mathbf{k} \circ \mathbf{k} - \hat{\epsilon}] = 0 \quad .$$

The three solutions are the **normal modes** of a cold plasma. We have defined the dielectric tensor

$$\hat{\epsilon} := \begin{pmatrix} \epsilon_0 & i\epsilon_1 & 0 \\ -i\epsilon_1 & \epsilon_0 & 0 \\ 0 & 0 & \epsilon_{\parallel} \end{pmatrix} :=$$

$$\begin{pmatrix} 1 - \sum_\alpha \frac{(\omega_p^\alpha)^2}{\omega^2 - (\Omega_z^\alpha)^2} & -i\sum_\alpha \frac{\Omega_z^\alpha(\omega_p^\alpha)^2}{\omega(\omega^2 - (\Omega_z^\alpha)^2)} & 0 \\ i\sum_\alpha \frac{\Omega_z^\alpha(\omega_p^\alpha)^2}{\omega(\omega^2 - (\Omega_z^\alpha)^2)} & 1 - \sum_\alpha \frac{(\omega_p^\alpha)^2}{\omega^2 - (\Omega_z^\alpha)^2} & 0 \\ 0 & 0 & 1 - \frac{(\omega_p)^2}{\omega^2} \end{pmatrix}$$

and the plasma frequencies for each species

$$(\omega_p^\alpha)^2 := \frac{4\pi q_\alpha^2 n_\alpha}{m_\alpha} \quad ,$$

$$\omega_p^2 := \sum_\alpha (\omega_p^\alpha)^2 \quad .$$

For quasi-parallel propagation, the three solutions are:

   1. Langmuir waves (electrostatic mode)

$$\omega = \omega_p$$

As the displacement current cancels the current produced by the oscillating particles, the wave magnetic field is zero. The mode is purely electric and longitudinal.

   2. Ordinary mode (o-mode)

$$\frac{c^2 k^2}{\omega^2} = 1 - \sum_\alpha \frac{(\omega_p^\alpha)^2}{\omega^2 - (\Omega_z^\alpha)^2} + \sum_\alpha \frac{\Omega_z^\alpha(\omega_p^\alpha)^2}{\omega(\omega^2 - (\Omega_z^\alpha)^2)}$$

The wave electric field oscillates perpendicular to the direction of wave propagation, thus the mode is called 'transverse'. The polarization of the wave, defined by the apparent motion of the electric field vector, is left circular.

   3. Extraordinary mode (x-mode)

$$\frac{c^2 k^2}{\omega^2} = 1 - \sum_\alpha \frac{(\omega_p^\alpha)^2}{\omega^2 - (\Omega_z^\alpha)^2} - \sum_\alpha \frac{\Omega_z^\alpha(\omega_p^\alpha)^2}{\omega(\omega^2 - (\Omega_z^\alpha)^2)}$$

The wave is similar to o-mode, but right-hand circularly polarized.

The dispersion relations for the two electromagnetic modes are

$$\frac{c^2 k^2}{\omega^2} = \left(\mathcal{N}_\circ\right)^2 \approx 1 - \left(\frac{\omega_p}{\omega}\right)^2 \left(1 \pm \frac{\Omega_e}{\omega}|\cos\theta|\right)^{-1} \quad .$$

For $\omega_p \to 0$, i.e. in vacuum, it becomes the well-known dispersion relation for propagating electromagnetic waves. The term including ω_p takes the dielectric properties into account arising from the oppositely charged particles, and the term involving the gyrofrequency $\Omega_e = eB/m_e c$ arises from the diamagnetic property of the gyrating electrons.

The change in dispersion relation by the ions and electrons partaking in the oscillation of the wave causes differences in the two modes, represented by the sign before the $\frac{\Omega_e}{\omega}$ term ($+$ for o-mode, $-$ for x-mode), respectively.

For $\omega \gg \Omega_e$, the group velocity becomes

$$v_{gr} = \frac{\partial \omega}{\partial k} \approx c(1 - \omega_p^2/\omega^2)^{1/2} < c \quad .$$

2.1 Beam Mode

Let $\mathbf{B}_0 = 0$, $\mathbf{V}_0^\alpha \neq 0$ and $\mathbf{E}_1 \parallel \mathbf{k} \parallel \mathbf{V}_1^\alpha$ and consider a wave with $\mathbf{B} = 0$. The plasma disturbances then are described by

$$-i\omega n_1^\alpha + V_0^\alpha ik n_1^\alpha + n_0^\alpha ik V_1^\alpha = 0$$

$$-i\omega \mathbf{V}_1^\alpha + \mathbf{V}_0^\alpha ik \mathbf{V}_1^\alpha = \frac{q_\alpha}{m_\alpha}(\mathbf{E}_1 + \frac{1}{c}\mathbf{V}_0^\alpha \times \mathbf{B}_1)$$

$$ik E_1 = 4\pi \sum_\alpha q_\alpha n_1^\alpha \quad .$$

Let us assume that the electrons are in motion, and the ions at rest ($\mathbf{V}_0^i = 0$, $\mathbf{V}_0^e \neq 0$).

$$1 = \frac{(\omega_p^i)^2}{\omega^2} + \frac{(\omega_p^e)^2}{(\omega - kV_0^e)^2} \quad .$$

This equation is of forth order. For sufficiently small $\mathbf{k}$ vectors, it has two real solutions (not of interest here) and two complex solutions,

$$\omega_3 = \omega_r - i\gamma_k \quad ,$$

$$\omega_4 = \omega_r + i\gamma_k \quad .$$

The solution ω_4 describes a wave with exponentially growing amplitude. It is called the *two-stream instability*. Thus we have shown that non-maxwellian velocity distributions constituting free energy can drive unstable the oscillations of an eigenmode. Usually the deviation from maxwellian is small and cannot be described in the cold plasma approximation. Nevertheless, the excited mode in hot plasma is, except for the imaginary part of the frequency, approximately identical to the cold plasma mode. It is therefore not surprising that a 'hot' beam, consisting of a bump on the tail of the thermal background velocity distribution, drives a similar instability (called 'bump-on-tail' instability) of Langmuir waves (for further information see Godfrey et al. 1975; Melrose 1986; Benz 1993).

3 Particle Beams

3.1 Plasma Emission

The Langmuir waves produced by an electron beam traversing the coronal plasma do not propagate from their region of origin. They cannot be observed from Earth unless they are converted into o-mode or x-mode waves. Two such processes, wave–particle scattering and wave–wave scattering, are here presented.

In the quantum mechanical picture an incident wave quantum (1) scatters into a quantum (2) of the secondary wave, changing the particle's energy and momentum by

$$\Delta\varepsilon = \hbar(\omega_2 - \omega_1) \quad,$$

$$\Delta\mathbf{p} = \hbar(\mathbf{k_2} - \mathbf{k_1}) \quad.$$

We define a spectral quantum density of bosons, $N(\mathbf{k})$, and their *effective wave temperature*, $T(\mathbf{k})$, using the Rayleigh - Jeans approximation,

$$N(\mathbf{k})\hbar\omega_\mathbf{k} := k_B T(\mathbf{k}) := W(\mathbf{k}) \quad.$$

A. Spontaneous Scattering off Ions

Let w be the rate at which a particle scatters a Langmuir wave into a transverse wave. The most efficient particles for this purpose are electrons, being far more agile than ions. However, the long wavelengths considered in radio astronomy require that Debye shielding is taken into account. The interacting electrons are not completely free, but are in such a close proximity of an ion that they move with the mean ion velocity. The Thomson scattering rate is reduced to

$$w \approx w_T/4$$

for primordial abundances, where the Thomson scattering rate per ion,

$$w_T = \frac{8\pi r_0^2 c \mathcal{N}}{3} \quad.$$

The scattering produces transverse photons, and the number density of photons increases with the rate

$$\frac{\partial N_t(\mathbf{k_t})}{\partial t} \approx \int d^3v \int \frac{d^3 k_L}{(2\pi)^3} w(\mathbf{v}, \mathbf{k_L}, \mathbf{k_t}) \, N_L(\mathbf{k_L}) \, f_i(\mathbf{v}) \quad.$$

We have used a differential scattering rate taking care of energy and momentum conservation (see below). The incident and emerging waves must obey the dispersion relations

$$\omega_L \approx \omega_p (1 + \frac{3k_L^2 v_{te}^2}{\omega_p^2})^{1/2} \quad,$$

including now thermal effects. Assuming $\omega \gg \Omega_e$, the transverse waves have

$$\omega_t \approx (\omega_p^2 + k_t^2 c^2)^{1/2} \quad .$$

If $\omega_L \approx \omega_t$, which will be shown below,

$$k_t \approx \sqrt{3}\, k_L \frac{v_{te}}{c} \quad .$$

The conservation equations require

$$\omega_L - \mathbf{k_L v} = \omega_t - \mathbf{k_t v} \quad .$$

The bandwidth of the scattered electromagnetic radiation becomes

$$\Delta\omega \approx |\mathbf{k_t} - \mathbf{k_L}|\, v_{ti} \approx k_L v_{ti} \quad .$$

Since $k_L v_{ti} \approx \omega_p v_{ti}/V_b \ll \omega_p$, the incident and scattered waves have practically the same frequency as assumed.

B. Induced Scattering

As shown already by Einstein, a spontaneous scattering process must be complemented by the inverse process (absorption) and by the corresponding induced process to satisfy the second law of thermodynamics. All these processes have the same scattering rates. Thus

$$\frac{\partial N_t(\mathbf{k_t})}{\partial t} = \int d^3 v \int \frac{d^3 k_L}{(2\pi)^3} w(\mathbf{v}, \mathbf{k_L}, \mathbf{k_t})[\{N_L(\mathbf{k_L}) - N_t(\mathbf{k_t})\} f_i(\mathbf{v}) -$$
$$N_L(\mathbf{k_L}) N_t(\mathbf{k_t}) \frac{\hbar(\mathbf{k_L} - \mathbf{k_t})}{m_i} \cdot \frac{\partial f_i}{\partial \mathbf{v}}] \quad ,$$

where the 3 terms in the bracket correspond to the 3 processes involved: The term which is linear in $N_L(\mathbf{k_L})$ accounts for spontaneous scattering of Langmuir waves into transverse waves weighted by the differential scattering rate $w(\mathbf{v}, \mathbf{k_L}, \mathbf{k_t})$ and the ion velocity distribution f_i. The term which is linear in $N_t(\mathbf{k_t})$ is negative, since it represents the inverse process: absorption. The term which is proportional to $N_L N_t$ describes the net effect of induced scattering in both directions differing by the ion distribution at the high and low energy levels. The difference can easily be evaluated using the approximation $f(\mathbf{v} - \hbar(\mathbf{k_L} - \mathbf{k_t})/m_i) \approx f(\mathbf{v}) - \hbar(\mathbf{k_L} - \mathbf{k_t})/m_i \cdot \partial f_i/\partial \mathbf{v}$.

C. Scattering off Other Waves

For scattering off a second wave (subscript 2), the conservation equations become

$$\omega_1 + \omega_2 = \omega_3 \quad ,$$

$$\mathbf{k_1} + \mathbf{k_2} = \mathbf{k_3} \quad ,$$

referred to as the *parametric conditions*. The growth rate of the transverse photon density is

$$\frac{\partial N_3(\mathbf{k_3})}{\partial t} = \int \frac{d^3 k_1}{(2\pi)^3} \int \frac{d^3 k_2}{(2\pi)^3} u^{123}(\mathbf{k_1}, \mathbf{k_2}, \mathbf{k_3})[N_1(\mathbf{k_1})N_2(\mathbf{k_2}) \\ - N_3(\mathbf{k_3})\{N_1(\mathbf{k_1}) + N_2(\mathbf{k_2})]$$

The scattering rate u^{123} can be evaluated from the linearized Maxwell's equations, from which the general wave equation for the fields of wave (3) is derived,

$$\frac{\partial}{\partial t}\left[\frac{|\mathbf{E_3}|^2}{8\pi} + \frac{|\mathbf{B_3}|^2}{8\pi}\right] + \nabla \cdot \left[\frac{c}{4\pi}\mathbf{E_3} \times \mathbf{B_3}\right] = -\mathbf{J}_{nl} \cdot \mathbf{E_3} \quad .$$

The non-linear current, J_{nl} includes the cross-products of the density and velocity disturbances of waves (1) and (2). The three-wave process saturates when the bracket in the rate equation in Section B vanishes, requiring

$$N_1(\mathbf{k_1})N_2(\mathbf{k_2}) = N_3(\mathbf{k_3})\{N_1(\mathbf{k_1}) + N_2(\mathbf{k_2})\} \quad .$$

Below we discuss two important examples.

a) Radio emission from Langmuir waves at the harmonic

For the process $L + L' \rightarrow t$ saturation occurs at a brightness temperature

$$T_{\max}(\mathbf{k_t}) \approx 2\frac{T_L(\mathbf{k_L})T_{L'}(\mathbf{k_L'})}{T_L(\mathbf{k_L}) + T_{L'}(\mathbf{k_L'})} \quad .$$

The emission is at $\omega_t \approx 2\omega_L \approx 2\omega_p$, giving rise to the factor of two.

Emission from ion acoustic waves

The coalescence of Langmuir waves and ion acoustic, $L + s \rightarrow t$, does not work efficiently in equilibrium plasma where ion acoustic waves are strongly damped. However, there are important non-equilibrium cases where ion acoustic waves reach a high energy density. If their phonon density $N_s \gg N_L$, they can make weak Langmuir waves observable at a brightness temperature of $T_t \lesssim T_L$. This may occur, for example, in currents (Section 5.3).

3.2 Radio Emission of Beams

Figure 2 shows the observation of a beam traversing the solar corona and escaping to interplanetary space. The beam excites Langmuir waves in the ambient plasma which are scattered into propagating transverse waves having roughly the local plasma frequency. Since the plasma frequency decreases with electron density, the emitted radiation has a frequency that also decreases with time. This drift of the peak emission is clearly visible in Fig. 2. Such rapidly drifting events are called type III bursts.

Radio signatures of electron beams (type III bursts) are usually observed to occur in groups. The number of elements per group increases with frequency, and above 300 MHz it can reach a hundred or more within a few minutes. Groups of type III

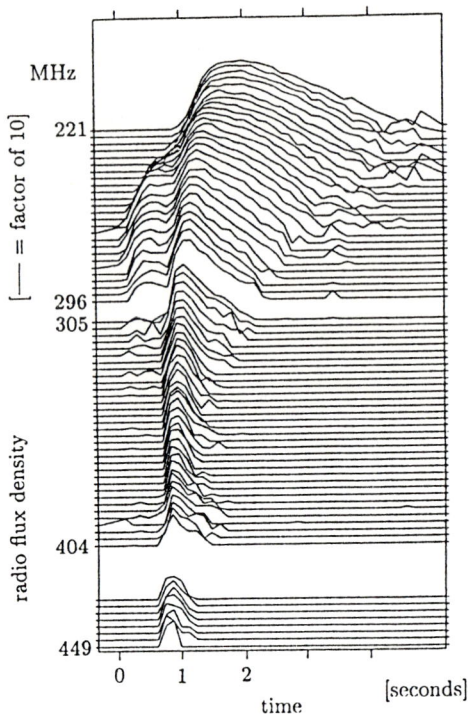

MHz

[——] = factor of 10]

radio flux density

221

296
305

404

449

0   1   2   [seconds]

time

Fig. 2: Radio emission of an electron beam traversing the solar corona. The flux density is shown on a logarithmic scale in different channels, the frequencies of which are indicated at the left (courtesy M.R. Perrenoud).

bursts are often associated with flares (seen in X-rays or Hα), particularly at high frequencies ($\gtrsim$ 300 MHz).

3.3 Drift of Electron Beams

The drift rate of the peak of emission is given by

$$\frac{d\nu}{dt} = \frac{d\nu}{dn_e}\frac{dn_e}{dh}\cos\phi\frac{ds}{dt} \quad .$$

The derivative dn_e/dh is given by the density model, and ds/dt corresponds to the beam velocity v_s. The angle ϕ is the inclination of the path s from the vertical. The relativistic motion of the source yields a correction,

$$dt = \frac{ds}{v_s} - \frac{ds\cos\theta}{v_{gr}} \quad ,$$

where θ is the angle between the beam propagation path and the direction to the observer. Note that the source is the excited background plasma which is not moving. Thus

$$\frac{d\nu}{dt} \approx -\frac{\nu v_s \cos\phi}{2H_n(1 - \beta\cos\theta)} \quad .$$

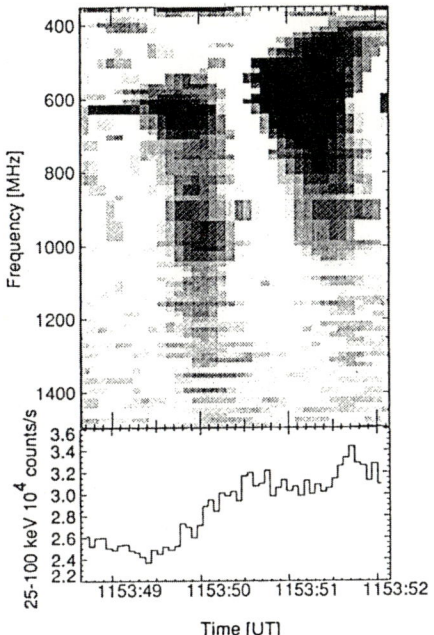

Fig. 3: Radio flux density of solar type III emission observed by a spectrometer at ETH Zurich. Two reversed slope bursts at high frequency ($\nu \gtrsim 600$ MHz) mark two down-going beams. The second one is accompanied by a simultaneous up-going beam ($\nu \lesssim 550$ MHz). *Bottom:* Hard X-ray counting rate measured by the BATSE experiment on the GRO satellite (Aschwanden et al. 1993).

The sign of the drift $\frac{d\nu}{dt}$ is determined by the sign of v_s, indicating whether the beam propagates up or down in the corona. The fraction of reverse drifting type III radio bursts (from low to high frequencies, indicating downward motion in the corona) increases with frequency and becomes dominant in microwaves ($\gtrsim 1$ GHz). Figure 3 shows a coincidence of the radio emission of down-going beams with peaks in hard X-ray emission. Moreover, the down-going beams are accompanied by upwards traveling beams. The separation of the two oppositely propagating type III bursts is at 600 MHz, corresponding to an electron density of about $4.5\ 10^9$ cm^{-3} (assuming fundamental emission). This seems to indicate the density in the acceleration region.

The electrons of one flare form a group of type III bursts. Thus, the acceleration in flares seems to be fragmented into many beams at low altitude, each producing a separate radio burst. The up-going fraction thereof later combines into one interplanetary type III burst observed also as a particle stream containing up to 10^{33} electrons at 1 AU. This is a small fraction ($\lesssim 10^{-3}$) of the typical numbers of energetic electrons deduced from the hard X-ray bremsstrahlung emission of flares. Apparently, only few of the accelerated electrons escape into interplanetary space. Nevertheless, even upward moving type III bursts and hard X-rays

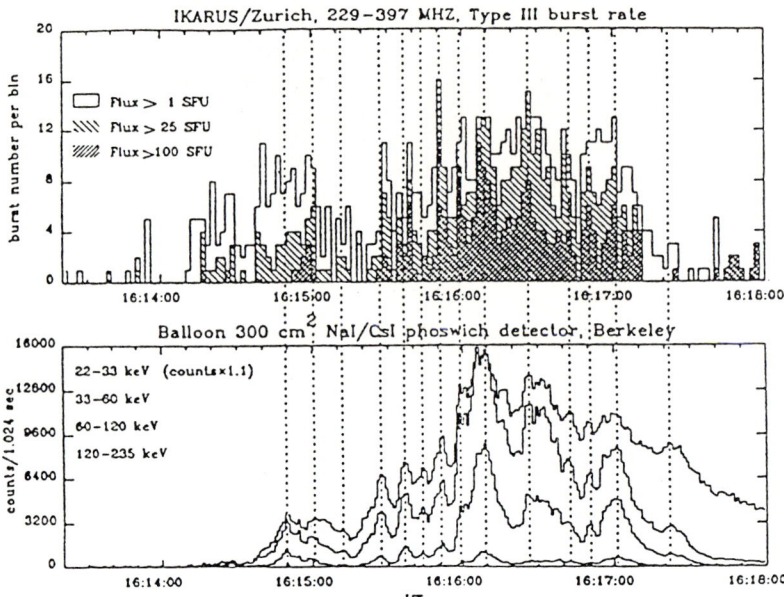

Fig. 4: Comparison of the number of beams visible as type III radio bursts with the flux of hard X-ray bremsstrahlung originating from electrons that have precipitated to high densities (Aschwanden et al. 1990).

sometimes correlate surprisingly well, although the two emissions do not involve the same electrons.

The flux density of type III bursts is the result of a beam-plasma instability growing exponentially in time. The emission process is coherent, indicating that the particles do not emit individually, but collectively. The brightness temperature can thus greatly exceed the particle temperature. As coherent radio emission is not proportional to particle number, its comparison with incoherent hard X-rays, which is proportional to particle number, requires special care. Often, one does not observe a one-to-one correspondence between individual hard X-ray peaks and down-going type III bursts. The down-going electrons presumably do not always develop a bump in velocity distribution and thus are not unstable toward growing electron plasma waves.

Figure 4 shows the *number* of type III bursts per unit time at different peak flux thresholds in a flare well observed in hard X-rays. The vertical dotted lines mark peaks in hard X-ray flux. They coincide with peaks in type III burst rate. The correlation of the rates of type III bursts and hard X-ray flux suggests that both upward and downward directed beams are generated. Much more importantly, the linear correlation suggests that the total number of electrons is roughly constant per type III producing beam. A flare seems to consist of hundreds of elementary processes of about equal size, each characterized by a type III burst, and each accelerating a similar number of hard X-ray producing electrons. In the case of Figure 4, the number is about 10^{32} electrons (>25 keV) per elementary event. It seems to be the result of fragmentation in the primary energy release of flares, consistent with the theoretical notion of small conversion sites.

Since beams of charged particles follow magnetic field lines, imaging radio telescopes may be used to trace the geometry of the field. Figure 5 shows an observation of an electron beam at three frequencies. The centroid of the sources outlines a loop having a radius that exceeds the photospheric radius of the Sun.

More frequently beam sources are observed to travel into interplanetary space. They can be used to map the interplanetary magnetic geometry.

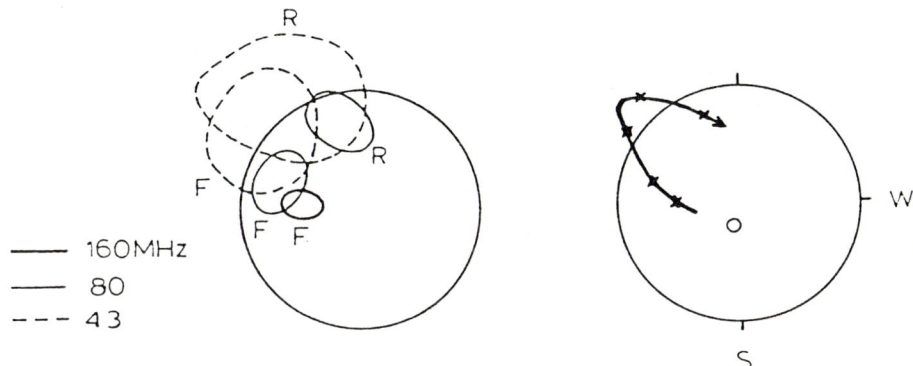

Fig. 5: *Left:* The half-power contours for three frequencies and at peak time trace the path of an electron beam in the solar corona (adapted from Suzuki 1978). The first source at a given frequency (upgoing) is labeled F, the second source is downgoing (R). *Right:* The observed centroids of each frequency (x) are connected to outline the path of the exciter. The photospheric disk is indicated for reference, and the site of the flare in H_α is shown by an open circle.

4 Trapped Particles

4.1 Conservation of Magnetic Moment

A collisionless particle experiencing a Lorentz force has the following equation of motion:

$$\frac{\partial(m\gamma\mathbf{v})}{\partial t} = \frac{q}{c}(\mathbf{v} \times \mathbf{B}) \quad .$$

In an inhomogeneous field the Lorentz force can be written as two components

$$\mathbf{F}_{L,r} = \frac{q}{c}(\mathbf{v}_\perp \times \mathbf{B}_z), \quad \mathbf{F}_{L,z} = \frac{q}{c}(\mathbf{v}_\perp \times \mathbf{B}_r) \quad ,$$

where the z-direction is parallel to the magnetic field line along which the particle moves. $\mathbf{B}_r$ is the radial field component at the particle orbit. The first component forces the particle into a gyrating orbit, the second component, parallel to the direction of cylindrical symmetry, constitutes a force in the direction opposite to which the field converges. This force slows down the parallel motion of a particle moving toward a converging field and eventually reflects the particle. The parallel force becomes in cylindrical coordinates

$$\frac{1}{r}\frac{\partial}{\partial r}(rB_r) + \frac{\partial B_z}{\partial z} = 0$$

or approximately

$$B_r \approx -\frac{1}{2}R\frac{\partial B_z}{\partial z} \quad ,$$

where R is the gyroradius. Let us define the magnetic moment

$$\mu := \frac{\pi R^2}{c} < I > = \frac{\frac{1}{2}mv_\perp^2}{B} \quad .$$

Thus

$$F_{L,z} = -\mu\frac{\partial B_z}{\partial z} \quad ,$$

The equation of motion in parallel direction is

$$mv_z\dot{v}_z = -\mu\frac{\partial B}{\partial z}\frac{dz}{dt} = -\mu\frac{dB}{dt} \quad .$$

The particle energy is conserved, since the Lorentz force is perpendicular to the velocity.

$$\frac{d}{dt}(\frac{1}{2}mv_z^2 + \frac{1}{2}mv_\perp^2) = 0$$

Inserting, one obtains

$$-\mu\frac{dB}{dt} + \frac{d}{dt}(\mu B) = 0 \qquad \text{Since } dB/dt = 0 , \quad \frac{d\mu}{dt} = 0$$

In a stationary magnetic field the magnetic moment is conserved.

4.2 Magnetic Trapping

A particle is reflected when all its energy is in transverse motion. The conservation of the magnetic moment requires that the magnetic field at the mirror point is

$$B_{mp} := B\left(\frac{v}{v_\perp}\right)^2 \quad .$$

at any given place along the loop with magnetic field B. If $B_{mp} \geq B_{tr}$, the field in the transition region, the particle penetrates the chromosphere and is lost from the trap due to collisions. The pitch angle of the particle orbit is defined by

$$\alpha := \arcsin(\frac{v_\perp}{v}) \quad .$$

If for example at the top of a loop

$$\alpha_{top} \leq \arcsin(B_{top}/B_{tr})^{1/2} \quad ,$$

the particle is trapped. The critical angle is

$$\alpha_c(B) = \arcsin \sqrt{\frac{B}{B_{\text{tr}}}}$$

for any location along the loop. Particles having $\alpha > \alpha_c$ are trapped, but particles with $\alpha < \alpha_c$ precipitate into the chromosphere and are missing in the velocity distribution of the trapped population in the corona, thus forming a loss-cone.

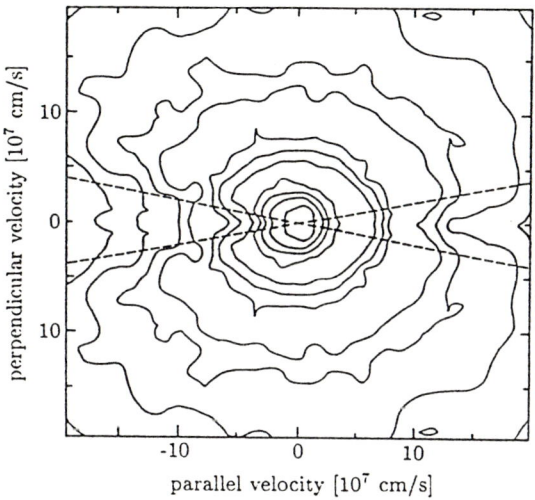

Fig. 6: Contour plot of the velocity distribution of protons trapped in the Earth's magnetosphere. The theoretical loss-cone angle is indicated by dashed lines (adapted from Boström et al. 1987).

4.3 Hot Plasma

In a loss-cone distribution (Fig. 6) particles near the parallel velocity axis are under-populated. This deviation from the equilibrium distribution constitutes free energy that can drive a maser instability. A hot plasma with a non-thermal velocity distribution is described by the Vlasov equation

$$\frac{\partial f}{\partial t} + \mathbf{v}\frac{\partial f}{\partial \mathbf{x}} + \frac{q}{m}\left(\mathbf{E} + \frac{1}{c}\mathbf{v} \times \mathbf{B}\right) \cdot \frac{\partial f}{\partial \mathbf{v}} = 0 \quad .$$

We assume a disturbance of the form

$$f(\mathbf{x}, \mathbf{v}, t) = f_0(\mathbf{v}) + f_1(\mathbf{x}, \mathbf{v}, t)$$

with

$$|f_1| \ll |f_0| \quad \text{for all} \quad \mathbf{x}, \mathbf{v}, t \quad .$$

Then the Vlasov equation becomes in first order

$$\left[\frac{\partial}{\partial t} + \mathbf{v}\cdot\nabla_{\mathbf{x}} + \frac{q}{mc}(\mathbf{v}\times\mathbf{B}_0)\cdot\nabla_{\mathbf{v}}\right] f_1 = -\frac{q}{m}(\mathbf{E}_1 + \frac{1}{c}\mathbf{v}\times\mathbf{B}_1)\cdot\nabla_{\mathbf{v}}f_0 \quad.$$

The formal solution is

$$f_1(\mathbf{x},\mathbf{v},t) = \frac{q}{m}\int_{-\infty}^{t} dt'(\mathbf{E}_1 + \frac{1}{c}\mathbf{v}'\times\mathbf{B}_1)\cdot\nabla_{\mathbf{v}'}f_0(\mathbf{v}') \quad.$$

The integral is along the undisturbed orbit of the particles. The Fourier transform can be carried out by assuming a wave of the form

$$f_1(\mathbf{x},\mathbf{v},t) = \bar{f}_1(\mathbf{v})\exp\left[i(k_z z + k_x x - \omega t)\right] \quad.$$

Similar to cold plasma, one finds another relation between $\mathbf{E}$ and f_1 through Ohm's law and Ampère's law. The dispersion relation becomes (e.g. for x-mode)

$$1 - \frac{c^2 k^2}{\omega^2} - \frac{2(\omega_p^e)^2}{\omega^2 n_e}\sum_l \int_{\mathcal{L}} d^3 v \left(\frac{l\Omega_e}{\gamma}\frac{\partial f_0}{\partial v_\perp^2} + k_z v_z \frac{\partial f_0}{\partial v_z^2}\right)\frac{v_\perp^2 J_l'(k_\perp R)^2}{k_z v_z + l\Omega_e/\gamma - \omega} = 0 \quad.$$

For the integration through the singularity at $\omega = k_z v_z + l\Omega_e/\gamma$ we use the Landau prescription indicated by the path integral $\mathcal{L}$. For $\omega_p^2 \ll \Omega_e^2$, the solutions have their real parts near the harmonics of the electron gyrofrequency:

$$\omega_l^r \approx l\Omega_e \quad l = 1, 2, ...$$

The growth rates are

$$\gamma_l = \int d^3 v A_l(\mathbf{v},\mathbf{k})\delta(k_z v_z + \frac{l\Omega_e}{\gamma} - \omega_l^r)[\frac{l\Omega_e}{\gamma}\frac{\partial}{\partial v_\perp} + k_z v_\perp\frac{\partial}{\partial v_z}]f_0(\mathbf{v})$$

where

$$A_l^z \approx \frac{\pi(\omega_p^e)^2 v_\perp}{4 n_e \omega_r}(\frac{l J_l}{R k_\perp})^2 \quad (l \gg 1, \; z-\text{mode})$$

The z-mode includes waves that are nearly perpendicular to the magnetic field. It is on the same branch of eigenmodes as the Langmuir waves for parallel propagation or unmagnetized plasma.

$$A_1^o \approx \frac{\pi(\omega_p^e)^2 v_\perp}{8 n_e \omega_r}(\frac{v_z}{c})^2 \quad (l = 2, \; o-\text{mode})$$

$$A_1^x \approx \frac{\pi(\omega_p^e)^2 v_\perp}{8 n_e \omega_r} \quad (l = 1, \; x-\text{mode})$$

(Wu & Lee 1979; Winglee & Dulk 1986). The above three loss-cone instabilities produce high-frequency waves at nearly perpendicular angles to the magnetic field.

4.4 Resonance Condition

The singularities of the dispersion relation define the waves that grow. They describe a resonance between the wave, Doppler shifted to the rest frame of a particle with velocity v_z, and the harmonic l of the (relativistic) gyration frequency Ω_e/γ of the particle.

$$\omega - k_z v_z = \frac{l\Omega_e}{\gamma}$$

In the semi-relativistic approximation ($\gamma \approx 1$) it becomes

$$v_z - \frac{v^2}{2c^2}\frac{l\Omega_e}{k}\frac{k}{k_z} + \frac{l\Omega_e}{k_z} - \frac{\omega}{k_z} \simeq 0 \quad ,$$

which is the equation of a circle. It is the location in velocity space of rapid energy exchange between the wave, given by ω and $\mathbf{k}$, and particles with velocity $\mathbf{v}$.

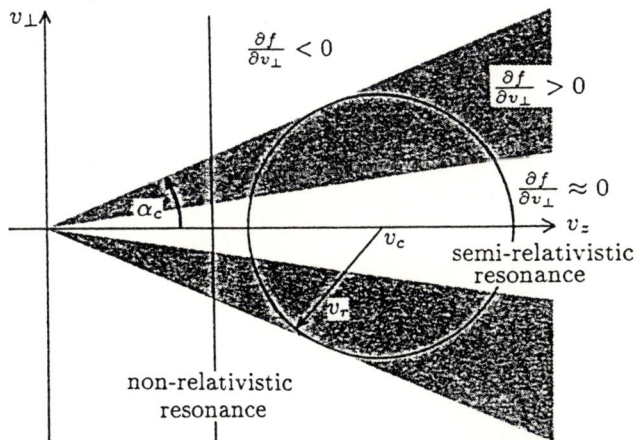

Fig. 7: Loss-cone (α_c) with ramp (shaded) velocity distribution of trapped particles. Curves of gyroresonance for two approximations are indicated.

Figure 7 shows schematically a loss-cone velocity distribution and curves of resonance on which electrons strongly interact with waves of a given l. Depending on whether the particles lose or gain energy in these interactions, the waves grow or are damped. It is given by the sign of the growth rate and determined by the sign of the derivatives in velocity space. The resonance circle (semi-relativistic) for a given v_c and maximum wave growth is drawn in Fig.7.

For further information on stellar observations and interpretations of loss-cone emissions, the reader is referred to the reviews by e.g. Kuijpers (1983), Bastian (1990) and Lang (1994).

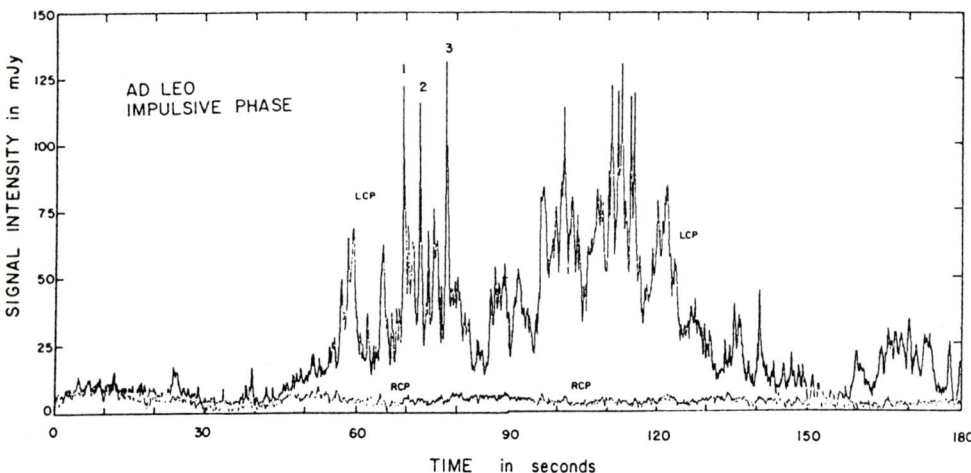

Fig. 8: The figure presents the radio emission of a flare on AD Leonis as an example. The emission is highly polarized characteristical for a loss-cone instability directly driving the o-mode or x-mode. Since one mode at one harmonic of the gyrofrequency grows fastest over many e-folding times, the emission is predominantly in this mode (Lang 1994).

5 Currents

5.1 Electric Conductivity

In a plasma carrying a stationary current, electrons and ions are in force equilibrium. In the rest frame of the electrons the ions form a beam with velocity V_d, and the equilibrium between the electric force on the ions and their friction with the electrons requires

$$q_i E = \frac{m_i V_d}{t_{\text{coll}}^{i,e}} \quad ,$$

where $t_{\text{coll}}^{i,e}$ is the ion-electron collision time. The theory of far-field interactions (Fokker-Planck equation) yields

$$q_i E = m_i (1 + \frac{m_i}{m_e}) \frac{A_d}{2(v_{te})^2} G(v/d_{te}) \quad .$$

A_d is a constant, and $G(v/v_{te})$ is a modified error function defined as

$$A_d := \frac{8\pi n_F q_T^2 q_F^2 \ln \Lambda}{(m_T)^2} \quad ,$$

$$G(x) := \frac{\Psi(x) - x\Psi'(x)}{x^2} \quad ,$$

$$\Psi(x) := \sqrt{\frac{2}{\pi}} \int_0^x \exp(-\frac{y^2}{2})dy \quad .$$

The indices F and T refer to the field particles and the test particle, respectively (Benz 1993).

For $v_z \ll v_{te}$,

$$\frac{1}{t_s^{i,e}} = \sqrt{\frac{2}{\pi}} \frac{m_e}{m_i} \frac{\ln \Lambda}{\Lambda} \omega_p^e \quad ,$$

and we can define a conductivity

$$\sigma := \frac{q_i \, n_i \, V_d}{E} = \frac{(\omega_p^i)^2 \, t_s^{i,e}}{4\pi} \approx 3.22 \cdot 10^6 \, T_e^{3/2} \frac{20}{\ln \Lambda} \quad [\text{s}^{-1}] \quad .$$

5.2 Runaway Electrons

Electrons have a much wider spread in velocity (cf. Fig. 9). Their friction with the ions depends on their velocity, decreasing the faster the electron. Therefore a velocity v_r exists at which friction equalizes the electric force of a single electron,

$$eE = \frac{m_e v_r}{t_{\text{coll}}^{e,i}(v_r)} \quad ,$$

where

$$v_r = \left(\frac{m_e A_d}{2eE} \right)^{1/2} \quad .$$

Electrons with $v > v_r$ accelerate and run away.

Runaway acceleration is an effective way to produce a large number of non-thermal electrons in coronal current sheets, possibly occurring in flares. The rate of runaway electrons, however, cannot exceed the current. At maximum, all the current is in the runaway electrons. Thus runaway acceleration rate is limited by

$$\dot{N} = \dot{n} \, V \leq \frac{2dwJ}{e} \approx \frac{cwB}{4\pi e} \quad [\text{electrons s}^{-1}] \quad .$$

5.3 Current Instabilities

If the current density increases, the drifting populations of ions and electrons form a beam situation that may become unstable. The nature of this instability and the type of growing waves both depend on the ratio of electron and ion temperatures, T_e/T_i. For currents parallel to the magnetic field the following three instabilities, given below with their thresholds, are the main contenders:

A. *Ion Cyclotron Instability*

$$V_d \gtrsim 15 \frac{T_i}{T_e} v_{ti}$$

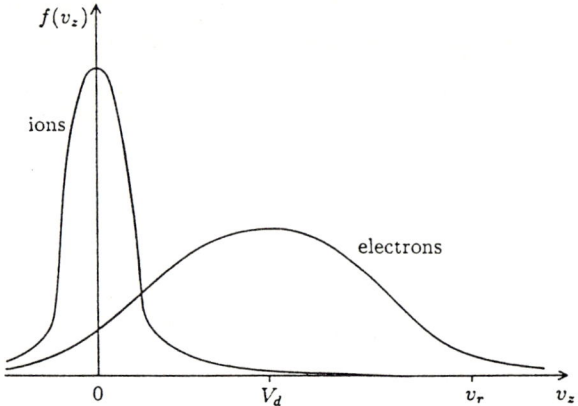

Fig. 9: The electron and ion velocity distributions are separated by the drift velocity V_d due to a parallel electric field.

B. *Buneman Instability*

$$V_d \gtrsim 1.7 \, (v_{te} + v_{ti})$$

C. *Ion Acoustic Instability*

$$V_d \gtrsim 2c_{is}, \quad \text{for } T_e > T_i,$$

where $c_{is} := \sqrt{\frac{k_B T_e}{m_i}}$ is the ion sound velocity (cf. Papadopoulos 1977).

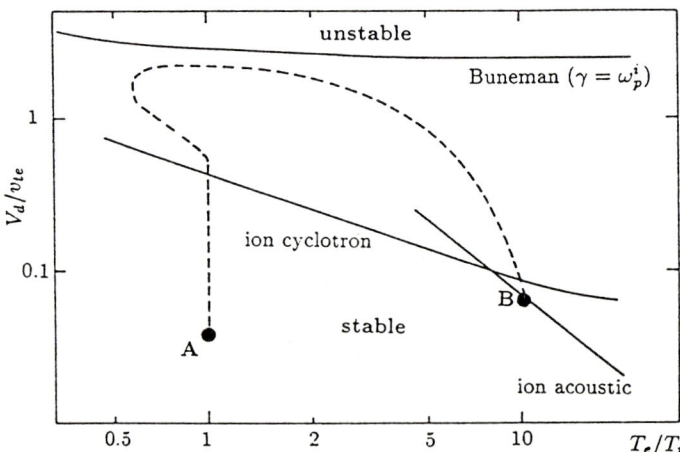

Fig. 10: The threshold drift velocity V_d of a parallel current is presented for the three instabilities discussed in the text. It depends strongly on the temperature ratio of electrons and ions. A likely evolutionary scenario for a current system with increasing V_d is indicated by the dashed curve from A to B (Benz 1993).

The occurrence of instability effects the current sheet in several ways:

1. The energy loss to the waves causes additional friction and thus enhanced electric resistivity.

2. Since the current density changes slowly, the enhanced resistivity causes an increase in the parallel electric field by Ohm's law. This may accelerate additional runaway electrons.

3. Enhanced resistivity increases Ohmic heating.

4. The enhanced resistivity may cause the current sheet to expand, reducing the current density and quenching the instability.

The dynamics of unstable currents in space is not well-known. A series of short pulses in accelerated electrons and low-frequency waves is most likely.

5.4 Radio Signatures of Unstable Currents

Narrowband, short peaks of metric radio bursts from the Sun are well-known to accompany the evolution of active regions (Fig. 11). They are not connected to flares, but may occur for hours and days. Noise storms or type I bursts, as they are called by the specialists, may be radio signatures of current instabilities in reconnecting current sheets. One of the still speculative models (Benz and Wentzel, 1981) suggests the following scenario:

1. New emerging magnetic flux pushes against preexisting magnetic fields and forms a current sheet.

2. As the current increases and becomes unstable, ion acoustic waves grow and electrons are accelerated.

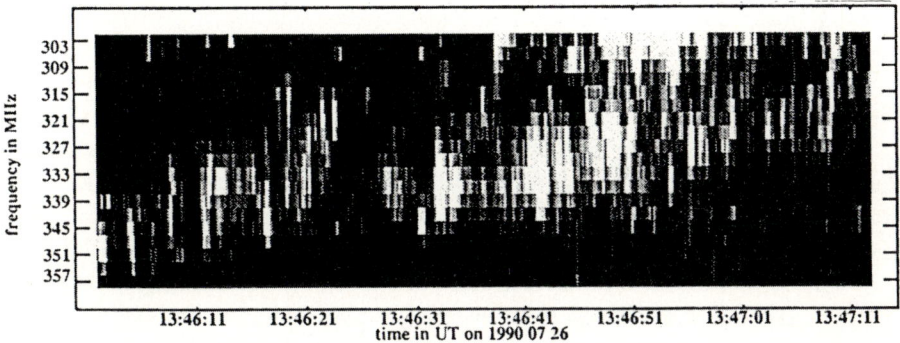

Fig. 11: A so-called 'chain' of hundreds of type I bursts traverses a spectral band of a radio spectrometer (PHOENIX instrument of ETH Zurich). White is enhanced emission.

3. The electrons form beams or traps and excite high-frequency electrostatic waves (Langmuir waves, if parallel to **B**).

4. The ion sound waves and Langmuir waves scatter into radio emission.

5. As the current sheet expands, the ion acoustic waves disappear and the radio emission is a short, narrowband peak emitted near the local plasma frequency.

This model is currently being tested using soft X-ray images.

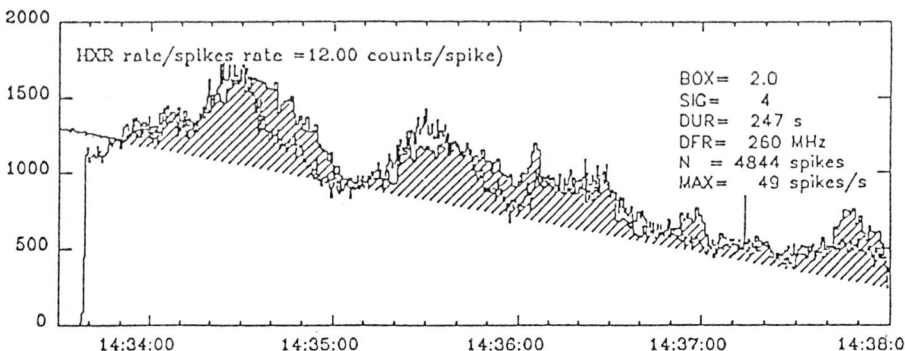

Fig. 12: Correlation of spike rate (shaded) with hard X-ray count rate. A total of 4844 spikes have been counted with a maximum rate of 49 spikes/s (Aschwanden & Güdel 1992).

A different kind of solar narrowband spikes also occur at decimetric and even centimetric wavelengths, but only during flares and at a much higher rate. Since their duration decreases with wavelength and is only a few tens of millisecond at 10 cm, they are often called 'millisecond spikes'. They may represent an even higher fragmentation in the energy release region than the beam radiation (type III). Generally their polarization is high. Spikes correlate with hard X-rays, suggesting a close relation between the spike source and the flare energy release. The correlation between spikes and hard X-rays improves when the spike rate is used (and not the spike flux density, Benz & Kane 1986). Furthermore, the duration of the spike activity is linearly correlated with the duration of the associated hard X-ray peak. A correlation coefficient of 0.78 for 84 events indicating a very high level of confidence demonstrates that the two emissions are physically related (Güdel et al. 1991).

The correlation between spikes and hard X-rays tends to be better than that of the type III bursts. Out of 84 spike events (groups) 40 had unambiguous hard X-ray correlation, and only 18 had better hard X-ray correlation with type III bursts. Figure 13 shows an example of the 26 cases with about equally good correlations between the three emissions.

A detailed analysis of spike *rate* and hard X-rays was recently made (Aschwanden & Güdel 1992), showing close correlation in 22% of all cases, and intermittent correlation in 52% of the cases. The ratio of hard X-ray counts per spike was found to be between 0.1 and 10, where one count by the HXRBS instrument on the SMM satellite corresponds to about an electron beam of 10^{24} erg hitting a thick

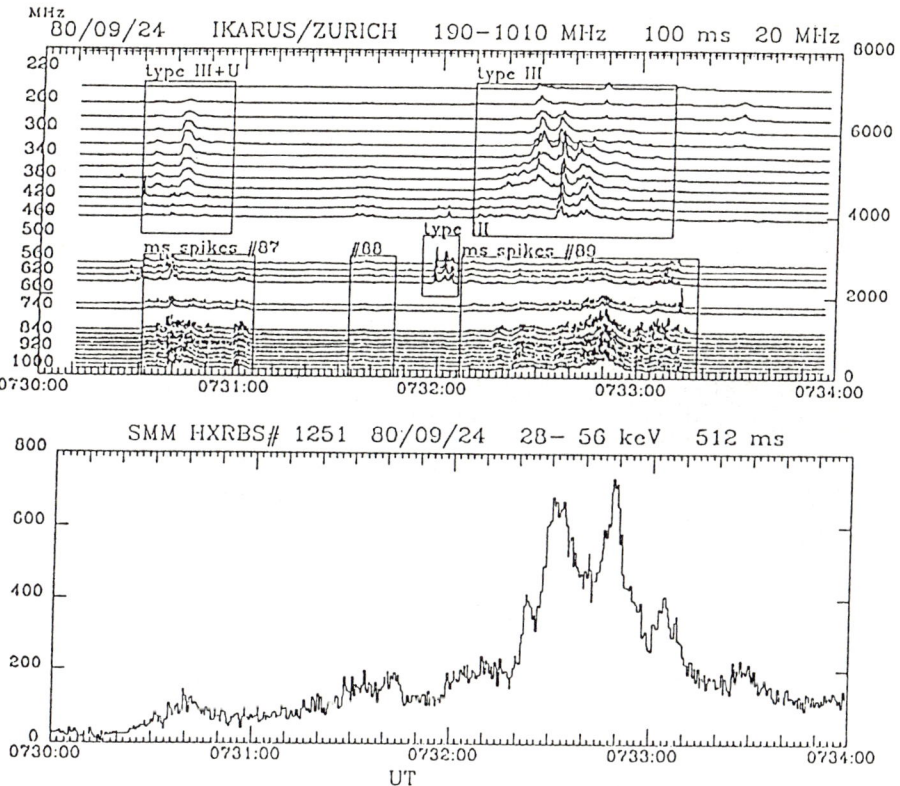

Fig. 13: Radio flux (top) in several channels (frequency in MHz given at left) versus time, and hard X-ray counts for comparison. Type III and spikes (frames #87, #88, and #89) are indicated (Güdel et al. 1991).

target. In Figure 12 an uncorrelated hard X-ray background has been subtracted. The ratio between spike rate and hard X-ray flux is then nearly constant.

Surprisingly, there is an apparent delay of the spike rate in relation to the associated hard X-ray peak. Other events showed delays between 0 and 8 seconds. The delay tends to increase with higher hard X-ray flux. The delay may simply reflect that the there is an important difference in the emission processes involved. The hard X-ray emission, being bremsstrahlung of individual electrons, is an incoherent process, strictly proportional to the number of electrons hitting the dense target region. Spikes, having brightness temperatures sometimes exceeding 10^{15}, are coherent radiations. Their production may be controlled by ambient plasma parameters or energetic particle density, and thus vary during the flare.

6 Collisionless Shocks

Figure 14 shows an example of a slowly drifting metric solar radio burst. In standard coronal models the drift rate corresponds to about the Alfvén velocity. The emission consists of two components: an intense, patchy, sometimes extremely narrow band and type III-like extension shooting out of the band and having a fast drift rate. These phenomena are generally agreed to be caused by the passage of a flare induced shock through the corona. It is called a type II radio burst.

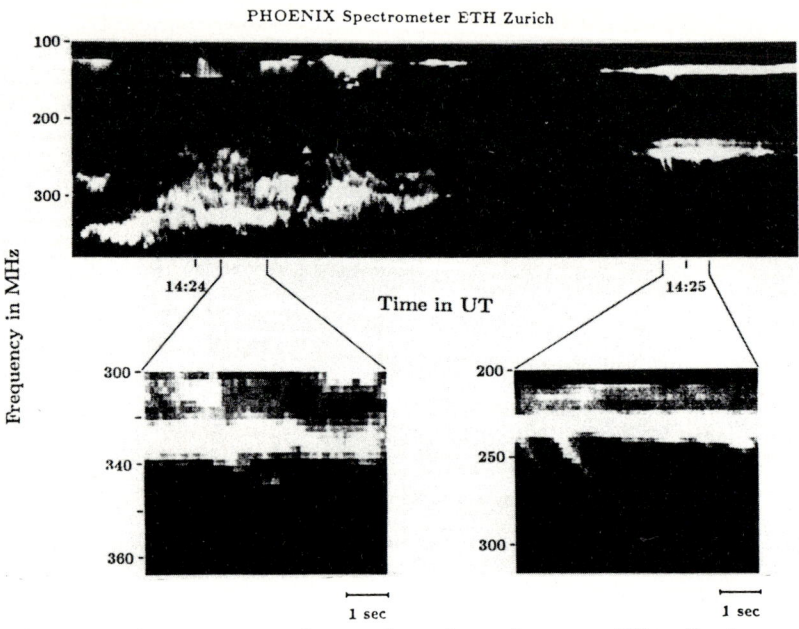

Fig. 14: Spectrogram of a section of a solar type III radio burst. The bright features indicate enhanced radio flux composed of a narrow band and type III-like extensions.

The acceleration processes on collisionless shocks strongly change from shocks hitting the upstream magnetic field in perpendicular or parallel direction. At the moment it is not clear which situation applies to the radio emission. To give an example, we follow the models by Holman & Pesses (1983) and Benz & Thejappa (1987) assuming a quasi-perpendicular shock.

6.1 Electron Acceleration

As in all models electrons are accelerated at the shock and form beams causing the jet-like fast drift radio bursts. In quasi-perpendicular shocks electrons are accelerated by single reflection (shock drift acceleration). This may be a magnetic mirroring similar to traps. However, for a correct treatment the electric field in the upstream flow, $E_1 = -\frac{1}{c} V_1 \times B_1$ (index $_1$ denotes upstream values), must be eliminated by a suitable coordinate transformation.

In the de Hoffmann-Teller frame of reference the shock is at rest and $V_1 \parallel B_1$, so the upstream electric field cancels. In this frame the upstream inflow velocity is

$$V_1^{HT} = \frac{V_1}{\cos\theta_1} \quad ,$$

where θ_1 is the angle of the upstream magnetic field in the normal incidence frame. Particles with a pitch angle

$$\alpha > \alpha_c := \arcsin(B_1/B_2)^{1/2}$$

are reflected with the inverted velocity V_1^{HT}. When this velocity is transformed back to the laboratory frame, it reaches a high value for θ_1 close to 90°.

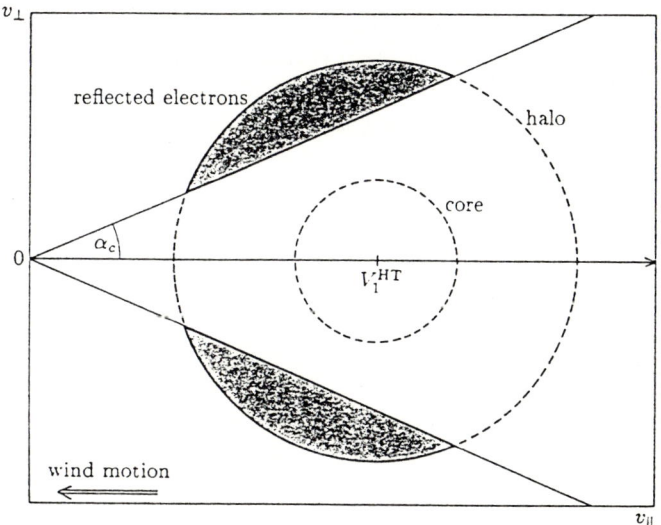

Fig. 15: Schematic velocity distribution of upstream electrons in the de Hoffmann-Teller frame of reference. The coordinates are chosen parallel and perpendicular to the upstream magnetic field, respectively. The initial distribution (not shown) is assumed to have two Maxwellian populations forming a core and a halo. The reflected particles are shaded. The reflection is only possible for incoming particles with large transverse velocities. This practically excludes the ions. The reflected and accelerated electrons have a displaced loss-cone distribution (Fig. 5) prone to instabilities.

1.   A loss-cone instability will occur and fill up the distribution near the parallel axis. This is a localized process and gives rise to radio emission by scattering z-mode waves (or similar) into harmonic radio waves or, scattering on low-frequency waves (such as lower hybrid), into fundamental radio waves.

2.   The resulting electrons will continue as a beam and produce Langmuir waves by the bump-on-tail instability.

The radio emission of shocks thus consists of two components as observed (Fig. 14). The loss-cone driven emission moves slowly with the speed of the shock. It is much stronger and more frequent than the type III - like ejections produced by the beam driven emission. Both radio emissions yield information on electron acceleration and on plasma waves in the upstream medium.

7 Conclusions

In late-type stars, where hard X-ray emission has not yet been detected, radio observations provide the most direct information on non-thermal electrons in flares and in the corona. The identification of the radio emission process by gyrosynchrotron mechanism in weakly polarized, incoherent radio emission allows, for example, to estimate the high-energy component of the flare and compare it to the total flare energy. The fraction of primary energy released into energetic electrons so determined appears to be large and similar to solar flares. The weakly polarized radio emission may serve as a proxy for hard X-ray signatures of relativistic electrons in stars.

Coherent radio emissions in the decimetric wavelength range of the Sun and stars yield informations on the region of energy release in flares, on electron acceleration, on ejected particle beams and on particle traps, and on shock waves propagating through the corona. Contrary to incoherent radio emission, the driving electrons do not have to be mildly relativistic to produce observable radiation. A non-thermal distribution in velocity space, such as a beam or a loss-cone, is sufficient. Furthermore, radio emissions give the source density (for plasma emission of electron beams) or the magnetic field (gyroresonance emission of trapped electrons having a loss-cone velocity distribution). The frequency range from 0.3 to 3 GHz corresponds to a plasma density of 10^9 to 10^11 cm^{-3} or 100 to 1000 gauss, respectively.

Until now, most the useful coherent emissions have been solar type III bursts, signatures of electron beams in the solar corona. The simple reason is that we understand (at least partially) their emission process. Since each burst requires an individual beam, the large number of type III bursts observed in flares and subflares suggests fragmentation of the flare energy release. Furthermore, since many type III are observed with no hard X-ray and even Hα counterparts, electron acceleration seems to take place also in smaller events and at higher altitude, where hard X-ray emission is less effective.

The narrowband millisecond spikes are very interesting for their correlation with hard X-rays, suggesting a close association with energy release. However, the spikes are still of less value, since their emission process is unknown. Nevertheless, some similarity with type III bursts in the association with hard X-rays may be taken as suggestion of an even higher fragmentation of the flare energy release. This remains a hypothesis until more is known about the spike emission process.

Loss-cone emissions are potentially more efficient, having been observed to convert up to 0.1% of the total non-thermal particle energy into radiation in the Earth's magnetosphere. They are most important in coronae with a strong magnetic field, and in particular for a high ratio of the gyrofrequency to the plasma frequency.

This may be the case in active dwarf M stars, which are very productive in producing highly polarized and extremely intense radio flares.

References

Aschwanden, M.J., Benz, A.O., and Schwartz, R.A.: 1993, 'The Timing of Electron Beam Signatures in Hard X-Ray and Radio Solar Flare Observations by BATSE/CGRO and PHOENIX', *Astrophys. J.* **417**, 790.

Aschwanden, M.J., Benz, A.O., Schwartz, R.A., Lin, R.P., Pelling, R.M., and Stehling, W.: 1990, 'Flare Fragmentation and Type III Productivity in the 1980 June 27 Flare', *Solar Phys.* **130**, 39.

Aschwanden M.J. and Güdel M.: 1992, 'The Coevolution of Decimetric Millisecond Spikes and Hard X-Ray Emission during Solar Flares', *Astrophys. J.* **401**, 736.

Bastian, T.S.: 1990, 'Radio Emission from Flare Stars', *Solar Phys.* **130**, 265.

Benz, A.O.: 1986, 'Millisecond Radio Spikes', *Solar Phys.* **104**, 99.

Benz, A.O.: 1993, *Plasma Astrophysics (Kinetic Processes in Solar and Stellar Coronae)*, Kluwer Acad. Publ., Dordrecht, Holland.

Benz A.O. and Kane S.R.: 'Electron Acceleration in Flares inferred from Decimetric Radio and Hard X-Ray Emissions', *Solar Phys.* **104**, 179.

Benz, A.O. and Thejappa, G.: 1988, 'Radio Emission of Coronal Shock Waves', *Astron. Astrophys.* **202**, 267.

Benz, A.O. and Wentzel, D.G.: 1981, 'Coronal Evolution and Solar Type I Radio Bursts: An Ion-acoustic Wave Model', *Astron. Astrophys.* **94**, 100.

Benz, A.O. and Güdel, M.: 1994, 'The Soft X-Ray/Microwave Ratio of Solar and Stellar Flares and Coronae', *Astron. Astrophys.* **285**, 621.

Boström, R., Koskinnen, H., and Holback, B.: 1987, 'Low-Frequency Waves and Small-Scale Solitary Structures Observed by Viking', *ESA SP* **275**, 185.

Gary, D.E., Linsky, J.L.: 1981, 'First Detection of Nonflare Microwave Emission from the Coronae of Single Late-type Dwarf Stars', *Astrophys. J.* **250**, 284.

Godfrey, B.B., Shananhan, W.R., and Thode, L.E.: 1975, 'Linear Theory of Cold Relativistic Beam Propagating along an External Magnetic Field', *Phys. Fluids* **18**, 346.

Güdel M., Aschwanden M.J., Benz A.O.: 1991, 'The Association of Solar Millisecond Radio Spike with Hard X-Ray Emission', *Astron. Astrophys.* **251**, 285.

Holman, G.D. and Pesses, M. E.: 1983, 'Solar Type II Radio Emission and the Shock Drift Acceleration of Electrons', *Astrophys. J.* **267**, 837.

Kuijpers, J.: 1989, 'Radio Emission from Stellar Flares', *Solar Phys.* **121**, 163.

Lang, K.R.: 1994, 'Radio Evidence for Nonthermal Particle Acceleration on Stars of Late Spectral Type', *Astrophys. J. Suppl. Ser.* **90**, 753.

Melrose, D.B.: 1986, *Instabilities in Space and Laboratory Plasmas*, Cambridge University Press, Cambridge, England.

Narain, U. and Ulmschneider, P.: 1991, 'Chromospheric and Coronal Heating Mechanisms', *Space Sci. Rev.* **57**, 199.

Pallavicini, R.: 1987, 'Solar and Stellar Coronae', *Lecture Notes in Physics* **292**, 98.

Papadopoulos, K.: 1977, 'A Review of Anomalous Resistivity for the Ionosphere', *Rev. Geophys. Sp. Sci.* **15**, 113.

Suzuki, S: 1978, 'On the Coronal Source Regions of U-Bursts', *Solar Phys.*, **57**, 415.

Winglee, R.M. and Dulk, G.A.: 1986, 'The Electron-Cyclotron Maser Instability as a Source of Plasma Radiation', *Astrophys. J.* **307**, 808.

Wu, C.S. and Lee, L.C.: 1979, 'A theory of the Terrestrial Kilometric Radiation', *Astrophys. J.* **230**, 621.

Spectroscopic Diagnostics of Astrophysical Plasmas

Roberto Pallavicini

Osservatorio Astrofisico di Arcetri, Florence, Italy

Abstract: This paper summarises a series of lectures on spectroscopic diagnostics of astrophysical plasmas given to graduate students. It focusses on optically thin thermal plasmas and on X-ray spectroscopy. The basic diagnostic techniques are first discussed in a general way and then applied to specific astrophysical cases, including the solar corona, stellar coronae, supernova remnants and clusters of galaxies. The limitations of current spectroscopic data in terms of both resolution and sensitivity are emphasised, and future prospects are briefly mentioned. These lectures are intended as an introduction to the field rather than as a survey of the most recent literature. The emphasis therefore is on basic concepts and the cited bibliography is limited mainly to review papers where further details and references to the original works can easily be found.

1. Introduction

At all wavelengths throughout the electromagnetic spectrum, spectroscopy is the best way to derive physical information about cosmic sources. By observing lines and continua with adequate resolution and sensitivity, it is possible to extract quantitites such as temperatures, densities, elemental abundances, magnetic field intensities, and flow velocities. The techniques used in different parts of the spectrum, or for different types of sources, differ from one case to the other, depending on the physical nature of the source and the quality of the available data. In all cases, the final result is a much deeper understanding of the physical properties of the objects under investigation.

The subject of plasma spectroscopy in astrophysics is so vast and complex that it is impossible to cover it in a comprehensive way in just a few lectures. Some selection needs to be made. I have chosen to focus on hot optically thin thermal plasmas, i.e. on plasmas which radiate by thermal processes, have typical temperatures in the range $\approx 10^6$ to 10^8 K, and have sufficiently low densities as to be transparent to their own radiation. These plasmas emit most of their radiation in the soft X-ray part of the spectrum, between ≈ 0.1 and 10 keV (i.e. over the wavelength range ≈ 1 to 100 Å). The physical conditions in optically thin thermal plasmas can differ enormously from one type of source to the other: the density can be as low as $\approx 10^{-3}$ cm^{-3} in the teneous gas pervading the space between galaxies in clusters of galaxies and as high as $\approx 10^{12}$ cm^{-3} in solar flares. Examples of hot thin plasmas are solar and stellar coronae, supernova remnants (SNR's), the hot interstellar medium (ISM), and the hot gas in clusters of galaxies and in galactic halos.

In these lectures I will summarise the basic diagnostic techniques for hot thin plasmas and the current understanding of cosmic sources as has been derived from X-ray spectroscopy. I will limit myself to the most important aspects of the subject while referring to the literature for details and complete lists of references. These lectures are organised as follow. First, I will discuss radiation processes from hot thin plasmas and spectroscopic diagnostic techniques in a general way. Then, I will present applications to individual classes of cosmic sources, with emphasis on the solar corona, stellar coronae, SNR's and clusters of galaxies. Non-thermal and/or optically thick sources (such as isolated white dwarfs, accretion-powered X-ray binaries and active galactic nuclei) will be excluded from these lectures.

2. Radiation Processes in Hot Thin Plasmas

2.1 The coronal model

In contrast to the more familiar case of high density plasmas like for instance stellar photospheres, high temperature low density plasmas are not in Local Thermodynamic Equilibrium (LTE). This means that each microscopic process is not balanced by its reverse (as would occur under LTE conditions by virtue of the *principle of detailed balance*). Instead, only a statistical balance of all relevant processes exists under steady conditions. For instance, collisional ionizations are balanced by radiative recombinations, rather then by 3-body recombinations, and photoionization processes are usually neglibible. As a consequence of this lack of detailed balance between opposite microscopic processes, we cannot use the Boltzmann equation to describe the population of atomic levels, neither the Saha equation to describe the populations of different ionization stages. The radiation field is not described locally by the Planck function, but must be calculated on the basis of the individual radiation processes.

A model which is often used to calculate the radiation of hot thin plasmas is the *coronal model*, first developed to describe the solar corona, but applicable also to other situations like stellar coronae, SNR's and clusters of galaxies. In this model, Maxwellian velocity distributions are assumed for both electrons and ions, the plasma is supposed to be of sufficiently high temperature ($\geq 10^6$ K) and low density ($\leq 10^{12}$ cm^{-3}) to be optically thin to its own radiation (by both free-free and Compton scattering processes), and the presence of a significant radiation field capable of photoionizing and photoexciting the plasma is neglected. Usually, steady conditions are also assumed, although this breaks down in young SNR's (see later) and sometimes in solar and stellar flares. The characteristic time for electro-electron collisions is $\tau_{ee} \approx 0.01\ T_e^{3/2}\ n_e^{-1}$ sec, while the characteristic time for ion-ion collisions is $\sqrt{m_i/m_e}\ \tau_{ee} \approx 43\ \tau_{ee}$; these times are sufficiently short as to ensure that both particle distributions are Maxwellian. The characteristic time for electron-ion collisions is however considerably longer ($\tau_{ei} = \tau_{ee}\ m_i/m_e \approx 10\ T_e^{3/2}\ n_e^{-1}$ sec) and transient conditions may therefore exist in which the two particle distributions have different temperatures (for instance in solar flares and in young SNR's). When photoionization and photoexcitations are not negligible, as may occur for accretion powered binaries and AGN's, the coronal approximation is no longer valid and a different model (the *nebular model*) must be considered (see, e.g, McCray 1987, Liedahl et al. 1990, Kahn and Liedahl 1991).

The ionization state and the radiation of hot thin plasmas under coronal conditions have been discussed in a number of review papers (e.g. Mewe 1988a,b,c,d, 1990, 1991a; Raymond 1988, 1990) to which we refer for full details and references to the original literature.

2.2 Ionization equilibrium

Let us indicate with Z^{+m} the element of atomic number Z m-times ionized (where m is an integer ranging from 0 to Z-1). Under non-LTE conditions the ionization state of the plasma is determined by a set of Z +1 coupled equations of the form:

$$\frac{1}{n_e}\frac{d}{dt}N_{Z,m} = N_{Z,m-1}\ S_{Z,m-1} - N_{Z,m}\ (S_{Z,m} + \alpha_{Z,m}) + N_{Z,m+1}\ \alpha_{Z,m+1} \quad (1)$$

where $S_{Z,m}$ is the total ionization rate from the ionization state m to $m+1$, and $\alpha_{Z,m}$ is the total recombination rate from state m to $m-1$. For steady conditions (i.e. when there is no dependence on time), the ratio of populations of two successive states of ionization is given simply by

$$\frac{N_{Z,m+1}}{N_{Z,m}} = \frac{n_e\ S_{Z,m}(T)}{n_e\ \alpha_{Z,m+1}(T)} = \Psi(T) \quad (2)$$

where $\Psi(T) = (S_c + S_{au})/(\alpha_r + \alpha_{di})$ is to first approximation a function of electron temperature only and depends on the detailed ionization and recombination processes (respectively, collisional ionization and autoionization, and radiative and dielectronic recombination). Note that, in contrast to the Saha equation valid under LTE conditions, the ionization state of the plasma is virtually independent of density (a slight dependence on density remains at relatively high densities through the dielectronic recombination rate).

The collisional ionization rate S_c, i.e. the ionization due to electron-ion impacts in the plasma, is determined from atomic calculations and/or fits to experimental data. An often used analytical approximation (*Lotz formula*) is

$$S_c \sim T^{1/2}\ \chi^{-1}\ \exp -(\chi/kT) \quad (3)$$

where χ is the ionization potential and k is the Boltzmann constant (here and in the following I indicate with T the electron temperature and I do not write down explicitly the numerical value of the proportionality factors).

Generally, autoionization is much less important than collisional ionization except for specific atomic configurations such as that of Na-like ions $1s^2\,2s^2\,2p^6\,3s$ as in Fe XVI. The autoionization rate is given by

$$S_a \sim T^{-1/2}\,E_0^{-1}\,\exp-(E_0/kT) \qquad (4)$$

where E_0 is the excitation energy of the autoionizing level. The latter is a double excited state above the first ionization limit of the ion Z^{+m}.

The radiative recombination rate (i.e. the capture of an electron by the ion Z^{m+1} with emission of a photon) is given by

$$\alpha_r \sim z_e^2\,T^{-1/2} \qquad (5)$$

where z_e is the effective nuclear charge ($= m+1$ for hydrogenic ions). Although this was the only recombination process considered up to 1964, it is usually less important in coronal conditions than dielectronic recombination, especially for lower values of Z and T. In the latter process, the capture of an electron by the Z^{m+1} ion results in a double excited state of the ion Z^{+m} which decays by the emission of a *satellite line* slightly shifted to the long-wavelength side of the parent transition in the Z^{m+1} ion. As we will see later, these satellite lines have great diagnostic potential for electron temperature and ionization state. For the moment, it may suffice to say that the dielectronic recombination rate via a double excited level s is given by

$$\alpha_{di} \sim T^{-3/2}\,\frac{g_s}{g_1}\,\frac{A_{as}A_{rs}}{A_{as}+A_{rs}}\,\exp-(E_s/kT) \qquad (6)$$

which depends on the statistical weight of level s and on the branching ratio of autoionization vs. radiative decay from this level (g_1 is the statistical weight of the ground level). Note that the double excited state s is collisionally coupled to the continuum at high densities, and this introduces a slight dependence on density in the ionization balance $\Psi(T)$.

The ionization equilibrium of high temperature low density plasmas has been computed by several authors over the past years (e.g. Jacobs et al. 1977, Raymond and Smith 1977; Mewe and Gronenschild 1981; Arnaud and Rothenflug 1985, Arnaud and Raymond 1992). Since the typical uncertainties on the ionization and recombination rates are of the order of $\approx 20 - 40\%$, the ionization balance is not known better than a factor $\approx 1.5-2$. This is demonstrated by a comparison of the different calculations, which shows shifts by ≈ 0.1 dex in T and differences by up to $\approx 50\%$ in the peak values (we recall that the ratio of the population of a given state of ionization of an element to the total abundance of that element is a single-peaked function of temperature, usually rather narrow in T). The agreement is much better for H-like and He-like ions, but much worse for lower ionization states (owing particularly to uncertainties on the dielectronic recombination rate). These uncertainties must be taken into account when deriving physical quantities from the observed spectra.

2.3 Emitted spectrum

Since the optical thickness $\tau \ll 1$, the emitted spectrum must be calculated by taking into account the individual line + continuum emission processes, i.e.: a) free-free continuum (*bremsstrahlung*); b) free-bound continuum (*recombination*); c) two-photon continuum; d) bound-bound transitions (i.e. *line emission*). For an isothermal source at distance D from the detector (with the latter outside the Earth's atmosphere), the observed flux over the spectral band $\Delta\lambda$ is given by:

$$f_{\Delta\lambda}(T, EM, A_{ab}) = \frac{1}{4\pi D^2} \exp -(\sigma_o N_H) \int P(\lambda, T, A_{ab}) \, d\lambda \int n_e^2 \, dV \quad (7)$$

where T is the electron temperature, $EM = \int n_e^2 \, dV$ is the volume emission measure, n_e is the electron density, A_{ab} are the elemental abundances, σ_o is the effective cross-section for photoelectric absorption, and N_H is the Hydrogen column density of the interstellar medium (plus any absorbing circumstellar material, if present). The function $P(\lambda, T, A_{ab})$ is the spectrum of the radiation per unit emission measure emitted by the plasma, and is a function of temperature T, elemental abundances A_{ab} and wavelength (or energy). The integral is made over the wavelength range $\Delta\lambda$. Eq.(7) assumes isothermal conditions, but can easily be extended to non-isothermal sources.

The radiative function $P(\lambda, T, A_{ab})$ has been computed by several authors and is continuously updated to take into account progress in atomic data calculations. Some popular plasma codes are those of Raymond and Smith (1977 and updates), Mewe et al. (1985 and updates), and Landini and Monsignori Fossi (1990 and updates). Although they basically consider the same processes, these codes differ considerably in the ionization equilibrium adopted, the number of lines included, the approximations used, and the wavelength coverage. Comparison of the different codes shows significant differences especially in the absolute values of line fluxes, although these differences tend to cancel out when the integrated emission over broad bands is considered. There are no *a priori* reasons to choose one spectral code rather than another; however, the significant differences that appear to be present between different codes are a matter of concern and a serious (but often neglected) source of errors when applying diagnostic techniques to the observed spectra.

More specifically, the spectrum of free-free emission from an isothermal source of temperature T per unit volume and unit frequency interval is given, as a function of frequency, by:

$$P_{ff}(\nu, T) \sim g_{ff}(\nu, T) \, n_e^2 \, T^{-1/2} \exp -(h\nu/kT) \quad (8)$$

or equivalently, as a function of wavelength, by

$$P_{ff}(\lambda, T) \sim g_{ff}(\lambda, T) \, n_e^2 \, T^{-1/2} \, \lambda^{-2} \, \exp -(hc/k\lambda T) \quad (9)$$

where g_{ff} is the average Gaunt factor.

The continuum due to recombination (per unit volume and unit frequency interval) is given by:

$$P_{fb}(\nu, T) \sim n_e \, n_H \, \frac{\sum N_{Z,m}}{n_H} \, \frac{N_{Z,m+1}}{\sum N_{Z,m}} \, T^{-3/2} \, \exp -[(h\nu - \chi)/kT] \qquad (10)$$

at energies $h\nu \geq \chi$ where χ is the ionization potential of the recombining ion. The total recombination spectrum is thus characterized by the presence of recombination edges corresponding to the ionization potential of the most abundant elements. In the equation above, n_H is the proton density, $\sum N_{Z,m}/n_H$ is the abundance of element Z with respect to Hydrogen, and $N_{Z,m+1}/\sum N_{Z,m}$ is the fractional abundance of ion Z^{m+1} with respect to the total number of ions of element Z. For a plasma of solar composition the product $n_e n_H \approx 0.8 \, n_e^2$.

The two-photon continuum, which is usually less important than the above mentioned processes, except for certain wavelengths and/or temperatures, is due to the decay of collisionally excited metastable levels of H-like and He-like ions by the emission of two photons. It contributes $\approx 10 - 20\%$ of the total continuum emission from these ions.

Finally, line emission is due to spontaneous decay of collisionally excited levels (usually from the ground level). The collisional rate is given by

$$C_{ex} \sim \frac{\Omega}{\omega \sqrt{T}} \, \exp -(E_o/kT) \qquad (11)$$

where ω is the statistical weight of the ground level and Ω is the collisional strength defined as

$$\Omega = \frac{8\pi}{\sqrt{3}} \, \frac{E_H}{E_o} \, \omega \, f \, g \qquad (12)$$

where f is the absorption oscillator strength and $g \approx 1$ for transitions with $\Delta n = 0$ and ≈ 0.2 for transitions with $\Delta n \neq 0$, where n is the principal quantum number. Ω is derived from atomic calculations and may have large uncertainties especially for forbidden and/or weak lines. Among the thousands of lines emitted by a hot thermal plasma, a special place is occupied by the so called satellite lines which are produced by dielectronic recombination and inner-shell excitation in Z^{+m} ions and which are located on the long-wavelength side of the resonance lines of $Z^{+(m+1)}$ ions. An example are the Li-like satellite lines adjacent to the resonant lines of He-like ions like for instance Ca XIX and Fe XXV.

Fig. 1 shows the X-ray spectrum between 1 and 100 Å from an isothermal plasma at $T = 2 \times 10^6$ K, as computed by Mewe and Gronenschild (1981). The contributions of the different continuum processes are indicated separately. The ions responsible for the most prominent emission lines are also indicated. By integrating the spectrum over a given spectral band, one gets the total radiative losses over that spectral band as a function of temperature (an important parameter for energy balance models). At temperatures less than $\approx 10^6$ K, most of the emission is in the EUV and UV parts of the spectrum, rather than in X-rays. In the latter range, the contribution of line emission to the total radiative emission

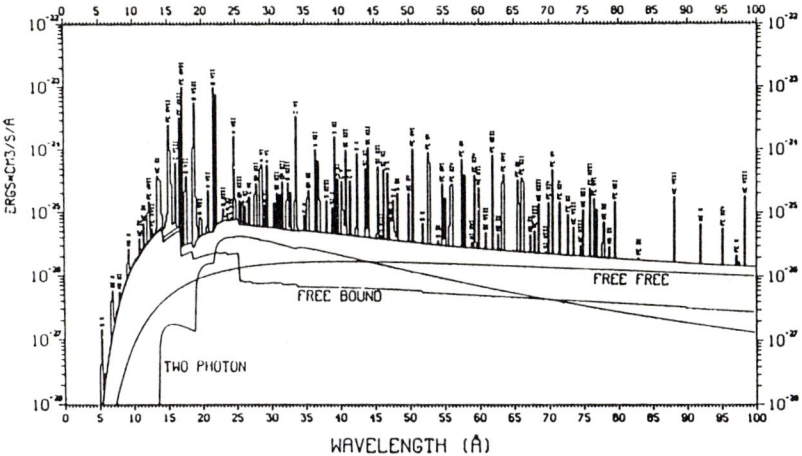

Fig. 1. Computed $1 - 100$ Å spectrum of an otically thin thermal plasma at $T = 2 \times 10^6$ K and unit emission measure (from Mewe and Gronenschild 1981)

becomes predominant at temperatures less than $\approx 10^7$ K: for soft sources like solar and stellar coronae, SNR's and cooling flows in clusters of galaxies, most of the physical information is indeed contained in lines that need to be resolved to infer the properties of the source.

3. Plasma Diagnostics Techniques

3.1. Atomic physics

In order to understand plasma diagnostic techniques, some knowledge is required of atomic physics. I refer to Herzberg (1944) for an elementary introduction to the subject. In low density plasmas, all ions are in the ground state. The electron configuration of the ground state in ions of increasing atomic number Z consists of shells with a maximum allowed number of electrons determined by the Pauli exclusion principle. For increasing values of the principal quantum number n, the shells are indicated by the letters K ($n = 1$), L ($n = 2$), M ($n = 3$), etc. Within each shell, there are subshells according to the azimuthal quantum number l, where l can take all integer values from $l = 0$ to $l = n - 1$. The subshells with $l = 0, 1, 2, \ldots$ are indicated respectively as s, p, d, etc. Thus, the ground state of, e.g., Oxygen (Z=8) is $1s^2\, 2s^2\, 2p^4$, that of Silicon (Z=14) is $1s^2\, 2s^2\, 2p^6\, 3s^2\, 3p^2$, that of Iron (Z=26) is $1s^2\, 2s^2\, 2p^6\, 3s^2\, 3p^6\, 3d^6\, 4s^2$, and so on. In addition to Hydrogen and Helium, the most important elements in X-ray emitting plasmas are those with even atomic number Z, like Carbon (Z=6), Oxygen (Z=8), Neon (Z=10), Magnesium (Z=12), Silicon (Z=14), Sulphur (Z=16), Argon (Z=18), Calcium (Z=20), Iron (Z=26) and Nickel (Z=28).

One of the simplest atomic configurations is that of the Lithium atom whose ground state is $1s^2\,2s$, i.e. a closed K-shell plus an outer electron. The diagram of the energy levels is thus similar to that of Hydrogen and H-like ions. We can specify the various energy levels by using the quantum numbers S, L and J, where for Li $S = 1/2$, $L = l = 0, 1, 2, \ldots, n-1$, and J takes the integer values $J = (L+S), (L+S-1), (L+S-2), \ldots, |L-S|$. Since $S = 1/2$, the energy levels consist of closely spaced doublets, and the energy levels corresponding to $L = 0, 1, 2, \ldots$ are indicated respectively as $S, P, D, \ldots$ terms. For $L = 0$, the quantum number J can take only the value $J = 1/2$, while, for $L = 1$, J can take the values $1/2$ and $3/2$, for $L = 2$ the values $3/2$ and $5/2$, and so on. Each atomic configuration, i.e. each energy term, can thus be indicated in a unique way by a somewhat cryptic notation which specifies the quantum numbers S, L and J. For instance, the ground state of the Li atom $1s^2\,2s$ will be indicated as $^2S_{1/2}$, where $S_{1/2}$ indicates an S term ($L = 0$) with $J = 1/2$ and the superscript to the left indicates a doublet state ($S = 1/2$); the excited state $1s^2\,2p$ will be indicated with $^2P_{3/2}$ or $^2P_{5/2}$ according to the value of J. Allowed transitions are those who satisfy the selection rules $\Delta L = \pm 1$, $\Delta J = 0, \pm 1$, except for the transition $J = 0$ to $J = 0$ which is not allowed.

Another interesting case is that of the Helium atom, or of He-like ions such as O VII (i.e. O^{+6}), Ne IX, Mg XI, Si XIII, S XV, Ar XVII, Ca XIX, Fe XXV, etc. Since in this case we have two outer electrons which may have either parallel or antiparallel spins, the quantum number S can take the values 0 and 1. The case $S = 0$ corresponds to singlet states, that with $S = 1$ to triplet states. The quantum number L is now given by the integer values $L = (l_1 + l_2), (l_1 + l_2 - 1), (l_1 + l_2 - 2), \ldots, |l_1 - l_2|$ where l_1 and l_2 are the azimuthal quantum numbers of the two electrons. We have again S, P, D, $\ldots$ terms, according to whether $L = 0, 1, 2, \ldots$. For singlets, $J = L$ and hence there is only one term for any value of L; instead, for triplets, $J = 1$ for $L = 0$, but $J = 0, 1, 2$ for $L = 1$, $J = 1, 2, 3$ for $L = 2$, and so on. The ground state of the He atom $1s^2$ is $^1S_0$, while the excited state $1s\,2p$ can both be single or triplet. For instance, the triplet state with $J = 1$ will be indicated as $^3P_1$. The selection rules are $\Delta L = 0, \pm 1$, $\Delta J = 0, \pm 1$, except for the transition $J = 0$ to $J = 0$ which is not allowed. In addition, there is the selection rule $\Delta S = 0$, i.e. transitions between singlet and triplet states are not allowed.

After recalling these basic concepts of atomic physics, we turn now our attention to diagnostic techniques. These have been discussed for instance by McWhirter (1975), Bely-Dubau (1988), Mason (1990), Jordan (1988), Gabriel (1992a), Mason and Monsignori Fossi (1994).

3.2. Temperature diagnostics

We have seen in section 2. that the ratio of the abundance of a given ionization state of an element to the total abundance of that element is a single-peaked function of temperature. In many cases, especially at relatively low temperatures, it is also a rather narrow function, with the peak temperature located at higher temperatures for higher ionization states. Thus the mere presence of lines of a given ionization state of an element is an indication, albeit crude, of the plasma temperature. A somewhat better diagnostics is the ratio of lines of the *same* element in two or more successive states of ionization. In fact, the intensity of a collisionally excited line will be proportional to the product of the relative abundance of the ion $N_{Z,m}/\sum N_{Z,m}$ and of the collisional excitation rate C_{ex}; both quantities are a function of temperature. The dependence on temperature, however, is much stronger for the ionization term and this allows the derivation of the plasma temperature by comparison with ionization balance calculations. The use of lines of the same element eliminates errors which may arise from uncertainties in the abundances of different elements. Note that temperatures derived in this way depends heavily on the accuracy of ionization balance calculations and this diagnostics can be applied in the semplified form discussed above only if the plasma is isothermal. In the presence of a temperature distribution, a Differential Emission Measure (DEM) analysis must be carried out as I will discuss later.

Electron temperatures can also be derived from the ratio of lines of the same ion Z^{+m}, thus eliminating the uncertanties related to the ionization balance. In order to understand this, let us consider a simple three-levels atom, with the ground level (g), and two excited levels (1 and 2) with excitation energies E_1 and E_2 with respect to the ground level. Since both upper levels are populated by collisional excitation from the ground level, the ratio of the intensities I_1 and I_2 of the two lines produced by spontaneous decay from levels 1 and 2 to the ground level is given by:

$$\frac{I_2}{I_1} = \frac{N_g \; n_e \; C_{g2}}{N_g \; n_e \; N_{g1}} = \frac{C_{g2}}{C_{g1}} \tag{13}$$

where the collisional excitation rate to the generic level i is given by

$$C_{gi} \sim T^{-1/2} \; \exp{-(E_i/kT)} \tag{14}$$

Thus, the ratio of the two lines will be given by

$$\frac{I_2}{I_1} \sim \exp{-[(E_2 - E_1)/kT]} \tag{15}$$

which is sensitive to temperature provided $(E_2 - E_1) \gg kT$. It is obvious that this method requires lines well separated in λ (with the associated difficulty of calibrating accurately the spectrometer over a large wavelength range); in addition, these lines will often have quite different intensities. For instance, this method has been used to derive temperatures from line ratios of the Li-like ion

O VI, by using the line $1s^2\, 3p\; {}^2P$ to $1s^2\, 2s\; {}^2S$ at 150 Å and the line $1s^2\, 2p\; {}^2P$ to $1s^2\, 2s\; {}^2S$ at 1032 Å.

To derive temperatures from lines which are closely spaced in wavelength and belong to the same element, one can use the ratio of collisionally excited resonant lines of He-like ions ($1s\, 2p \rightarrow 1s^2$) to adjacent satellite lines ($1s\, 2s\, 2p \rightarrow 1s^2\, 2s$) produced by dielectronic recombination in the Li-like ion of the same element. The temperature sensitivity derives from the different dependences on temperature of the resonant and satellite lines. In fact, if the subscripts r and s indicate the He-like resonant line and the Li-like satellite line, respectively, the intensity of the two lines is given by:

$$I_r \sim n_i\, n_e\, T^{-1/2}\, \exp -(E_r/kT) \tag{16}$$

$$I_s \sim n_i\, n_e\, T^{-3/2}\, A_r\, \exp -(E_s/kT) \tag{17}$$

where A_r is the transition probability for radiative decay from the double excited level produced by dielectronic recombination. The ratio of the two lines is thus:

$$\frac{I_s}{I_r} \sim T^{-1}\, A_r\, \exp\left[(E_r - E_s)/kT\right] \tag{18}$$

which depends on temperature as $\sim T^{-1}$ since $(E_r - E_s) \ll kT$. Since $A_r \sim Z^4$, this method is particularly important for high Z ions, and hence for high temperatures, since high ionization states like the He-like one require increasingly higher temperatures in higher Z ions. This method, for instance, has been applied to derive electron temperatures from Ca XIX and Fe XXV lines and their associated satellite lines in solar flares (e.g. Antonucci 1989).

3.3. Density diagnostics

Line ratios can also be used as density diagnostics. To illustrate this point, let us consider a two-level atom where 1 indicates the lower level and 2 the upper level. In steady conditions, collisional excitations to the upper level must be balanced by collisional plus radiative depopulations, i.e.:

$$n_1\, n_e\, C_{12} = n_2\, n_e\, C_{21} + n_2\, A_{21} \tag{19}$$

where C_{ij} is the collisional excitation (or depopulation) rate and A_{ij} is the rate of radiative decay. In a low-density plasma, the first term on the right-hand side is negligible. However, it may not be negligible for metastable levels and/or forbidden transitions. In this case, A_{21} is small and may become comparable to $n_e C_{21}$. If this occurs, only a part of the excitation energy of the upper level contributes to the emitted line, and the fraction which is emitted in the line depends on density (which control the population of the upper level). If we indicate with the subscripts a and m the intensity of an allowed transition and, respectively, of a forbidden transition from a metastable level, the ratio of the two lines is:

$$\frac{I_m}{I_a} = \frac{C_{1m}}{C_{1a}}\, \frac{A_{m1}}{A_{m1} + n_e \sum C_{dep}} \tag{20}$$

where $\sum C_{dep}$ is the collisional depopulation rate to all other levels. It is clear that the ratio will depend on the density n_e, but the details depend on the particular atomic configuration for the specific line. The ratio, in fact, can either increase or decrease with density: what is important in all cases is that there is a critical density over which (or under which) the line ratio becomes density sensitive. There are many transitions in the X-ray and EUV part of the spectrum which are density sensitive, like a number of low Z He-like ions over the range $\approx 10 - 20$ Å, the Fe XXI-XXII lines in the range $\approx 100 - 140$ Å, and a number of other lines in the EUV region between ≈ 200 and 900 Å.

A particularly interesting case is that of the triplets of He-like ions (for instance O VII at ≈ 20 Å). An He-like triplet consists of the resonance (R) line $1s\ 2p\ ^1P \rightarrow 1s^2\ ^1S$, the intercombination (I) line $1s\ 2p\ ^3P \rightarrow 1s^2\ ^1S$, and the forbidden (F) line $1s\ 2s\ ^3S \rightarrow 1s^2\ ^1S$ (these lines are often indicated also as w, y and z, respectively). The ratio of the F to the I line is densitive sensitive because the populations of the upper levels of these lines are collisionally coupled. In the low-density regime (which is of interest here), we have a transfer of electrons from the upper level of the forbidden line to the upper level of the intercombination line. Hence, F will decrease and I will increase, while the resonance line R remains approximately constant. For O VII, the F and I lines start to become density sensitive for $n_e > 10^{10}$ cm^{-3} and can be used as density diagnostics up to $\approx 10^{13}$ cm^{-3} (at higher densities, the F line disappears and the ratio R/I increases with density).

For He-like ions in high-temperature low-density plasmas, the critical density above which the F/I ratio becomes density sensitive increases with the atomic number Z. It is about 3×10^{10} cm^{-3} for O VII (formed at $\approx 10^6$ K), but becomes $\approx 6 \times 10^{11}$ for Ne IX (at $\approx 4 \times 10^6$ K) and $\approx 5 \times 10^{12}$ for Mg XI (at $\approx 6 \times 10^7$ K). For Ca XIX and Fe XXV, which form at temperatures higher than $\approx 10^7$ K, the critical densities are extremely large ($\geq 2 \times 10^{16}$ and 7×10^{17} cm^{-3}, respectively), so these He-like ions cannot be used for density diagnostic in hot plasmas at $T > 10^7$ K. There are other lines, however, at longer wavelengths, which can be used at least in principle for determining the density of very hot plasmas. Note also that in He-like ions, the ratio R/(F+I) is slightly temperature sensitive.

3.4. Other diagnostics

3.4.1. Emission measure distribution and abundances

The intensity of a collisionally excited line can be expressed as:

$$I_\lambda = h\nu \int N_{Z,m}\ n_e\ C_{ex}\ dV = h\nu \int \frac{N_{Z,m}}{\sum N_{Z.m}} \frac{\sum N_{Z,m}}{n_H} \frac{n_H}{n_e}\ n_e^2\ C_{ex}\ dV \quad (21)$$

where the first term under the latter integral is given by the ionization balance, the second one represents the chemical abundance of the element, and C_{ex} is the collisional excitation rate given by Eq.(14). For a fully ionized isothermal plasma of solar abundances for which $n_H/n_e \approx 0.8$, this can be rewritten as:

$$I_\lambda = 0.8 \ h\nu \ A_{ab} \ \frac{N_{Z,m}}{\sum N_{Z.m}} \ C_{ex} \int n_e^2 \ dV = 0.8 \ h\nu \ A_{ab} \ G_i(T) \ EM \qquad (22)$$

where EM is the volume emission measure and $G_i(T)$ is a function of temperature which represents the ionization and excitation properties of the line.

If the plasma is not isothermal, one can introduce a differential emission measure $Q(T)$ defined as

$$Q(T) \ dT = n_e^2 \ dV \qquad (23)$$

and the intensity of the line can be rewritten as

$$I_\lambda \sim A_{ab} \int G_i(T) \ Q(T) \ dT \qquad (24)$$

If we have N lines and/or narrow continuum bands, there are N equations of the above form which can be inverted to give $Q(T)$ i.e. the distribution of the emitting material with temperature. Obviously, a good reconstruction of the Differential Emission Measure (DEM) can be obtained only if the functions $G_i(T)$ for the various lines are narrow and the total temperature interval is large. Usually, one starts with an approximate expression for $Q(T)$ and try to reach the best fit to the observed line intensities by means of an iterative procedure.

The same technique can also be used to derive relative abundances for a multi-temperature plasma. In fact, if one uses many lines of different elements which form over a wide range of temperatures, the different lines should give the same $Q(T)$. Systematic differences due to different elements are produced by relative differences in abundances, which can thus be corrected for. Equivalently, *relative* abundances can be determined from line ratios of different elements with similar ionization potentials (which form over the same temperature range). Instead, *absolute* abundances can be determined from line to continuum ratios.

3.4.2. Out-of-equilibrium conditions

The ionization equilibrium computed as described in the previous section assumes steady conditions. This assumption breaks down when the relevant time scales are shorter than the characteristic times for ionization and recombination. Examples are transient events like solar and stellar flares, and dynamic situations like young SNR's. The presence of out-of-ionization equilibrium conditions can be inferred from the analysis of the spectra.

For instance in solar flares, model fits to the He-like resonance lines and to the associated Li-like satellite lines allows the abundance ratio of these two successive stages of ionizations to be derived in a way independent of ionization equilibrium calculations. On the other hand, the ratio of a satellite to a resonance line is also a measure of the electron temperature and this results in a value for the ratio of He-like to Li-like ions given by ionization equilibrium calculations. The comparison of these two abundances provides a diagnostics of out-of-equilibrium conditions.

In young SNR's, the plasma is underionized and there is an overall shift of the lines to longer wavelengths with respect to what expected from ionization equilibrium at the given temperature. Even if the lines are not fully resolved, a shift of line complex centroids to longer wavelengths is a good diagnostics of out-of-equilibrium conditions. These and other diagnostics of non-equilibrium ionization in SNR's, such as the R/(F+I) ratio of He-like triplets, will be discussed in more detail in section 6.

3.4.3. Velocity fields

The Doppler broadening of a spectral line is given by

$$\Delta\lambda_D = \frac{\lambda_o}{c} \sqrt{\frac{2kT_i}{m_i + v_{tur}^2}} \tag{25}$$

where T_i is the temperature of the ion emitting the line. If $T_i = T_e$ (as we have assumed implicitely up to now), and if we can determine T_e from line ratios, we can estimate the velocity of turbulent motions. If on the contrary turbulent motions are negligible, the observed broadening of the line can be used to estimate T_i and thus check whether the ion temperature is equal or not to the electron temperature T_e (as could occur for instance in young SNR's and in the early phases of solar flares).

If there are bulk motions rather than turbulent motions, the entire line will be shifted in wavelength by an amount given by

$$\Delta\lambda_{shift} = \frac{\lambda_o}{c} \, v_{bulk} \tag{26}$$

Examples are the blueshifted components observed in solar flares (with $v \approx 100 - 400$ km/s), the red and blue-shited lines expected from the two components of stellar binary systems (with $v_{orb} \geq 100$ Km/s) and the differential radial velocities in the expanding shells of SNR's (with $v_{exp} \approx 10^3$ Km/s).

3.5. Effects of finite resolution and sensitivity

So far in discussing diagnostics techniques I have assumed infinite resolving power $\lambda/\Delta\lambda$ and infinite signal to noise ratio S/N. Unfortunately, real data have both finite resolution and limited sensitivity, and we have to cope in practice with observational limitations. Since high dispersion and high sensitivity are mutually exclusive, some trade-off is necessary, and the applicability of diagnostic techniques will depend largely on the type of astrophysical source we are dealing with.

In order to have some appreciation of the kind of spectral resolution required for spectroscopic diagnostics in X rays, we may note that the typical separation of H-like and He-like ions of some of the most abundant elements ranges from ≈ 3 Å for Oxygen at ≈ 20 Å to ≈ 0.5 Å for Silicon at ≈ 6.5 Å. Resolving powers ≥ 20 are thus required to separate H-like from He-like lines in X-rays. For O VII, the separation of the components of the He-like triplet is ≈ 0.2 Å; we need resolving powers ≥ 200 to apply density diagnostics at the O VII lines. A similar resolution is also needed to separate the many L-shell lines of Fe which are present in the spectral interval $\approx 10 - 15$ Å. A much higher resolution is needed to study satellite lines adjacent to the He-like lines of Fe XXV at 1.8 Å (the typical separation is ≈ 0.002 Å which requires resolving powers in excess of 2,000). Similarly, a resolving power $\geq 3,000$ is required to measure velocities ≤ 100 Km/s. While it is possible to reach such resolutions in the X-ray spectral band (see lectures by Schnopper 1995 elsewhere in this volume), there are usually not enough photons available for most cosmic sources. As a result of this, many diagnostics techniques of the type described above have been applied so far only to the Sun, or to a few very bright SNR's. In most other cases, only observations at much lower resolution have so far been possible.

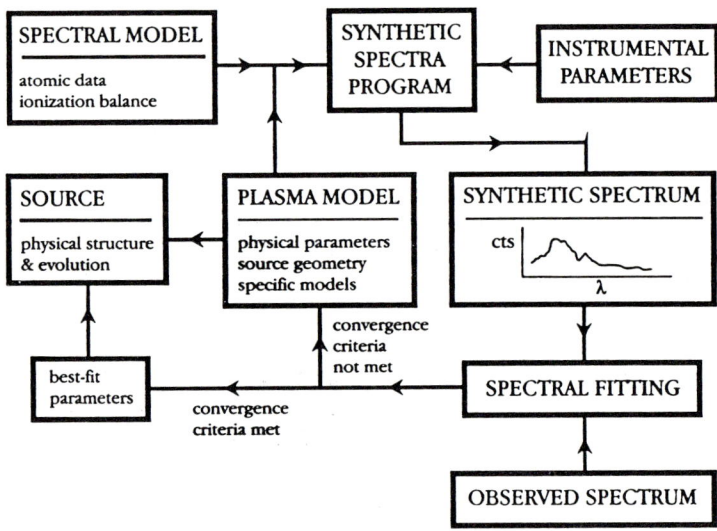

Fig. 2. Block diagram showing the various steps to be followed for deriving physical information about cosmic sources using data of limited spectral resolution (from Mewe and Kaastra 1994).

When we have data of low resolution (and often also low S/N), much of the physical information contained in the spectra is lost. In many cases, the observational data available are X-ray spectra where the lines are usually blended and severely degraded by the low resolution of the istrument. Even in this case, however, it is possible to derive some limited information on the source by model fitting techniques. The way of doing this is illustrated in Fig. 2. One starts with a plausible spectral model (for instance the coronal model discussed in section 2) which includes all atomic parameters and a computed ionization balance (steady-state or time-dependent). This spectral model is then applied to the specific source, assuming a set of values for the relevant physical parameters (e.g. temperature, emission measure, interstellar absorption, etc.), and then is folded with the instrument response. The resulting synthetic spectrum is than compared with the observed one and the source model parameters are modified by an iterative procedure until an acceptable fit is obtained by minimizing the χ^2. The best fit parameters are then used to describe the physical properties of the source. It is clear that this procedure depends crucially on the accuracy of the atomic data and of the spectral model assumed. A good fit does not guarantee that the best fit model is also a good physical description of the source.

In the rest of these lectures, I will discuss how the general techniques discussed above have been used in practice to derive information on various classes of optically thin plasmas observed in astrophysics.

4. The solar corona

4.1. Generalities

The corona of the Sun has been the first cosmic source to be detected in X-rays (since 1948). Although it is an intrinsically weak X-ray source (with an X-ray luminosity $L_x \approx 10^{27}$ erg/s), it is extremely bright because of its proximity; moreover, it is an extended source and in fact the Sun is the only star for which we can obtain spatially resolved observations. For this reasons, it occupies a special place in the field of X-ray spectroscopy.

The solar coronal plasma has a typical temperature of $\approx 1.5 \times 10^6$ K and a density of $\approx 10^8$ cm^{-3} in quiet areas. In active regions (i.e. regions where the surface magnetic field is stronger) the temperature is somewhat higher ($\approx 2 - 4 \times 10^6$ K) and the density can reach values as high as $\approx 10^9 - 10^{10}$ cm^{-3}. Still higher temperatures ($\approx 20 - 40 \times 10^6$ K) and densities ($\approx 10^{11} - 10^{12}$ cm^{-3}) are reached in solar flares.

Spatially resolved X-ray images of the solar corona obtained with *Skylab* and *Yohkoh* (see also Fig. 3) have shown that the coronal plasma is highly structured, with the observed features strongly related to the presence and strength of dynamo-generated magnetic fields. Most X-ray emission comes from closed loop like structures which connect regions of opposite magnetic polarities (Vaiana and Rosner 1978). Very weak X-ray radiation is emitted by open field regions (*coronal holes*) which are the source of high speed wind streams (with a typical mass loss rate of $\approx 10^{-14}$ M$_\odot$/yr and velocities of $600-800$ Km/s).

The high brightness of the solar corona in X-rays allows the acquisition of high-resolution high S/N spectra and the application of sophisticated plasma diagnostic techniques for temperature, density, elemental abundances and velocity. Virtually all diagnostic techiques currently used for X-ray plasmas have originally been developed in the context of solar physics and in close connection with studies of laboratory plasmas and atomic physics. On the other hand, the highly inhomogeneous structure of the solar corona requires spectrometers with a sufficiently narrow field of view so as to allow the application of plasma diagnostics to individual coronal features. This is an important difference with

Fig. 3. X-ray image of the solar corona in the Fe XVI line at 63.5 Å taken by a rocket-borne normal incidence X-ray telescope with multilayer coating (from Golub 1991).

respect to spectroscopy of other stars, for which only integrated observations over the entire stellar surface are possible.

4.2. Solar X-ray spectroscopy

X-ray spectra of the solar corona have been obtained since the early '70 with rockets and satellites (see reviews by Doschek 1988 a,b,c,d; Gabriel 1992b, Mason 1990, Mason and Monsignori Fossi 1994). The early solar spectrometers were uncollimated, and could give only an average description of the physical properties of the corona. Much effort therefore was devoted to the development of collimated spectrometers that could isolate single active regions from the rest of the corona. For transient events like flares, the technological problem was simplified by the fact that a single solar flare may be as bright in X-rays or brigher than the entire corona. These spectra clearly showed the importance of H-like and He-like transitions of O, Ne, Mg and Si in the range $\approx 5-25$ Å, the presence of many L-shell transitions of Fe in the range $\approx 10-15$ Å, and the diagnostic importance of the region $\approx 90-170$ Å, a spectral range which has remained poorly explored up to now. The He-like triplets of He-like ions like O VII, Ne IX and Mg XI were clearly resolved, showing the R, I and F lines which are important density diagnostics. The comparison of spectra obtained during solar

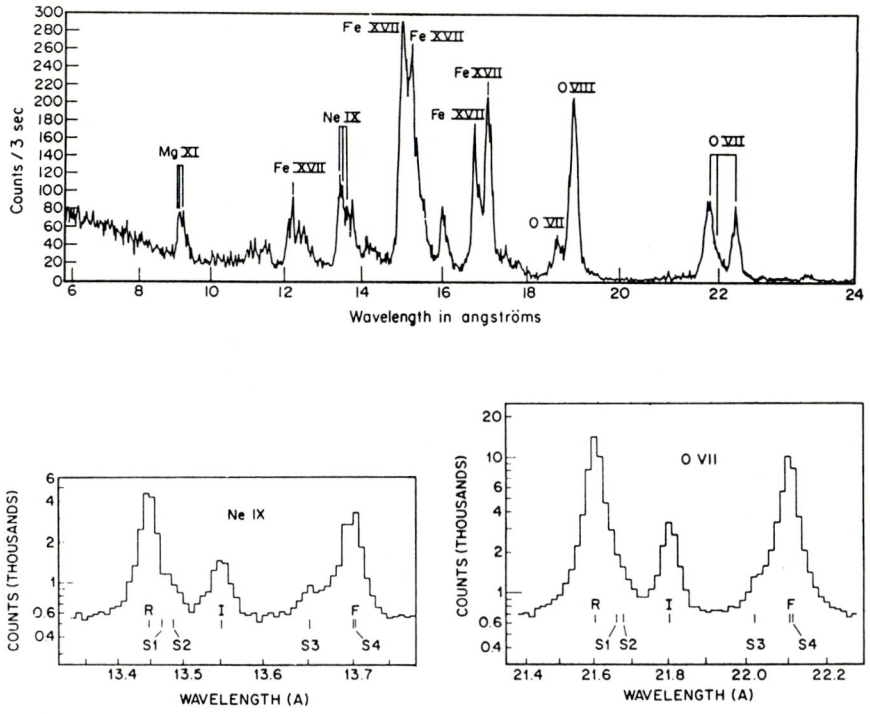

Fig. 4. X-ray spectra of the solar corona obtained in the early seventies showing H-like and He-like lines of various ions as well as L-lines of Fe. The lower panels show the triplet structure of the He-line ions Ne IX and O VII (from Culhane 1977).

flares with those obtained for isolated active regions and quite areas revealed spectral changes due to the different temperatures of the various features (see Fig. 4 for examples of these early spectra).

More recently, high resolution flare spectra of the Ca XIX and Fe XXV He-like ions with associated satellite lines have been obtained with Bragg spectrometers on satellites like *P78-1, SMM, Hinotori* and *Yohkoh*. Examples of these spectra are shown in Fig. (5). The resolution (several thousands) and S/N are sufficiently high to allow the application of diagnostic techniques for temperature, out-of-ionization equilibrium and velocity fields. Particularly interesting has been the detection of non-thermal broadenings in the early phases of flares (deduced from the apparent difference between the ion and the electron temperature) and the discovery of blue-shifted components that have been attributed to plasma moving upwards along coronal loops in response to a sudden deposition of energy at the onset of the flare.

Spectroscopic observations in the EUV (100 - 900 Å), FUV (900 - 1200 Å) and UV (1200 - 3000 Å) spectral ranges are particularly useful to investigate the

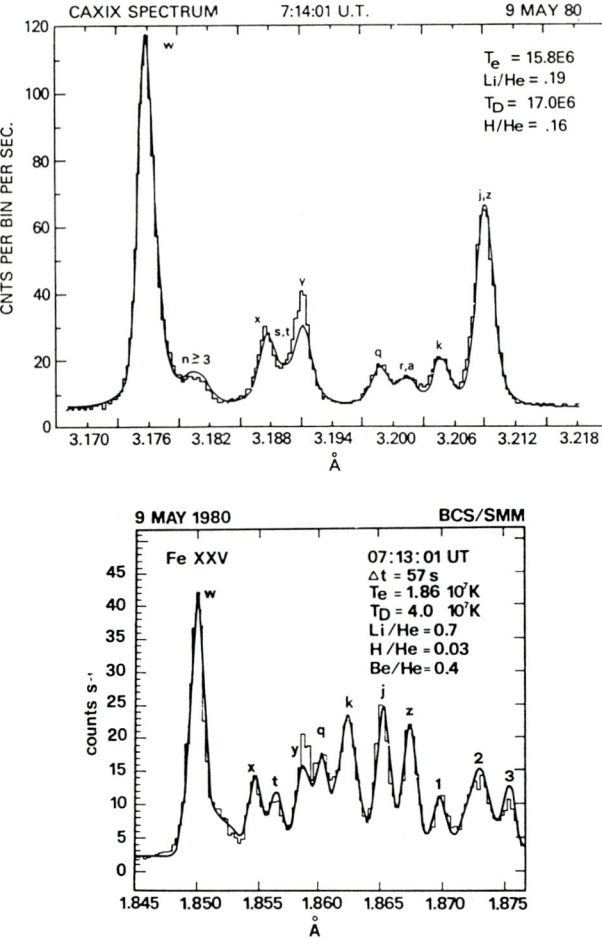

Fig. 5. High resolution spectra of solar flares showing Ca lines around 3.20 Å and Fe lines around 1.86 Å. The He-like lines of Ca XIX and Fe XXV, and their associated satellite lines, are clearly resolved (from Antonucci 1989).

temperature structure of the narrow transition region between the high temperature corona ($T \geq 10^6$ K) and the lower temperature chromosphere ($T \approx 10^4$ K). Since the ionization state of the plasma depends strongly on temperature, each ionization state forms over a narrow range of temperatures. By observing many lines of different elements in different ionization states, it is possible to derive the variation of temperature with height and the amount of material present at each temperature (i.e. the differential emission measure distribution). For the Sun, this can be done for spatially resolved regions; spectroheliograms can also be obtained which show monochromatic images of the Sun in specific spectral lines. These spectroscopic observations are the primary data for investigating the structure, energy balance and dynamics of the outer atmosphere of the Sun Similar techniques can also be used for other stars, although in this case the spatial information is totally lost, and spatially averaged quantities can only be derived (Jordan 1988).

4.3. Loop models

Although plane-parallel models can be used to obtain a rough description of the lower solar atmosphere (possibly taking into account the cell/network structure caused by the presence of strong photospheric magnetic fields at the boundaries of the supergranular cells, cf. Gabriel 1992b), magnetically confined loops are clearly required to model coronal features. A simple model for a coronal loop can be obtained by assuming stationary conditions and absence of plasma flows. For a semicircular loop of varying cross section $A(s)$, confined by the magnetic field (i.e. $\beta = p_g/p_B = 8\pi p_g/B^2 \ll 1$), and in hydrostatic equilibrium, the energy equation reads:

$$\frac{1}{A(s)} \left[\frac{d}{ds} \left(A(s) \ K \ T^{5/2} \ \frac{dT(s)}{ds} \right) \right] = n_e^2 \ P[T(s)] - E_H(s) \qquad (27)$$

where s is the coordinate along the loop, $K \ T^{5/2} \ dt/ds$ is the conductive flux, $n_e^2 \ P(T)$ are the radiative losses and $E_H(s)$ is the heating term (the physical nature of which is still poorly known).

By solving the energy equation with appropriate boundary conditions, one can derive the variation of temperature and density along the loop. Spectral observations can than be used to check the reliability of the model and derive the appropriate parameters for individual solar features. For instance, the energy equation allows the prediction of the differential emission measure distribution for a given loop geometry, and a given heat deposition. By comparing it with the observed DEM (as derived from spatially resolved observations of various lines and narrow continuum bands) one can thus derive parameters like the pressure and density of the emitting structure (e.g. Pallavicini et al. 1981, Mewe 1991c).

It is interesting to note that, as a consequence of the balance between the various gain and loss terms in the energy equation, simple scaling laws exist between the global parameters of the loop. For a loop of constant cross section, constant heating and constant pressure (i.e. loops that are small with respect to the pressure scale-height), the scaling law reads (Rosner et al. 1978):

$$T_{max} \approx 10^3 \ (pL)^{1/3} \qquad (28)$$

or, equivalently,

$$T_{max} \approx 55 \ L^{4/7} \ E_H^{2/7} \qquad (29)$$

where T_{max} is the maximum temperature at the top of the loop, p and E_H are the pressure and heat deposition (assumed constant along the loop) and L is the loop semilength. The first of these scaling laws shows that, for a given loop geometry, the maximum temperature and base pressure cannot vary independently, while the second scaling laws indicates the obvious fact that a larger heating deposition results in a higher loop temperature. Comparison of these scaling laws with spatially resolved spectroscopic observations of quiescent solar features shows a good agreement between observations and static loop models (e.g. Pallavicini et al. 1981).

For transient situations like flares, the energy equation must be solved together with the momentum and continuity equations for a prescribed dependence of the heating deposition on space and time. Even in this case, spectroscopic observations of lines fluxes at different times during the flare are the primary observational data to check the model and derive the flare parameters. By obtaining a good fit between spectroscopically measured quantities and model predictions, one can get information on the energy release during the transient event (e.g Pallavicini et al. 1983, Peres et al. 1987).

5. Stellar coronae

5.1. Generalities

Normal stars of nearly all spectral types and luminosity classes are surrounded by high-temperature low-density coronae which emit by thermal processes predominantly in the X-ray part of the spectrum. Apparently, the only stars with no significant coronal emission are late-type giants and supergiants and, possibly, A-type dwarfs. Since coronae are observed for both early and late-type dwarfs, which have completely different internal structure and atmospheric properties, it is unlikely that a single mechanism of coronal formation is at work in all types of stars.

The X-ray luminosities of stellar coronae range from less than 10^{27} erg/s to nearly 10^{34} erg/s. By comparison, the solar corona with an average X-ray luminosity of $\approx 10^{27}$ erg/s is one of the weakest coronal sources. If all stars were as weak as the Sun, only a handful of coronal sources would have been revealed, even with the most sensitive X-ray instruments flown up to now. For instance, if the Sun were located at 10 pc, its X-ray flux would be $\approx 1.7 \times 10^{-13}$ erg cm^{-2} s^{-1}, which is the typical sensitivity of the pointed observations carried out with the *Einstein* satellite and of the full-sky survey carried out by *ROSAT*. Before *Einstein*, in fact, only a few coronal sources were known, and most of them were emitting at levels several orders of magnitude higher than the solar corona. At any rate, stellar coronae are intrinsically weak X-ray sources, much weaker than other X-ray emitting sources like accretion-powered binaries, SNR's and clusters of galaxies.

Among normal stars, the brightest ones are those of early spectral type (O-type) with typical X-ray luminosities of $\approx 10^{31}$ to 10^{34} erg/s. Their X-ray luminosity is strongly correlated with the bolometric luminosity ($L_x \sim 10^{-7} L_{bol}$), which suggests a causal relationship with the strong radiation-driven winds of these stars. On the contrary, the X-ray luminosity of late-type stars is independent of bolometric luminosity (except under conditions of saturated activity), while being strongly dependent of the rotation rate ($L_x \sim v_{rot}^{\alpha}$ with α of the order of 2). Since late-type stars are characterised by the presence of an outer convection zone, where the convective motions interact with rotation to produce differential rotation and dynamo-generated magnetic fields, it is usually assumed that the X-ray emission of late-type stars originates from magnetic processes, as is believed to occur for the Sun.

The X-ray luminosities of late-type stars are in the range $\approx 10^{27} - 10^{31}$ erg/sec, and are virtually independent of effective temperature. Instead, at each colour, there is a large range of coronal emission levels: the stronger coronal sources are either younger or more rapidly rotating stars. Usually, for single stars, rapid rotation and age go together, since late-type stars are spun-down in the course of their evolution by the braking action of magnetised stellar winds. In close binaries, however, tidal interaction can maintain rapid rotation: RS CVn binaries, for instance, are among the brightest coronal sources of late spectral type, in spite of the fact that they usually contain an evolved (class IV) late-type component and thus are not young.

The number of coronal sources detected with the *Einstein* Observatory is $\approx 2,000$. The *ROSAT* all-sky survey has increased this number by at least one order of magnitude, and many more stellar coronal sources have been detected with *ROSAT* pointed observations. However, the number of sources for which spectroscopic observations have so far been possible is orders of magnitude less, and only for a very few of them the data are of sufficient quality as to allow some limited plasma diagnostics. Only with the powerful X-ray missions (*XMM* and *AXAF*) planned for the end of this century, it will be possibly to apply to other stars most of the diagnostic techniques that are currently used for the study of the solar corona.

In this section, I will limit myself to X-ray spectroscopy of coronal sources, while referring to a number of review papers for a general discussion of stellar coronae (see e.g. Stern 1983; Rosner et al. 1985; Linsky 1985, 1990a; Haisch 1986; Vaiana and Sciortino 1987; Serio 1987; Pallavicini 1988, 1989, 1992a,b; Schmitt 1988, 1990a, 1992a,b, 1993; Vaiana 1990; Vaiana et al. 1992; Sciortino 1993; Fleming 1993). The most up-to-date reference on the physics of stellar coronae (including the Sun) are the proceedings of the Vaiana Memorial Symposium edited by Linsky and Serio (1993). Spectroscopy of stellar coronae has been discussed in particular by Schmitt (1990b), Linsky (1990b), Mewe (1991a,b; 1992;1993).

5.2. X-ray spectroscopy of stellar coronae

Most observations of stellar coronal sources have been obtained so far with proportional counters with very limited spectral resolution. For instance, the IPC on *Einstein* had a resolving power $E/\Delta E \approx 1$ and the PSPC on *ROSAT* had a resolving power only a factor 2 higher. This resolution is insufficient to show lines or line complexes, but can only reveal the overall shape of the spectrum, thus allowing a determination of the average temperature and emission measure of the emitting plasma. For spectra of sufficiently high S/N, and sources with a multi-temperature structure, the best we can do is to derive a two-temperature model fit.

Somewhat better information have been provided by the Solid State Spectrometer (SSS) on *Einstein* which had a resolving power of 2 to 20. Although lines were not resolved even in this case, there was at least a hint of their presence in the spectra, thus allowing checking whether elemental abundances were consistent or not with solar abundances. Probably, the best coronal spectra obtained so far are those provided by the Transmission Grating Spectrometer (TGS) on *EXOSAT* which clearly showed the presence of lines complexes over the spectral band $\approx 1 - 200$ Å. Fig. 6 shows one of such spectra which has a resolution of $\approx$ 50. Unfortunately, only for three coronal sources (Capella, σ^2 CrB and Procyon) were spectra of this type obtained, and these too are of much inferior quality than typical solar spectra. The many L-shell transitions of Fe over the range $\approx 10 - 15$ Å appear as a single feature, and the same occurs for lines at longer wavelengths, which are all heavily blended.

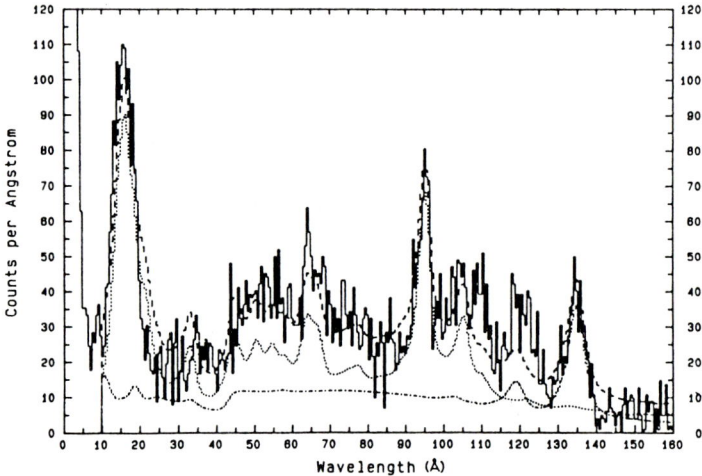

Fig. 6. *EXOSAT* TGS spectrum of Capella compared with the best fit two-temperature component model (from Mewe 1992).

The analysis of the *EXOSAT* TGS and *Einstein* SSS spectra (together with the analysis of lower resolution spectra from the *Einstein* IPC, *EXOSAT* ME and *ROSAT* PSPC) has shown that stellar coronae are best fitted by 1- or 2-temperature models (the latter case usually occurs for higher S/N spectra, a clear indication that a multitemperature structure is probably present in all cases). In the case of 2-temperature fits, the lower component is typically of a few million degrees, while the higher component is around 20×10^6 K. Since these spectra refer to the integrated emission from the whole star, it is still unclear whether the 2-temperature model arises from a continuous emission measure distribution within a single family of coronal loops or rather from two spatially distinct coronal regions (or equivalently two different families of loops). Eclipse observations of the type discussed below seem to favor this second possibility; however, it must be taken into account that the vast majority of sources for which reasonably good spectra have so far been obtained are RS CVn binaries, and it is not yet clear whether the available sample is representative of stars in general.

For a spatially unresolved observation, the X-ray luminosity (which results from the integrated emission of a large number of spatially distinct structures) will be given by

$$L_x \sim P(T) \, n_e^2 \, V \sim P(T) \, \frac{p^2}{T^2} \, N \, A_\circ \, L \tag{30}$$

where $P(T)$ are the total radiative losses over the observed spectral band, n_e is the electron density, and the total emitting volume V has been expressed as the sum of N equal loops of semilength L and constant cross-section $A_\circ$. Using the scaling law (28), we can eliminate L from the above equation, and hence the flux at the star surface can be expressed as

$$F_x = \frac{L_x}{4\pi R_\star^2} \sim \Phi(T, A_f, p) \tag{31}$$

where A_f is the fraction of the stellar surface covered by X-ray emitting features. By fitting the observed spectrum with a model spectrum, we can derive the temperature T and a normalization factor related to the amount of emitting material (i.e. to the total emission measure). However, it is usually not possible to separate the pressure p from the coverage factor A_f. Even neglecting the fact that the derived parameters refer in any case to *average* conditions over the star, one needs additional constraints to infer the spatial distribution and density of stellar features.

A way of solving the above difficulty is to measure the density from density-sensitive line ratios. So far, this has been possible in the EUV and UV parts of the spectrum (which refer to regions at temperatures $T \approx 10^4 - 10^5$ K, cf. Jordan 1988), but may become possible also in X-rays with future missions; the other is to observe eclipsing binary systems, and to use spectral observations at different phases to disentagle the spatial from the temperature structure. An example are the extensive observations of the RS CVn eclipsing binary AR Lac carried out with *Einstein*, *EXOSAT*, *ROSAT* and *ASCA* which have suggested the presence in this system of both compact loops and of more extended structures, possibly embracing the whole system.

For early-type stars, an important diagnostics is the presence of Oxygen absorption edges around 0.5 keV. Such edges are predicted by models which attribute the X-ray emission to a shallow corona at the base of the massive winds of early-type stars. The fact that these absorption edges have not been observed (at least as prominent as expected) has been taken as evidence that coronal emission in early-type stars does not originate at the base of the wind, but rather throughout it. Current models attribute the coronae of early-type stars to shock heating against high density blobs which form throughout the wind by radiative instabilities. In Wolf-Rayet + O star binary systems, the heating of the plasma may also result from colliding winds, a model that could be tested by observing eclipsing binary systems at different phases.

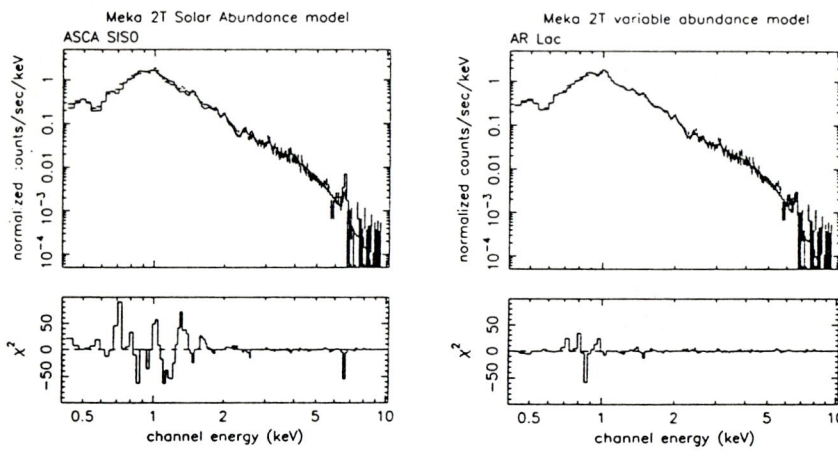

Fig. 7. *ASCA* SIS spectrum of AR Lac fitted with a two-temperature model with solar abundances (left) and with variable abundances (right). The residuals are plotted in the lower panels (from White et al. 1994).

Spectra of significantly better quality at energies > 1 keV have been obtained recently with CCD detectors on the Japanese satellite *ASCA*. These spectra have a resolving power of 10 to 50 over the range 1 to 10 Å, which is sufficient to resolve the H-like and He-like lines of the most abundant elements. Not only the temperature structure can be determined much better than with previous non dispersive spectrometers, but also the elemental abundances can be investigated with much higher accuracy. Particularly puzzling is the finding that variable non-solar abundances are required to fit the observed spectra, although the observed pattern remains unclear. Fig. (7) shows the spectrum of AR Lac obtained by ASCA.

Although brief, the above considerations should have made it clear that plasma diagnostics are far from being fully exploited in the case of the coronae of other stars. With the development of new dispersive spectrometers for *XMM* and *AXAF*, this situation should change drastically. The spectra of stellar coronae, like the spectrum of the solar corona. are very rich in emission lines: when these lines will be properly resolved, it will be possible to derive temperature, densities, elemental abundances, flow velocities and circumstellar absorbing components for a large number of stars of different spectral types, luminosity classes, ages, and binary status. Only at that stage, it will be possible to determine the physical properties of stellar coronae in a truly quantitative way.

6. Supernova Remnants

6.1. Generalities

Supernova remnants (SNR's) are the debris of supernova explosions. There are two types of SNR's: *crab-like* and *shell-type*. The crab-like SNR's (of which the best known example is the Crab Nebula) are non-thermal sources in which the rotational energy of the central engine (the pulsar) is converted into accelerated particles and radiation. These non-thermal sources (which show a continuum spectrum due to synchrotron emission) will be excluded from the following discussion. The other type of SNR's are instead optically-thin thermal plasmas produced by a blast wave heating the ISM (which produces a high temperature component) and by a reverse shock heating the stellar ejecta (thus giving rise to a low-temperature component). The expanding shell can shine for thousands of years after the explosion of the supernova, providing an ideal enviromnent for the application of diagnostic techniques. For general reviews on SNR's see for instance Danziger and Gorenstein (1983), Roger and Landecker (1988) and Cioffi (1990). X-ray observations of SNR's have been discussed by Aschenbach (1988a,b; 1993), Canizares (1990a,b.c), Bleeker (1990), Kaastra and Bleeker (1991).

The typical parameters for the hot plasma in shell-type SNR's are electron temperatures of $\sim 1 \times 10^6 - 5 \times 10^7$ K, electron densities of ~ 0.1 to 10 cm^{-3}, expansion velocities of ~ 1000 to 5000 Km/s, and X-ray luminosities of $\sim 10^{34}$ to 10^{38} erg/s. There are about 50 X-ray emitting SNR's in our Galaxy, with distances ranging from ~ 500 pc to ~ 3 kpc, but there are also several known SNR's in the Magellanic Clouds. The sizes of galactic SNR's range from a few arcmin to a few degrees (from a few parsec to about 50 parsec), and their ages range from ~ 300 yrs to more than 20,000 yrs. As expected from an expanding shell, the size is larger for older systems. In any case, SNR's are extended sources with spatial structures observable down to a scale of about 1 armin. The spatial structure observed in X-rays bears often complicated relationships with the structures observed at optical and radio wavelengths.

The youngest SNR's in our Galaxy have ages of $\sim 300 - 400$ yrs and angular diameters of ~ 5 to 25 armin: these include Cas A (probably due to the explosion of a supernova in the year 1680) and the well known historical supernovae Tycho (SN 1572) and Kepler (SN 1604). At intermediate ages (about one to a few thousand years) there is Pup A, with an angular diameter of ~ 50 arcmin. Finally, there are old SNR's with ages exceeding $\sim 10^4$ years, and angular diameters of several degrees, like the Cygnus Loop and Vela (the latter also houses a pulsar at the centre). The so-called North Polar Spur, which is a prominent feature in the soft X-ray background which extends for more than 100 degrees across the

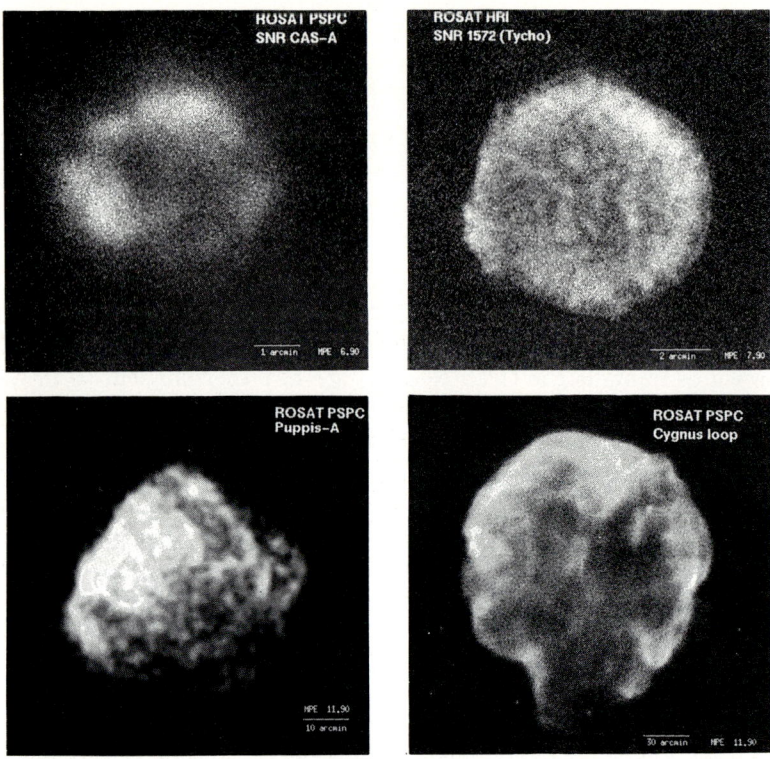

Fig. 8. *ROSAT* PSPC images of supernova remnants (courtesy of B. Aschenbach).

sky, is a very old SNR with an age of $\sim 10^5$ years. Fig. 8 shows some examples of SNR's of various ages observed with *ROSAT*.

The spectroscopic diagnostics which can be applied to SNR observations are analogous to those for other X-ray emitting optically thin plasmas, but with some important differences. The first is that we have an extended source for which spatially resolved spectroscopic observations can (and should) be obtained. The other is that SNR's, particularly the youngest ones, are sources which are not in ionization equilibrium, since the relevant ionization and recombination processes are often much longer than the age of the system. Time-dependent ionization equilibrium calculations must be performed and applied to the interpretation of the observations. Since the evolution of SNR's is a complicated hydrodynamical process, non-equilibrium ionization (NEI) calculations must be coupled to hydrodynamical codes to fit the observed spectra. Direct evidence for NEI conditions and a multi-temperature structure is provided by the X-ray observations, as I will discuss in the next section.

6.2. Non-equilibrium ionization

An important parameter while discussing SNR's is the so-called ionization parameter $\tau = n_e\,t$, i.e. the time integral of the electron density. Since $\tau \approx 10^4/n_e$ yrs is larger than the typical ages of young SNR's, the plasma is underionized with respect to collisional equilibrium ionization (CEI). As τ increases, the plasma becomes progressively more ionized and all lines shift to shorter wavelengths as higher ionization stages become more important. If the lines are not resolved (as often occurs in X-rays) the centroid of line complexes shifts to shorter wavelengths, thus providing an observational test for the occurrence of NEI conditions. This shift can amout to a few tenths of a keV. Typically, CEI conditions are reached for values of the ionization parameter of $\sim 10^{12} - 10^{13}$ cm^{-3} sec, i.e. over a time scale of $\sim 10^5$ yrs for a typical density of 1 cm^{-3}. For lower densities and/or younger ages, NEI conditions are clearly important. Shifts of X-ray line complexes indicative of NEI conditions have been measured with *EXOSAT* and *TENMA*. A value $\tau = 10^{11}$ cm^{-3} s has been inferred for Cas A, and $\tau = 10^{10}$ for Tycho.

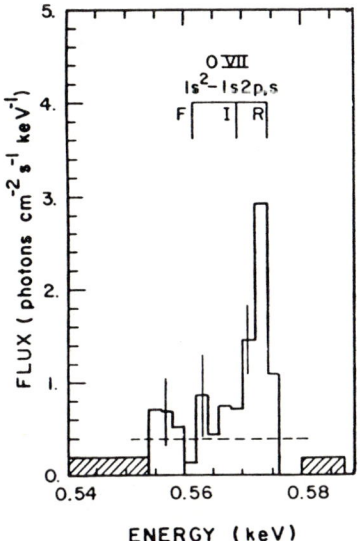
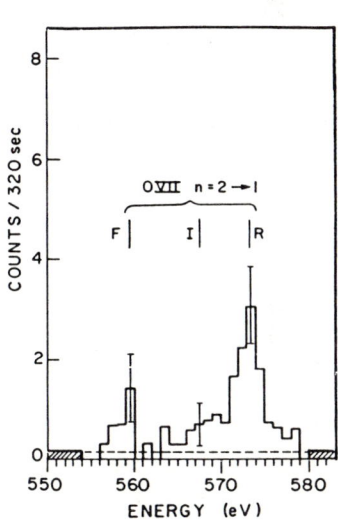

Fig. 9. *Einstein* FPCS spectra of the He-like triplet of O VII in Puppis A (right) and in the Cygnus Loop (left) (from Canizares 1990b).

Another way to infer the presence of NEI conditions is to use the F/R ratio of the forbidden to resonance line in He-like ions. As discussed above, in a low-density plasma in collisional equilibrium (like a stellar corona) the ratio F/R (or more correctly the ratio (F+I)/R where I is the intercombination line of the He-like triplet) is of the order of 1 (and decreases slightly for increasing temperatures). At densities larger than a critical value (which is of the order of

$\sim 10^{10}$ cm^{-3} for O VII), this ratio decreases as the density increases. However, this very high densities do not occur in the case of SNR's, and hence the ratio F/R should be ~ 1 under conditions of collisional ionization equilibrium. The observations show instead that this ratio (or better the ratio (F+I)/R when the three lines are all resolved) is significantly less than one, and the difference is larger for increasing values of the electron temperature T_e. This is a clear indication of NEI conditions, and is due to the fact that the upper $1s\,2s\ {}^3S$ level of the F and I transitions are not populated anymore by cascade following radiative recombination (a process which is more important for triplet states than for singlet states like the upper levels of the R transition). High resolution spectra of the OVII triplet in Pup A and in the Cyg Loop observed by the Focal Plane Crystal Spectrometer (FPCS) on *Einstein* show that the F/R ratio is indeed much less than 1 (see Fig. 9), indicating the occurrence of NEI conditions.

More quantitatively, while in CEI conditions there is a unique relationship between the line ratio of two successive stages of ionization of the same element (e.g. O VIII and O VII) and the F/R ratio of the He-like triplet, a family of solutions exists under NEI conditions depending on the two parameters T_e and τ. By comparing the observed line ratios to the predictions of NEI calculations, it is thus possible to infer the values of the electron temperature T_e and of the ionization parameter τ which fit simultaneously the O VIII/OVII and the F/R ratios. If more than two diagnostically important line ratios are available, one can determine the region in the parameter space T_e vs. τ allowed by all line ratios. For instance, various line ratios observed with the FPCS on *Einstein*, including O VII Lyα to Lyβ, O VII F to R, O VIII Lyβ to O VII R and O VII $(1s-3p)$ to O VIII Lyα, were used to derive the allowed values of T_e and τ for Pup A (see Canizares 1990c).

6.3. Spectra of SNR's

The detailed diagnostics discussed above require very high resolution to allow the study of individual lines. Only a few such spectra have been obtained so far. For instance, the high-resolution spectrum of Pup A obtained by the FPCS on *Einstein* is still the best so far obtained for an extrasolar source; its quality approaches that of solar spectra (see Fig. 10). Unfortunately, owing to the very low efficiency of Bragg spectrometers, only a few such spectra were obtained during the *Einstein* mission, and they required extremely long exposure times. More typical spectra have lower resolution such as those of Tycho shown in Fig. 11. They were obtained, respectively, by the Solid State Spectrometer (SSS) on *Einstein* and by the Gas Scientillation Proportional Counter (GSPC) on *EXOSAT*. Fig. 12 shows the spectrum of the SNR W49B obtained recently with a CCD detector on board *ASCA*. The H- and He-like ions of important elements such as Si, S, Ar, Ca and Fe are clearly evident in this spectrum. By measuring the strengths of these lines, one can determine elemental abundances and infer the mass of the emitting material. This in turn gives information on the nucleosynthesis in the progenitor star and on the relative importance of stellar ejecta with respect to swept-up interstellar matter.

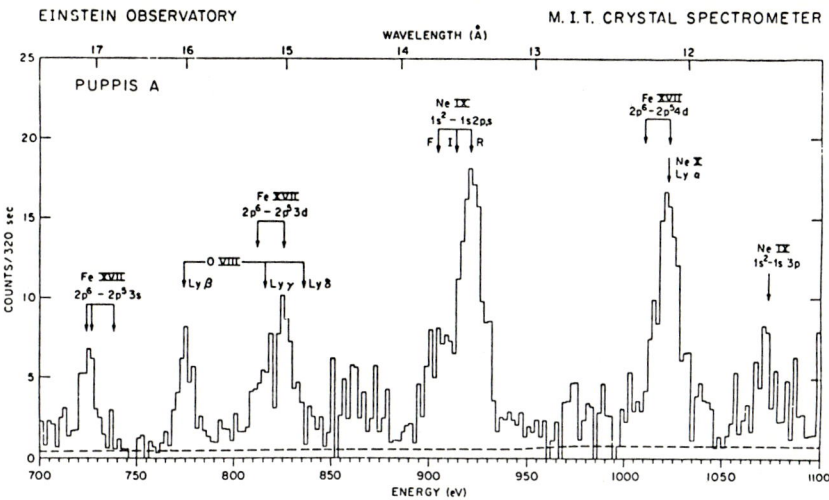

Fig. 10. High-resolution spectrum of Puppis A observed with the FPCS on *Einstein* (from Winkler et al. 1981).

The spectra of SNR's show typically stronger lines of O, Si, Ca and other heavy elements (except Fe) than solar spectra. The interpretation of these line strengths in terms of elemental abundances, however, is complicated by the presence of NEI effects, which have often been neglected in the earlier analyses. It is not surprising therefore that quite different results have been reported by different authors, both in terms of elemental abundances and of mass determinations. By including NEI effects, one generally gets lower abundances and lower masses than for CEI conditions. Even so, there is clear evidence in SNR spectra of an excess of nucleosynthesis products with respect to solar abundances, although there are still large uncertainties on the precise values of these abundance excesses. Fe, instead, is underabundant, by about a factor of 2.

Another important observational limitation is the fact that many observations of SNR's, especially those obtained in the past, had no, or only very limited, spatial resolution: the contribution of different parts of the remnant, including the emission of the swept-up interstellar mass and of the stellar ejecta heated by the reverse shock, are thus combined together, not to mention additional complications like departures from spherical symmetry caused by inhomogeneities in the ISM. To infer the properties of the remnant is thus a complicated process, which requires a detailed hydrodynamic modelling of the expansion.

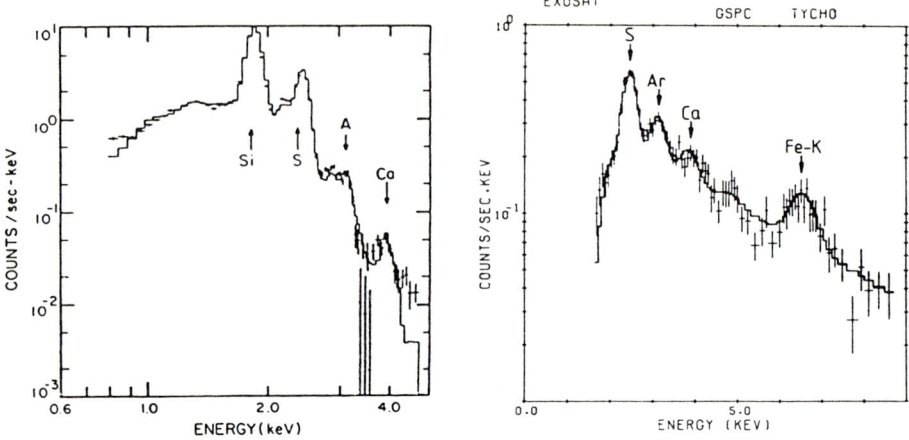

Fig. 11. X-ray spectra of the SNR Tycho obtained with the SSS on *Einstein* (left) and the GSPC on *EXOSAT* (right) (from Aschenbach 1988).

6.4. Hydrodynamic modelling of SNR

From what I have discussed above, it is clear that the volume-integrated X-ray luminosity of a SNR's under NEI conditions will have the form:

$$L_x \sim \int n_e^2 \ P(\lambda, T, A_{ab}, \tau) \ dV \qquad (32)$$

where τ is the ionization parameter, A_{ab} are the elemental abundances and T is the electron temperature (note that the assumption of equal electron and ion temperatures is not necessarily true in SNRs, since the characteristic time for electron-ion collisions, $\tau_{ei} \approx 10^4 \ (kT)^{3/2} n_e^{-1}$ s, with kT in keV, is pretty long at the low densities of SNR's).

As SNR's are extended sources with a high degree of spatial inhomogeneities, the temperature, density, ionization parameter and expansion velocity are all dependent on the position within the remnant: spatially resolved observations and modelling of the expansion process are therefore required for a proper analysis of the observed spectra. Although there are still considerable uncertainties about the physics of the expansion process, two models have often been assumed: an adiabatic expansion (*self-similar Sedov solution*) and an adiabatic expansion plus a reverse shock (*Chevalier solution*). With these models, one can predict the dependence of temperature, density, ionization parameter and expansion velocity as a function of the radial coordinate (assuming spherically symmetric conditions). This, together with a time-dependent ionization code (e.g. Kaastra and Jansen 1993), allows the expected spectrum to be predicted. By comparing it

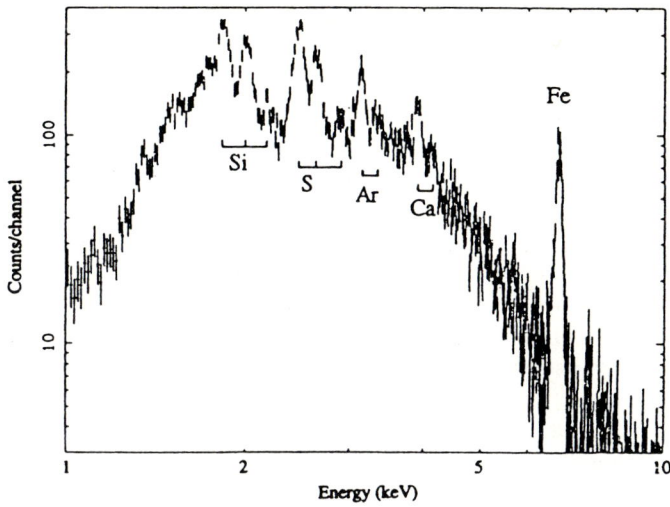

Fig. 12. X-ray spectrum of the supernova remnant W49B obtained with the SIS detector on *ASCA* (from Tanaka et al. 1994).

with the observed spectrum one can in turn derive the parameters of the source and/or improve the dynamical model.

As an example of this approach, I refer to *EXOSAT* observations of Cas A obtained with the PSD and the GSPC (Jansen et al. 1988). The spectrum could not be fitted by a two temperature model in CEI conditions. However, it could be fitted by a two-temperature bremsstrahlung (with T of, respectively, 0.6 and 3.7 keV) and by a number of individual spectral lines with adjustable intensities (so as to simulate the occurrence of NEI conditions). This suggests the presence of an expanding shell responsible for the high-temperature component, plus a reverse shock responsable for the low-temperature component. More detailed modelling using a NEI code has shown in fact that the Sedov adiabatic expansion, while fitting well the high energy part of the spectrum. predicts much less flux at low energy. On the contray, a satisfactory fit at both low and high energies can be obained by adopting the Chavalier's model of adiabatic expansion plus reverse shock (Bleeker 1990, Kaastra and Bleeker 1991). The derived masses are ~ 5 $M_\odot$ for the swept-up ISM heated by the blast wave and ~ 9 $M_\odot$ for the stellar ejecta heated by the reverse shock. This is consistent with a Type II supernova explosion.

7. Clusters of Galaxies

7.1. Generalities

Clusters of galaxies are the largest organised structures in the Universe. Typically, a cluster of galaxies contains hundreds of galaxies in a region of a few Mpc and has a total mass (including galaxies, hot gas and dark matter) of the order of $\sim 10^{15}$ $M_\odot$. There are about 4,000 clusters listed in the Abell's catalog. These objects are strong X-ray sources with X-ray luminosities of 10^{43} to 10^{45} erg/s. The X-ray emission is extended and non-variable and the X-ray spectra at energies ≥ 2 keV are consistent with thermal bremsstrahlung emission from a low-density ($n_e \sim 10^{-4} - 10^{-2}$ cm^{-3}) high-temperature ($T \sim 10^7 - 10^8$ K) plasma. The X-ray emission originates from a hot Intra Cluster Medium (ICM) with a total mass comparable to the mass of the galaxies in the cluster. Both masses are only a small fraction ($\sim 0.1 - 0.3$) of the gravitational binding mass of the cluster, indicating the presence of a large amount of unseen dark matter.

In addition to continuum emission, the X-ray spectra of clusters show the presence of lines, including the Fe K-line at 6.7 keV, the Fe L-lines around 1 keV and lines of O, Mg, Si and S. This is a clear evidence that the emission is due to a hot thermal plasma. The observed low-temperature lines ($\leq 10^7$ K) indicate the presence of a significant amount of relatively cool gas, which is now attributed to cooling flows at the center of the cluster. The Fe lines indicates Fe abundances of ~ 0.3 solar, which implies that the emitting gas must have been processed in stars and stripped, at least partly, from the galaxies in the cluster.

Fig. 13 shows examples of X-ray clusters observed by *ROSAT*. In all cases, the emission is extended and embraces the whole optical cluster: however, there are significant differences in the overal X-ray morphology from one cluster to the other. The Coma cluster contains some gigantic elliptical galaxies, but the cluster emission is not centered around any of them. On the contrary, the Virgo cluster is strongly centered in X-rays on the M87 galaxy, although the emission is extended and embraces also the other galaxies in the cluster. Similarly, the Perseus cluster is strongly centered on the active galaxy NGC 1275, which is significantly diplaced from the center of the optical cluster. NGC 1265, the well known head-tail radio galaxy (another indication for the presence of an intra-cluster medium) is a member of this cluster. Finally there are cases like Abell 1367 with no central dominant galaxy (similarly to the Coma cluster) and a very irregular X-ray morphology (unlikely Coma).

The bewildering variety of X-ray cluster morphologies has been investigated by Forman (1988), Jones and Forman (1990, 1992) and Jones et al. (1990) using the spatially resolved images provided by the IPC detector on *Einstein*. The presumption was that the different morphologies could be a clue to the dynamical evolution of clusters. To this aim, they introduced a bidimensional classification, based on the overall X-ray shape (regular, irregular) and the presence or not of a giant dominant galaxy at the cluster centre. The prototypes are the well known clusters mentioned above: Coma (regular with no dominant galaxy at the centre), Perseus (regular with a dominant galaxy, NGC 1275), A1367 (irregular with no dominant galaxy), Virgo (irregular, but centered on the dominant galaxy

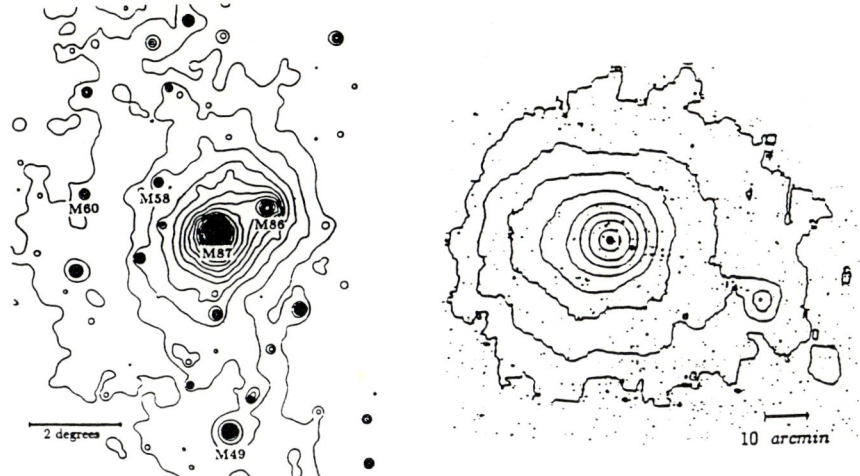

Fig. 13. Contour plots of X-ray images of the Virgo Cluster (left) and of the Perseus cluster (right) obtained with ROSAT (from Börinher 1993).

M87). If these scheme reflects an evolutionary sequence, it could suggest an evolution towards more dynamically relaxed systems with or without a central condensation. However, the situation is far more complex, and the relationship between cluster morphology and dynamical evolution is still largely unclear.

The X-ray emission of clusters of galaxies has been reviewed in a comprehensive monograph (Sarazin 1988a) and additional information can be found in the conference proceedings edited by Fabian (1988a, 1992a). In the following sections I will discuss a few selected topics with emphasis on the diagnostic potential of X-ray spectra of clusters.

7.2. X-ray imaging of clusters

Besides providing information on morphology, spatially resolved images of clusters also give information on the mass and radial distribution of the hot gas in the cluster. We have already seen that the X-ray emission of clusters is extended with a typical core radius of $\sim 0.07 - 0.9\ h_{50}^{-1}$ Mpc (where h has been normalized to an Hubble constant $H_o = 50$ Km s^{-1} Mpc^{-1}). For a self-gravitating hydrostatic isothermal sphere, the surface brightness at the projected distance r from the center of the cluster will be given by

$$I(r) = I(0)\ [1 + (r/r_c)^2]^{-3\beta + 1/2} \tag{33}$$

where

$$\beta = \frac{\sigma_{gal}^2}{\sigma_{gas}^2} = \frac{\mu m_H \sigma_r^2}{k T_{gas}} = 0.76 \left(\frac{\sigma_r}{10^3 km/s}\right)^2 \left(\frac{T}{10^8 K}\right)^{-1} \tag{34}$$

In the above equation σ is the dispersion velocity for the galaxies and for the gas, rispectively, σ_r is the dispersion of the radial velocites of the galaxies, and T_{gas} is the temperature of the intracluster medium.

The distribution of galaxies is usually assumed to follow a King's approximation

$$\rho_{gal}(r) = \rho_{gal}(0) \left[1 + (r/r_c)^2\right]^{-3/2} \tag{35}$$

where r_c is the core radius for the galaxy distribution.

Under these assumptions, the gas distribution will be given by

$$\rho_{gas}(r) = \rho_{gal}(r)^\beta = \rho_{gas}(0) \left[1 + (r/r_c)^2\right]^{-3/2\beta} \tag{36}$$

By fitting the observed brightness distribution $I(r)$, it is possible to derive the density distribution of the gas as a function of radius, and hence estimate the total mass of the gas. The observations show that a hydrostatic-isothermal model fits well the observed surface brightness profiles for clusters with no central galaxies, but does not fit well clusters with a central dominant galaxy. In the latter clusters there is a clear emission excess at the centre (Forman 1988, Bohringer et al. 1992, Böhringer 1993). As we shall see later, this is one of the observational evidence for the existence of cooling flows.

7.3. X-ray spectra

X-ray spectra of clusters provide much information on their physical properties (Mushotzky 1988, 1992, Mushotzky and Szymkowiak 1988, Koyama 1991, Stewart 1992). At energies $\geq$ 2 keV, cluster spectra are well fitted by thermal bremsstrahlung emission at temperatures of $\sim 2 \times 10^7$ to 1×10^8 K. Since the very first spectral observations obtained with *HEAO-1*, it was clear that the Fe-K line at 6.7 keV was present in the spectra of the Coma and Perseus cluster, thus confirming the thermal nature of their spectrum. Subsequent observations with the SSS and FPCS on *Einstein* provided evidence for the presence also of lower temperature material (with $T \sim 10^6 - 10^7$ K emitting the Fe L lines as well as lines of O, Mg and Si). The H-like and He-like ions of these elements, as well as Fe L-lines like Fe XVII, are due to a plasma much cooler than the one emitting the continuum at energies $\geq$ 2 keV and the Fe K-line at 6.7 keV (Canizares et al. 1988). Spatially resolved spectra obtained recently with *ROSAT, BBXRT* and *ASCA* have clearly demonstrated that the cooler material is located close to the cluster core (Mushotzky 1992, Böhringer 1993, Fukazawa et al. 1994). The temperature distribution as a function of radius shows in fact a significant decrease towards the centre. This gives support to the interpretation of the cool material as due to a cooling flow. Fig. 14 shows an example of X-ray spectra of clusters.

A parameter which can easily be determined from the shape of the continuum at energies $\geq$ 2 keV is the electron temperature T. There is a clear relationship

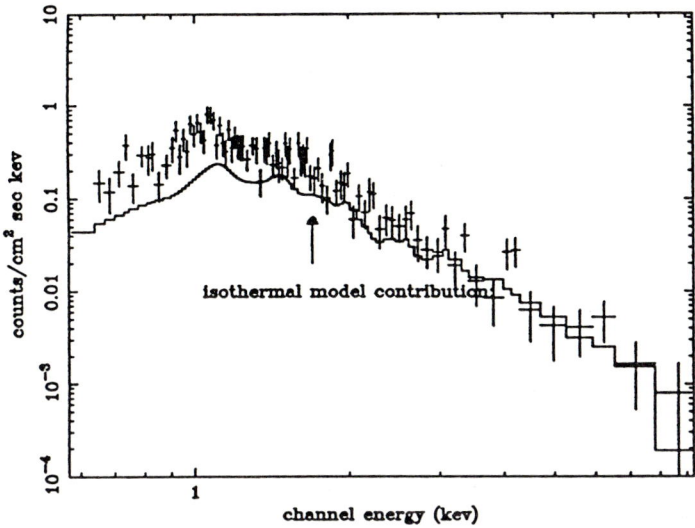

Fig. 14. X-ray spectrum of the central region of the cluster A262 obtained with *BBXRT*. The excess emission at low energies with respect to the model fit is due to the contribution of a cooling flow (from Mushotzky 1992).

between the total X-ray luminosity of a cluster and the temperature T, i.e. richer clusters have typically higher temperatures. Approximately, L_x scales as $\sim T^{5/2}$. A similar relationship is expected from a gravitationally bound cluster. In fact, the mass of the cluster (as estimated from the virial theorem) will be $M \sim R\sigma_{gal}^2/G$ which, for $\beta \sim 1$, scales as $M \sim RT$, where R is the radius of the cluster. On the other hand, the X-ray luminosity of an optically thin bremsstrahlung source is given by $L_x \sim n^2 P(T)V \sim M^2 R^{-3}T^{1/2}$, and hence $L_x \sim T^{5/2}$ if R is constant.

An important observational result derived from cluster spectra is the abundance of Fe, which is typically 0.3 times the solar value, as has been derived from many spectral observations obtained by *Einstein*, *EXOSAT*, *GINGA* and *ASCA*. If the ICM were of primordial origin, one would have expected no Fe at all. The fact that we observe a significant amount of Fe indicates that the ICM cannot be primordial but is material processed by thermonuclear reactions in stars. It must come, at least partly, from the galaxies. However, it cannot be ejected entirely from the cluster galaxies since its mass is comparable to and even larger than the total mass observed at present in the galaxies.

Spatially resolved spectroscopy of clusters provides a means to derive the gravitational mass of a cluster in a much better way than by the virial theorem (which depends exclusively on optically determined quantities). In fact, for a spherically symmetric gravitationally bound cluster in hydrostatic equilibrium

$$\frac{dP_{gas}}{dr} = -\frac{\rho_{gas}\ G\ M(r)}{r^2} \tag{37}$$

which, together with the perfect gas law

$$P_{gas} = \frac{\rho_{gas}}{\mu m_H}\ kT \tag{38}$$

leads by differentiation to

$$M(r) = -\frac{k\ T(r)\ r}{\mu\ m_H\ G}\left[\frac{dln\rho_{gas}}{dlnr} + \frac{dlnT}{dlnr}\right] \tag{39}$$

If $T(r)$ and $\rho(r)$ can be determined from spatially resolved spectroscopy, one can determine the total gravitational mass of the cluster. This in turns can be compared with the observed mass in the optical galaxies and with the mass of the X-ray emitting gas (see e.g. Sarazin 1988a,b, 1990, 1992). The gas mass M_{gas} is comparable to or somewhat larger than the mass in galaxies M_{gal}, with a tendency of the ratio M_{gas}/M_{gal} to increase for increasing gas temperatures. For instance, for the Perseus and Coma clusters with a gas temperature of $\sim 7 - 10$ keV, the gas to stellar mass ratio is about 3 to 4, while in cooler clusters with gas temperatures of 1-2 keV, this ratio is of the order of 1. On the other hand, the sum of the gas and stellar masses is only a small fraction ($\sim 0.1 - 0.3$) of the total gravitational binding mass of the cluster. Therefore, the largest part of the cluster mass must be in a form not observable both at optical and X-ray wavelengths. It has been speculated that this dark matter may be due to low mass stars generated in cooling flows.

7.4. Cooling flows

The concept of cooling flows has been much advocated in recent years both on theoretical grounds and to explain a number of observational facts, like the excess X-ray surface brightness and the presence of cool material in the core of many clusters (Fabian 1988b; 1990a,b; 1992b). The basic argument is simple. The characteristic cooling time of an optically thin plasma by bremsstrahlung (i.e. at energies ≥ 2 keV) is given by $\tau_{cool} \sim 7 \times 10^{10}\ T_8^{1/2}\ n_{-3}^{-1}$ yrs. In general, $\tau_{cool} \gg \tau_a$ where τ_a is the age of the cluster ($\sim H^{-1} \sim 2 \times 10^{10}$ yrs). However, in the core of a cluster (within a radius of about 200 Kpc), n is much higher and $\tau_{cool} \leq \tau_a$.

If $\tau_{grav} < \tau_{cool} < \tau_a$, where τ_{grav} is the gravitationally free-fall time, the temperature drops at the centre, the pressure also drops in the absence of a supporting force, and the plasma moves in to establish pressure equilibrium. This is the origin of cooling flows at the centre of clusters of galaxies. Observational support for the occurrence of cooling flows comes from the surface brightness excess in the core of clusters dominated by a central galaxy, and from the presence of cool plasma (as evidenced by low ionization lines) in the cluster core. Model simulations of cooling flows predict X-ray spectra which are qualitatively similar to those observed from clusters of galaxies, including the presence of a high temperature continuum emission, of a Fe-K line at 6.7 keV due to the hot plasma, and of O and Fe L-lines produced in the lower temperature plasma of the cooling flow (Sarazin 1990, Sarazin and Wise 1991).

For radial distances $r < r_{cool}$, where r_{cool} is the radius at which $\tau_{cool} = \tau_a$, the X-ray luminosity of the cooling flow (due to radiation of the thermal energy plus the work done) is given by:

$$L_{cool} = \frac{5}{2} \frac{M}{\mu m_H} kT \tag{40}$$

which implies a mass flow $M \sim 10 - 300$ $M_\odot$/yr at a typical inflow speed of $\sim$ 10 Km/s. The total mass in the cooling flow will be:

$$M \tau_a = 10^{12} \left(\frac{M}{100 M_\odot/yr} \right) \left(\frac{\tau_a}{10^{10} yr} \right) M_\odot \tag{41}$$

which is a significant fraction of the mass of the central galaxy. Where this cool material will eventually end up, it is still unclear. If it gives rise to star formation, it is likely to produce low mass stars. This can be an important source of the dark matter that appears to be needed to bind the cluster gravitationally.

8. Conclusions

In these series of lectures, I have presented a few examples of the power of X-ray spectroscopy to infer the properties of cosmic sources. The choice of topics is been highly selective. I have limited myself only to hot optically thin plasmas, but even within this more restricted area I have not discussed important topics like the hot phase of the interstellar medium or X-ray emission from normal galaxies. More importantly, I have completely neglected other classes of sources (either non-thermal or optically-thick or photoionized), in which spectroscopic diagnostic techniques play a fundamental role. These include, for instance, accretion-powered sources and active galactic nuclei. To include these topics would have required much more time, and many more pages, than those allowed to me.

However, it should be sufficiently clear at this point that X-ray spectroscopy is a powerful tool to infer the properties of virtually all types of cosmic sources, from the Sun to the most distant galaxies and clusters of galaxies. It should also be clear from the above sections that the diagnostic potential of X-ray spectroscopy has been hampered so far by the low flux of most extrasolar sources which has imposed severe limitations on the spectral, spatial and time resolution of X-ray observations. There are good hopes that this situation will change in the near future (e.g. Culhane 1990). The powerful X-ray missions which are under development for the end of this century (*AXAF, XMM, ASTRO-E*) will carry on board powerful spectroscopic devices which are expected to provide major advances in our understanding of cosmic sources. With these missions, many of the diagnostic techniques that have been discussed in these lectures will become possible for fainter and/or more distant objects. This is a good time for young researchers to enter the field.

References

Antonucci, E. (1989): "Solar Flare Diagnostics: Present and Future", in *Solar and Stellar Flares* (B.M. Haisch and M. Rodonò eds.), Dordrecht: Kluwer, p. 31.

Arnaud, M., and Raymond, J. (1992): "Iron Ionization and Recombination Rates and Ionization Equilibrium", *Ap. J.* **398**, 394.

Arnaud, M., and Rothenflug, R. (1985): "An Updated Evaluation of Recombination and Ionization Rates", *Astron. Astrophys. Suppl* **60**, 425.

Aschenbach, B. (1988a): "X-ray Observations of Hot Thin Plasma in Supernova Remnants", in *Hot Thin Plasmas in Astrophysics* (R. Pallavicini ed.), Dordrecht: Kluwer, p. 185.

Aschenbach, B. (1988b): "Recent X-ray Observations of Supernova Remnants and their Interpretation", in *Supernova Remnants and the Interstellar Medium* (R.S. Roger and T.L. Landecker eds.), Cambridge University Press, p. 99.

Aschenbach, B. (1993): "X-ray Emission from Supernova Remnants Observed with ROSAT", *Adv. Space Res.* **13**, (12)45.

Bely-Dubau, F. (1988): "Diagnostic Techniques for Hot Thin Astrophysical Plasmas", in *Hot Thin Plasmas in Astrophysics* (R. Pallavicini ed.), Dordrecht: Kluwer, p. 21.

Bleeker, J.A.M. (1990): "X-ray Observations of Supernova Remnants", *Adv. Space Res.* **10**, (2)143.

Böhringer, H. (1993): "ROSAT Observations of Galaxy Clusters and Cosmological Implications", *Adv. Space Res.* **13**, (12)181.

Böhringer, H., Schwarz, R.A., Briel, U.G., Voges, W., Ebeling, H., Hartner, G., and Cruddace, R.G. (1992): " ROSAT Observations of Clusters of Galaxies", in *Clusters and Superclusters of Galaxies* (A.C. Fabian ed.). Dordrecht: Kluwer, p. 71.

Canizares, C.R. (1990a): "X-ray Observations of Supernova Remnants", in *Physical Processes in Hot Cosmic Plasmas* (W. Brinkmann, A. C. Fabian and F. Giovannelli eds.), Dordrecht: Kluwer, p. 17.

Canizares, C.R. (1990b): "High Resolution Spectroscopy and Plasma Diagnostics of Supernova Remnants", in *High Resolution X-ray Spectroscopy of Cosmic Plasmas* (P. Gorenstein and M. Zombeck eds.), Cambridge University Press, p. 136.

Canizares, C.R. (1990c) "High Resolution X-ray Spectroscopy of Thermal Plasmas", in *Imaging X-ray Astronomy* (M. Elvis ed.), Cambridge University Press, p. 123.

Canizares, C.R., Markert, T.H., and Donahue, M.E. (1988): "X-ray Emission Lines from Cooling Flows", in *Cooling Flows in Clusters and Galaxies* (A.C. Fabian ed.), Dordrecht: Kluwer, p. 63.

Cioffi, D.F. (1990): "Supernova Remnants as Probes of the Interstellar Medium" in *Physical Processes in Hot Cosmic Plasmas* (W. Brinkmann, A. C. Fabian and F. Giovannelli eds.), Dordrecht: Kluwer, p. 1.

Culhane, J.L. (1977): "X-ray Astronomy", *Vistas in Astronomy*, **19**, 1.

Culhane, J.L. (1990): "Comments on the Future Observatories and their X-ray Spectroscopy Capability", in *High Resolution X-ray Spectroscopy of Cosmic Plasmas* (P. Gorenstein and M. Zombeck eds.), Cambridge University Press, p. 282.

Danziger, J., and Gorenstein, P. (eds.): 1983, *Supernova Remnants and their X-ray Emission*, Dordrecht: Reidel.

Doschek, G.A. (1988a): "Introduction to Solar Spectroscopy", in *Astrophysical and Laboratory Spectroscopy* (R. Brown and J. Lang eds.), Edinburg: SUSSP Publications, p. 237.

Doschek, G.A. (1988b): "UV Solar Spectroscopy", in *Astrophysical and Laboratory Spectroscopy* (R. Brown and J. Lang eds.), Edinburg: SUSSP Publications, p. 251.

Doschek, G.A. (1988c): "XUV Solar Spectroscopy", in *Astrophysical and Laboratory Spectroscopy* (R. Brown and J. Lang eds.), Edinburg: SUSSP Publications, p. 265.

Doschek, G.A. (1988d): "The Solar X-ray Spectrum", in *Astrophysical and Laboratory Spectroscopy* (R. Brown and J. Lang eds.), Edinburg: SUSSP Publications, p. 279.

Fabian, A.C. (ed.) (1988a): *Cooling Flows in Clusters and Galaxies*, Dordrecht: Kluwer.

Fabian, A.C. (1988b): "Theory of Intracluster Gas", in *Hot Thin Plasmas in Astrophysics* (R. Pallavicini ed.), Dordrecht: Kluwer, p. 293.

Fabian, A.C. (1990a): "Conduction and Turbulence in Intracluster Gas", in *Physical Processes in Hot Cosmic Plasmas* (W. Brinkmann, A. C. Fabian and F. Giovannelli eds.), Dordrecht: Kluwer, p. 271.

Fabian, A.C. (1990b): "Signatures of Cooling Flows", in *High Resolution X-ray Spectroscopy of Cosmic Plasmas* (P. Gorenstein and M. Zombeck eds.), Cambridge University Press, p. 219.

Fabian, A.C. (ed.) (1992a): *Clusters and Superclusters of Galaxies*, Dordrecht: Kluwer.

Fabian, A.C. (1992b): "Cooling Flows in Clusters of Galaxies", in *Clusters and Superclusters of Galaxies* (A.C. Fabian ed.), Dordrecht: Kluwer, p. 151.

Fleming, T.A. (1993): "Recent ROSAT Results in Stellar X-ray Astronomy", *Adv. Space Res.* **13**, (12)7.

Forman, W. (1988): "Clusters of Galaxies and Cooling Hot Gas", in *Cooling Flows in Clusters and Galaxies* (A.C. Fabian ed.), Dordrecht: Kluwer, p. 17.

Fukazawa, Y., Ohashi, T., Fabian, A.C., Canizares, C.R., Ikebe, Y., Makishima, K., Mushotzky, R.F., and Yamashita, K. (1994): "Metal Concentration and X-ray Cool Spectral Component in the Central Region of the Centaurus Cluster of Galaxies", *Publ. Astron. Soc. Japan* **46**, L55.

Gabriel, A.H. (1992a): "Spectroscopic Diagnostics", in *The Sun. A Laboratory for Astrophysics* (J.T. Schmelz and J.C. Brown eds.), Dordrecht: Kluwer, p. 261.

Gabriel, A.H. (1992b): "The Solar Corona", in *The Sun. A Laboratory for Astrophysics* (J.T. Schmelz and J.C. Brown eds.), Dordrecht: Kluwer, p. 277.

Golub, L. (1991): "Very High Resolution Solar X-ray Imaging", in *Mechanisms of Chromospheric and Coronal Heating* (P. Ulmschneider, E.R. Priest and R. Rosner eds.), Berlin: Springer, p. 115.

Haisch, B.M. (1986): "Coronae on Stars", *Irish Astron. J.* **17**, 200.

Herzberg, G. (1944): *Atomic Spectra and Atomic Structure*, New York: Dover.

Jacobs, V.L., Davis, J., Kepple, P.C., Blaha, M. (1977): "The Influence of Autoionization Accompanied by Excitation on Dielectronic Recombination and Ionization Equilibrium", *Ap. J.* **211**, 605.

Jansen, F., Smith, A., Bleeker, J.A.M., de Korte, P.A.J., Peacock, A., and White, N.E. (1988): "EXOSAT Observations of the Cassiopeia A Supernova Remnant", *Ap. J.* **331**, 949.

Jones, C., and Forman, W. (1990): "Clusters of Galaxies and the Hot Intracluster Medium", *Adv. Space Res.* **10**, (2)209.

Jones, C., and Forman, W. (1992): "Imaging the Hot Intracluster Medium", in *Clusters and Superclusters of Galaxies* (A.C. Fabian ed.), Dordrecht: Kluwer, p. 49.

Jones, C., Forman, W., and David, L. (1990): "Clusters of Galaxies and the Hot Intr-acluster Medium", in *High Resolution X-ray Spectroscopy of Cosmic Plasmas* (P. Gorenstein and M. Zombeck eds.), Cambridge University Press, p. 249.

Jordan, C. (1988): "Ultraviolet Stellar Spectroscopy", in *Hot Thin Plasmas in Astro-physics* (R. Pallavicini ed.), Dordrecht: Kluwer, p. 97.

Kaastra, J.S., and Bleeker, J.A.M. (1991): "Line Spectroscopy in Supernova Rem-nants", in *Iron Line Diagnostics in X-ray Sources* (A. Treves, G.C. Perola and L. Stella eds.), *Lecture Notes in Phys.* **385**, 35.

Kaastra, J.S., and Jansen, F.A. (1993): "A Spectral Code for X-ray Spectra of Super-nova Remnants", *Astron. Astrophys. Suppl.* **97**, 873.

Kahn, S.M., and Liedahl, D.A. (1991): "Iron L-shell Line Formation in Diverse As-trophysical Environments", in *Iron Line Diagnostics in X-ray Sources* (A. Treves, G.C. Perola and L. Stella eds.), *Lecture Notes in Phys.* **385**, 3.

Koyama, K. (1991): "Iron Line Emission Features in Clusters of Galaxies", in *Iron Line Diagnostics in X-ray Sources* (A. Treves, G.C. Perola and L. Stella eds.), *Lecture Notes in Phys.* **385**, 67.

Landini, M., and Monsignori Fossi, B.C. (1990): "The X-UV Spectrum of Thin Plas-mas", *Astron. Astrophys. Suppl.* **82**, 229.

Liedahl. D.A., Kahn, S.M., Osterheld, A.L., and Goldstein, W.H. (1990): "X-ray Spec-tral Signatures of Photoionized Plasmas", *Ap. J. Letters* **350**, L37.

Linsky, J.L. (1985): "Nonradiative Activity across the H-R Diagram: Which Types of Stars are Solar-like?", *Solar Phys.* **100**, 333.

Linsky, J.L. (1990a): "*Einstein* and Stellar Sources", in *Imaging X-ray Astronomy* (M. Elvis ed.), p. 39.

Linsky, J.L. (1990b): "Goals for the Applications of High Resolution X-ray Spec-troscopy to the Diagnosis of Stellar Coronal Plasmas", in *High Resolution X-ray Spectroscopy of Cosmic Plasmas* (P. Gorenstein and M. Zombeck eds.), Cambridge University Press, p. 94.

Linsky, J.L., and Serio, S. (eds.) (1993): *Physics of Solar and Stellar Coronae*, Dor-drecht: Kluwer.

Mason, H.E. (1990): "Spectroscopic Diagnostics for Ions Observed in Solar and Cosmic Plasmas", in *High Resolution X-ray Spectroscopy of Cosmic Plasmas* (P. Gorenstein and M. Zombeck eds.), Cambridge University Press, p. 11.

Mason, H.E., and Monsignori Fossi, B.C. (1994): "Spectroscopic Diagnostics in the EUV-UV for Solar and Stellar Plasmas", *Astron. Astrophys. Reviews* **6**, 123.

McCray, R.A. (1987): "Coronal Interstellar Gas and Supernova Remnants", in *Spec-troscopy of Astrophysical Plasmas* (A. Dalgarno and D. Layzer eds.), Cambridge University Press, p. 255.

McWhirter, R.W.P. (1975): "The Contribution of Laboratory Measurements to the Interpretation of Astronomical Spectra", in *Atomic and Molecular Processes in Astrophysics* (M.C.E. Huber and H. Nussbaumer eds.), Saas-Fee: Swiss Soc. Astron. Astrophys., p. 185.

Mewe, R. (1988a): "Ionization Balance in Low-Density Plasmas: Steady-State and Transient Case", in *Astrophysical and Laboratory Spectroscopy* (R. Brown and J. Lang eds.), Edinburg: SUSSP Publications, p. 129.

Mewe, R. (1988b): "Ionization Rate Coefficients", in *Astrophysical and Laboratory Spec-troscopy* (R. Brown and J. Lang eds.), Edinburg: SUSSP Publications, p. 141.

Mewe, R. (1988c): "Recombination Rate Coefficients", in *Astrophysical and Laboratory Spectroscopy* (R. Brown and J. Lang eds.), Edinburg: SUSSP Publications, p. 155.

Mewe, R. (1988d): "Satellite Lines as a Diagnostic", in *Astrophysical and Laboratory Spectroscopy* (R. Brown and J. Lang eds.), Edinburg: SUSSP Publications, p. 167.

Mewe, R. (1990): "Ionization of Hot Plasmas", in *Physical Processes in Hot Cosmic Plasmas* (W. Brinkmann, A. C. Fabian and F. Giovannelli eds.), Dordrecht: Kluwer, p. 39.

Mewe, R. (1991a): "X-ray Spectroscopy of Stellar Coronae", *Astron. Astrophys. Reviews* **3**, 127.

Mewe, R. (1991b): "X-ray Lines in Stellar Coronae", in *Iron Line Diagnostics in X-ray Sources* (A. Treves, G.C. Perola and L. Stella eds.), *Lecture Notes in Phys.* **385**, 13.

Mewe, R. (1991c): "Solar and Stellar Coronal Loops", *Adv. Space Res.* **11**, (1)127.

Mewe. R. (1992): "X-ray Spectroscopy of Solar and Stellar Coronae", in *Solar and Stellar Coronae* (R. Pallavicini ed.), *Memorie Soc. Astron. Ital.* **63**, 681.

Mewe, R. (1993): "Stellar Coronal X-ray Spectroscopy". in *Physics of Solar and Stellar Coronae* (J.L. Linsky and S. Serio eds.), Dordrecht: Kluwer, p. 225.

Mewe, R., and Gronenschild, E.H.B.M. (1981): "Calculated X Radiation from Optically Thin Plasmas. IV. Atomic Data and Rate Coefficients for Spectra in the Range 1-270 Å", in *Astron. Astrophys. Suppl.* **45**, 11.

Mewe, R., Gronenschild, E.H.B.M., and van den Oord, G.H.J. (1985): "Calculated X Radiation from Optically Thin Plasmas. V. Improved Atomic Data ...", in *Astron. Astrophys. Suppl.* **62**, 197.

Mewe, R., and Kaastra, J.S. (1994): "X-ray Spectral Modelling of Hot Astrophysical Plasmas", *EAS Newsletter* **8**, 3.

Mushotzky, R. (1988): "X-ray Observations of Clusters of Galaxies", in *Hot Thin Plasmas in Astrophysics* (R. Pallavicini ed.), Dordrecht: Kluwer, p. 273.

Mushotzky, R. (1992): "X-ray Spectral Images of Clusters", in *Clusters and Superclusters of Galaxies* (A.C. Fabian ed.), Dordrecht: Kluwer, p. 91.

Mushotzky, R.F., and Szymkowiak, A.E. (1988): "*Einstein* Observatory Solid State Detector Observations of Cooling Flows in Clusters of Galaxies", in *Cooling Flows in Clusters and Galaxies* (A.C. Fabian ed.), Dordrecht: Kluwer, p. 53.

Pallavicini, R. (1988): "Stellar Coronae", in *Solar and Stellar Coronal Structure and Dynamics* (R.C. Altrock ed.), Sunspot: National Solar Obs., p. 19.

Pallavicini, R. (1989): "X-ray Emission from Stellar Coronae", *Astron. Astrophys. Reviews* **1**, 177.

Pallavicini, R. (1992a): "Stellar Chromospheres, Coronae, and Winds", in *The Sun. A Laboratory for Astrophysics* (J.T. Schmelz and J.C. Brown eds.), Dordrecht: Kluwer, p. 313.

Pallavicini, R. (1992b): "X-rays from Normal Stars", in *Frontiers of X-ray Astronomy* (Y. Tanaka and K. Koyama eds.), Tokyo: Universal Academy Press, p. 265.

Pallavicini, R., Peres, G., Serio, S., Vaiana, G.S., Golub, L., and Rosner, R. (1981), "Closed Coronal Structures. III. Comparison of Static Models with X-ray, EUV, and Radio Observations", *Ap. J.* **247**, 692.

Pallavicini, R., Peres, G., Serio, S., Vaiana, G.S., Acton, L., Leibacher, J., and Rosner, R. (1983): "Closed Coronal Structures. V. Gasdynamic Models of Flaring Loops and Comparison with SMM Observations", *Ap. J.* **270**, 270.

Peres, G, Reale, F., Serio, S., and Pallavicini, R. (1987): "Hydrodynamic Flare Modelling: Comparison of Numerical Calculations with SMM Observations", *Ap. J.* **312**, 895.

Raymond, J.C. (1988): "Radiation from Hot Thin Plasmas", in *Hot thin Plasmas in Astrophysics* (R. Pallavicini ed.), Dordrecht: Kluwer, p. 3.

Raymond, J.C. (1990): "Emission Lines from Hot Astrophysical Plasmas", in *High Resolution X-ray Spectroscopy of Cosmic Plasmas* (P. Gorenstein and M. Zombeck eds.), Cambridge University Press, p. 1.

Raymond, J.C., and Smith, B.W. (1977): "Soft X-ray Spectrum of a Hot Plasma", *Ap. J. Suppl.* **35**, 419.

Roger, R.S., and Landecker, T.L. (eds.): 1988, *Supernova Remnants and the Interstellar Medium* Cambridge University Press.

Rosner, R., Golub, L., and Vaiana, G.S. (1985): " On Stellar X-ray Emission", *Ann. Rev. Astron. Astrophys.* **23**, 413.

Rosner, R., Tucker, W.H., and Vaiana, G.S. (1978): "Dynamics of the Quiescent Solar Corona", *Ap. J.* **220**, 643.

Sarazin, C.L. (1988a): *X-ray Emissions from Clusters of Galaxies*, Cambridge University Press.

Sarazin, C.L. (1988b): "Properties of Clusters of Galaxies", in *Cooling Flows in Clusters and Galaxies* (A.C. Fabian ed.), Dordrecht: Kluwer, p. 1.

Sarazin, C.L. (1990): "X-ray Spectra of Clusters of Galaxies", in *High Resolution X-ray Spectroscopy of Cosmic Plasmas* (P. Gorenstein and M. Zombeck eds.), Cambridge University Press, p. 209.

Sarazin, C.L. (1992): "The Intracluster Medium", in *Clusters and Superclusters of Galaxies* (A.C. Fabian ed.), Dordrecht: Kluwer, p. 131.

Sarazin, C.L., and Wise, M.W. (1991): "Iron Line Diagnostics in Elliptical Galaxies and Cluster Cooling Flows", in *Iron Line Diagnostics in X-ray Sources* (A. Treves, G.C. Perola and L. Stella eds.), *Lecture Notes in Phys.* **385**, 57.

Schmitt, J.H.M.M. (1988): "X-ray Emission from Normal Stars", in *Hot Thin Plasmas in Astrophysics* (R. Pallavicini ed.), Dordrecht: Kluwer, p. 109.

Schmitt, J.H.M.M. (1990a): "Stellar X-ray Astronomy", *Adv. Space Res.* **10**, (2)115.

Schmitt, J.H.M.M. (1990b): "X-ray Spectroscopy across the HR Diagram", in *High Resolution X-ray Spectroscopy of Cosmic Plasmas* (P. Gorenstein and M. Zombeck eds.), Cambridge University Press, p. 110.

Schmitt, J.H.M.M. (1992a): "ROSAT Observations of Late-type Stars", in *Cool Stars, Stellar Systems, and the Sun* (M.S. Giampapa and J.A. Bookbinder eds.), ASP Conference Series N. 26, p. 83.

Schmitt, J.H.M.M. (1992b): "First Stellar Results from the ROSAT XRT", in *Solar and Stellar Coronae* (R. Pallavicini ed.), *Memorie Soc. Astron. Ital.* **63**, 563.

Schmitt, J.H.M.M. (1993) "ROSAT Observations of Late-type Stars", in *Physics of Solar and Stellar Coronae* (J.L. Linsky and S. Serio eds.), p. 327.

Schnopper, H.W. (1995): "Hot Plasmas in Space: X-ray Diagnostic Instrumentation", this volume.

Sciortino, S. (1993): "Stellar Coronal Emission: What we Have Learned from pre-ROSAT Observations", in *Physics of Solar and Stellar Coronae* (J.L. Linsky and S. Serio eds.), p. 211.

Serio, S. (1987): "Theory of Stellar Coronae", in *Circumstellar Matter* (I. Appenzeller and C. Jordan eds.), p. 347.

Stern, R.A. (1983): "*Einstein* Observations of Cool Stars", *Adv. Space Res.* **2**, (9)39.

Stewart, G.C. (1992): "X-ray Emission from Clusters of Galaxies", in *Frontiers of X-ray Astronomy* (Y. Tanaka and K. Koyama eds.), Tokyo: Universal Academy Press, p. 447.

Tanaka, Y., Inoue, H., and Holt, S.S. (1994), "The X-ray Astronomy Satellite ASCA", *Publ. Astron. Soc. Japan* **46**, L37.

Vaiana, G.S. (1990): "X-ray Emission from Stars: A Sharper and Deeper View of Our Galaxy", in *Imaging X-ray Astronomy* (M. Elvis ed.), p. 61

Vaiana, G.S., Maggio, A., Micela, G., and Sciortino, S. (1992): "Coronal Emission and Stellar Evolution", in *Solar and Stellar Coronae* (R. Pallavicini ed.), *Memorie Soc. Astron. Ital.* **63**, 545.

Vaiana, G.S., and Sciortino, S. (1987): "Observations of Stellar Coronae", in *Circumstellar Matter* (I. Appenzeller and C. Jordan eds.), p. 333.

Vaiana, G.S., and Rosner, R. (1978): "Recent Advances in Coronal Physics", *Ann. Rev. Astron. Astrophys.* **16**, 393.

White, N.E., Arnaud, K., Day, C.S.R., Ebisawa, K., Gotthelf, E.V., Mukai, K., Soong, Y., Yaqoob, T., and Antunes, A. (1994): "An ASCA Observation of One Orbital Cycle of AR Lacertae", *Publ. Astron. Soc. Japan* **46**, L97.

Winkler, P.F., Canizares, C.R., Clark, G.W., Markert, T.H., Kalata, K., and Schnopper, H.W. (1981): "A Survey of X-ray Line Emission from the Supernova Remnant Puppis A", *Ap. J. Letters* **246**, L27.

Hot Plasmas in Space:
X-ray Diagnostic Instrumentation

Herbert W. Schnopper

Danish Space Research Institute, Gl. Lundtoftevej 7, DK-2800, Lyngby, Denmark; hs@dsri.dk
Laboratory for High Energy Astrophysics, NASA/GSFC, Greenbelt, MD 20971, USA

Abstract. A new era of X-ray astrophysics began when the imaging and spectral data from *ROSAT* and *ASCA* became available. The information content of cosmic X-ray plasmas is revealed when these data are combined with powerful plasma diagnostic codes. Plasmas as diverse as stellar coronal regions and cooling flows on massive members of clusters of galaxies begin to reveal their secrets. The next generation of X-ray missions will extend this progress by introducing medium- and high-resolution spectroscopy for routine observations. The present and future missions, their instrumentation and their sensitivities for spectral features are discussed.

Keywords. Hot plasmas, X-ray diagnostics, X-ray detectors, X-ray imagers, X-ray spectrometers, space instrumentation

1 Introduction

The hot plasmas encountered in the Universe are characterized by a large range of parameter space. They can range in size from a small active region on a star to the whole of the intergalactic medium in a large cluster of galaxies or even a supercluster. Their temperatures can range from a few hundred eV in the coronal regions of some stars to ten thousand eV in rich clusters of galaxies. Their densities can range from less than a few per cm^3 in clusters of galaxies to almost liquid consistency in an accreting compact object. Except in rare cases, the plasmas are not in thermal equilibrium and different temperatures must be used to characterize the motions of electrons and ions. In some sources, there is a distribution of temperatures for both motions. The X-ray spectrum of Centaurus X-3 (Fig. 1.1), recorded by the *CCD* detector on the Japanese satellite *ASCA*, is a good example of this complexity. After modest beginnings on early missions, the variety of diagnostic instrumentation on present and future X-ray satellites will provide data of a quality similar to that obtained routinely for Solar plasmas. Stimulated by this flow of data, those who seek an understanding of the underlying physical processes in cosmic X-ray sources will find their endeavors well rewarded.

X-rays with energies that are interesting for the diagnostics of cosmic plasmas are almost fully absorbed in air over distances from a few mm to several m. Elaborate vacuum systems are used to gain access to the X-rays produced by Tokamaks and laser produced plasmas. The enormous X-ray output from our most brilliant Sun is overwhelmed by the Earth's atmosphereand it cannot be seen from the ground. At balloon altitudes of 40-50 km, the X-ray diagnostic energy range is fully absorbed and only X-rays above 15 keV can be detected. Rockets bring instruments to heights of hundreds of km, but they remain there for only several hundred seconds observing only the brightest of extra-Solar sources. X-ray astronomy came of age with the first satellite, *UHURU* (1970), and reached maturity with the first X-ray telescope on *EINSTEIN* (1979). X-ray telescopes are now quite common.

The present discussion will start with an overview of missions, present and future, that have the capability to perform meaningful plasma diagnostics of cosmic X-ray sources. Their instruments will be discussed and finally, their sensitivities for the detection of spectral features will be presented. A list of acronyms used in the text is provided in Section 10.

1.1 Setting the Stage

From an X-ray astrophysicist's point of view, there is the Sun and then there is everything else. Its relative brilliance made it the trend setter in the development of X-ray diagnostic instrumentation. Solar physicists using rocket borne, Bragg spectrometers of the classical type were able to obtain spectra with high resolution and statistical significance while astrophysicists seeking extra-solar sources were barely able to find them using large area gas filled proportional counters with almost no spectral resolution.

NASA's Solar Maximum Mission *(SMM)* and Japan's *YOHKOH* have provided data with unprecedented spectral and imaging resolution of Solar active regions and flares. Various spectra of Solar flares recorded in the He-like wavelength range for Fe, Ca, S, Si and O are shown in Fig. 1.2 (Culhane *et al.* 1981; Harra-Murnion *et al.* 1995; Keenan *et al.* 1994)and in Fig. 1.3 (McKenzie *et al.* 1980). The resolution is sufficient to reveal satellite lines and contributions from Li- and Be-like charge states. These have great diagnostic significance. Perhaps the best non-Solar example is the *EINSTEIN* Bragg Spectrometer spectrum of the Puppis A SNR shown in Fig.1.4 (Winkler *et al.* 1981). The most meaningful plasma diagnostics, those that resolve the main features, are possible only for $\Delta E/E \gtrsim 10^{-3}$. This is the goal of some of the high resolution instruments described below. Only a small number of known cosmic sources, however, have fluxes sufficient to be studied with a high energy resolution at the modest sensitivity levels attainable with some of these instruments.

2 About Hot, X-ray Emitting, Plasmas

Following ground based observers, X-ray astrophysicists seeking to study faint sources use telescopes to collect a large cross section of incoming flux and concentrate it on a detector pixel of relatively small size. Usually, the telescope images as well as concentrates. Elements can be introduced into the optical path that will disperse the beam into its many spectral components and the instrument is transformed into a spectrometer. It is usually easier to design a spectrometer to study point (star-like) sources rather than extended ones.

The goal for the proper diagnostics of a cosmic plasma is to obtain spectra and to deduce from them the following parameters and their distribution in point- and extended sources:

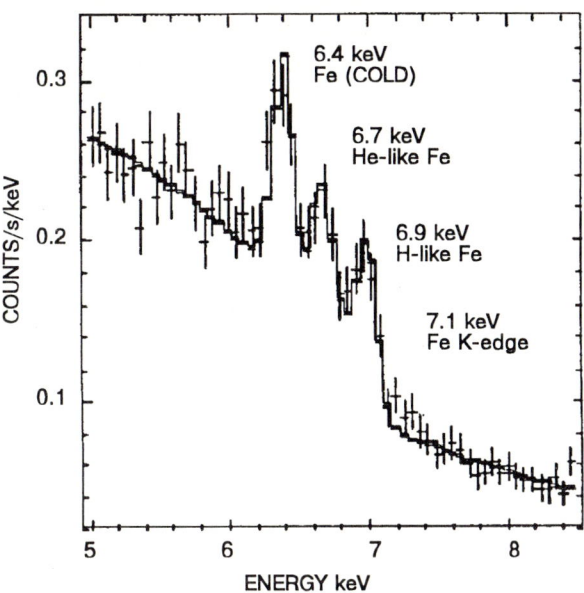

Fig. 1.1 The *ASCA* spectrum of CEN X-3. Hot plasma is responsible for the continuum and the He- and H-like Fe emission. A portion of the X-rays with energies above the Fe K-absorption edge at 7.1 keV is removed from the continuum by ionizing a gas cloud surrounding the central source. The thickness of the gas and its density determine the amount by which the continuum is lowered. The Fe $K\alpha_{1,2}$ radiation at 6.4 keV comes from cold material.

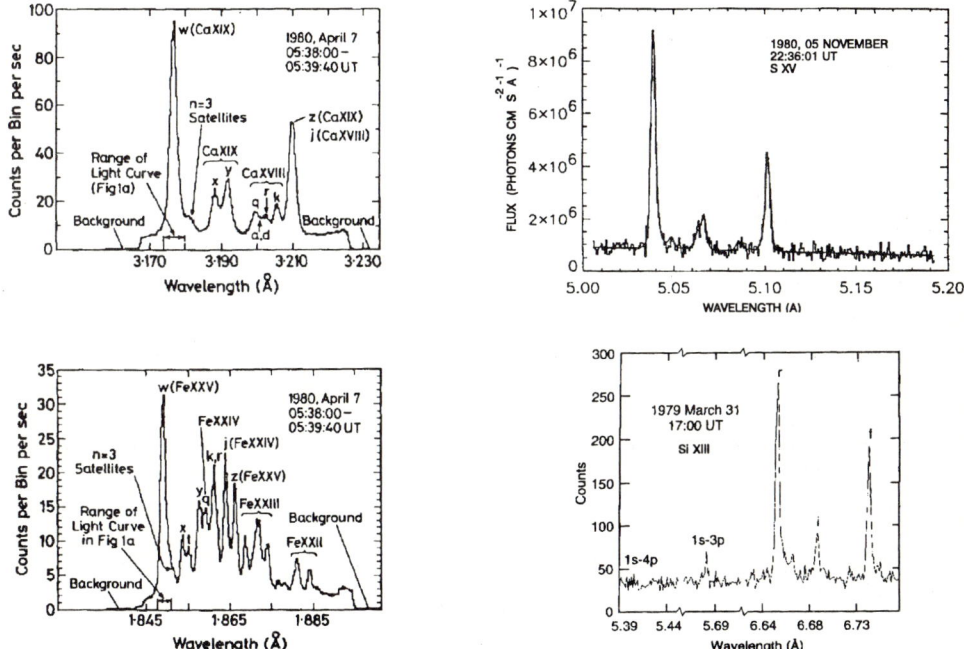

Fig. 1.2 Solar spectra in the He-like wavelength range for Fe, Ca, S, and Si taken by various missions.

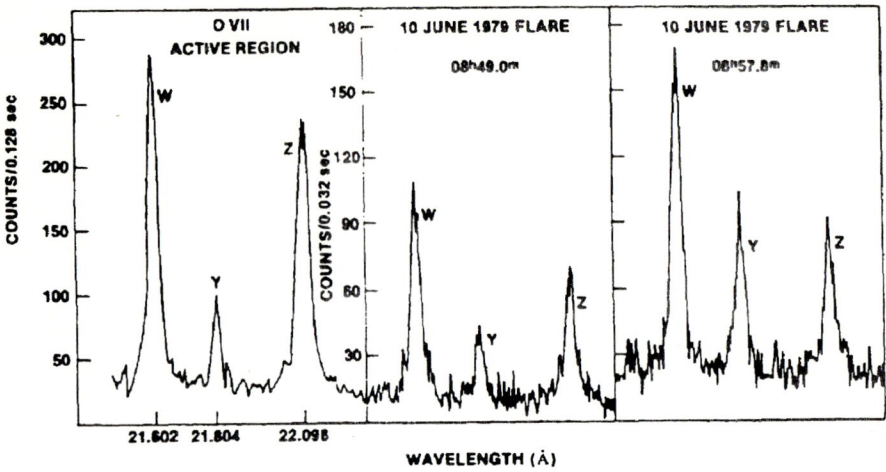

Fig. 1.3 The Solar spectrum in the He-like wavelength range for O. The line ratios depend upon plasma conditions and vary accordingly. Dramatic changes can happen within the time span of a single flare. Note the absence of major satellite structure

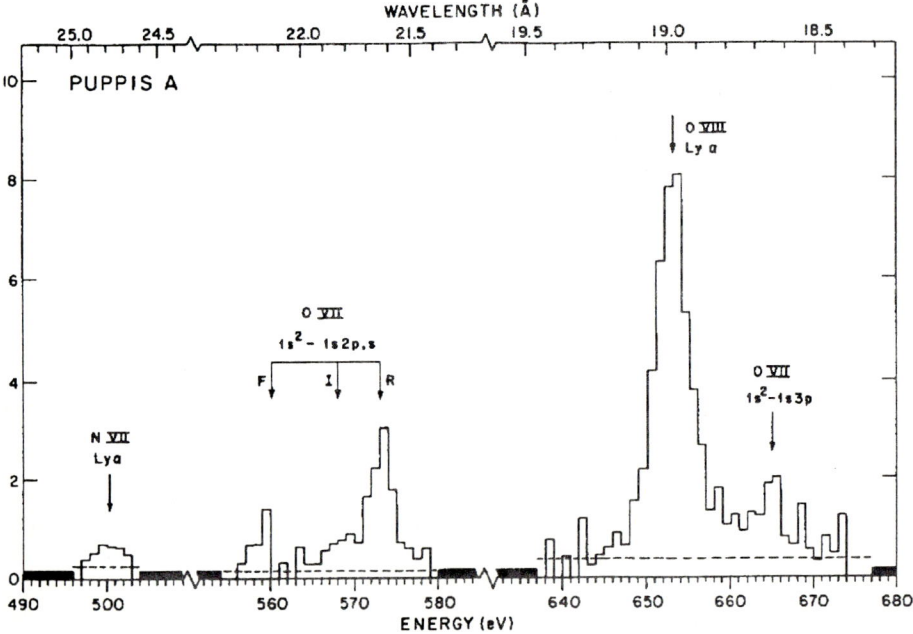

Fig. 1.4 Three segments of the spectrum of a portion of the Puppis A SNR. Compare the He-like O lines (OVII) with the Solar data. Note that the counting rates differ by a factor $>10^5$ in the two spectra.

♦ The electron temperature and density.
♦ The ion abundances, temperatures, charge states and densities.
♦ (The redshift distance of the source).

The information that enables the diagnostics comes in the form of broad- and narrow-band spectra over a wide range of energies. The 0.1 - 10 keV energy range, which is most useful for the diagnostics of cosmic X-ray plasmas, contains the K- and L-series emission lines and absorption edges for various states of ionization of the abundant ions from C through Fe.

Electron-proton collisions produce a thermal bremsstrahlung whose featureless form, in the ideal case, is defined by a single parameter, T_e, the electron temperature. Hot plasmas produce ions with charge states that depend upon the atomic number, Z, of the ion and T_e. Ions that are most useful for diagnostic studies are those with a few remaining electrons, the H-like and He-like states being the favorites. Collisions between the hot ions and electrons produce excited states that sometimes include an inner shell vacancy. The excited state subsequently decays either by the emission of radiation or as a result of a collision with one of the plasma electrons. If the vacancy was filled by a bound electron, this process leads to the emission of an X-ray line that is characteristic of: the ion; its initial charge state; its thermal motion defined by its temperature, T_i; its density, n_i; and its bulk motion.

The details of the observed spectrum are determined by a combination of the atomic properties of a single ion at rest and the collective properties of the plasma. For a population of isolated ions of atomic number Z, the strengths and widths of lines emitted from various excited states are determined by their lifetime. Put the same ions in motion in a collisional plasma and the line strengths and widths of the long- vs. short-lived states will change differently depending upon the density and temperature. X-ray emission from long lived, radiative states will not be as prominent when the channel for collisional decay opens. Line ratios are important diagnostics for the determination of temperatures and densities . The random thermal motion of the ions in the plasma can lead to line widths in excess of those determined by their lifetime and, if measurable, can be used to determine T_i. When $T_e \neq T_i$, the plasma is not in equilibrium.

If the plasma has a bulk motion e. g., an expanding shell of a supernova remnant or a jet escaping a quasar, then the whole of the emitted spectrum will be red/blue shifted according to the line of sight component of the motion. Cosmological red shifts produce the same effects on the spectrum. The emitted spectrum may also be modified as the X-rays pass through material along the line of sight to the observer e. g. X-rays emitted from the core of a compact object that pass through a cooler accretion disc. Absorption and radiatively excited emission features can be added in this way.

The observed spectrum usually consists of a continuum and a series of lines with different strengths and widths. The emitted spectrum has been modified by the properties of the detection system. The diagnostic information can be extracted from the observed spectrum if the spectrometer has been well calibrated and the processes taking place in the source are described by a well suited model. Various models appropriate to hot plasmas are discussed by Pallavicini (1995) elsewhere in this volume.

3 Missions: Present and Future

Five missions, *ROSAT, ASCA, SRG, AXAF* and *XMM* are discussed briefly. Parameters relevant to the telescopes and the instruments relevant to each are given later in Tables 4.4.1, 5.3.1 and 6.4.1. Auxiliary instruments, co-aligned with the X-ray instruments, are often present on X-ray missions. Star trackers provide accurate aspect information that can be used to map the pointing direction and roll angle of the X-ray telescope as the spacecraft drifts with time. Then, this information is used on the ground to accurately map the origin of each arriving X-ray photon on the celestial sphere. Co-aligned optical and UV monitors, both having broad band filters, provide complementary spectroscopic and timing data at other wavelengths. Their data, usually obtained with high angular resolution, can help identify previously unknown X-ray sources. These auxiliary instruments and non-spectroscopic X-ray detectors are not discussed further.

3.1 *ROSAT*

On 01 April 1990, *ROSAT* (Fig. 3.1) (Trümper unpublished) was launched from Cape Canaveral on board a *Delta II* launch vehicle. Its mission goals were: To obtain broad band fluxes and spectra from the first complete, high angular resolution, all-sky survey at X-ray and *XUV* energies, and; To obtain spectral, temporal and spatial properties of selected sources from pointed mode studies. *ROSAT* (Trümper, 1984) is a German mission with the X-ray telescope and multiwire proportional counter were provided by MPE, Garching, the high resolution channel plate imager by SAO/CFA and the wide field *XUV* telescope and camera by a consortium of British institutions led by the University of Leicester. (The high resolution imager and wide field *XUV* camera are not discussed in this paper.)

ROSAT completed the all-sky survey after the first six months of operation. Since then it has been used to examine a wide variety of sources at distances as close as the Moon and Jupiter and as distant as quasars. It has obtained valuable imaging and spectral data on the hot plasmas in stellar coronae, supernova remnants, galaxies, and clusters of galaxies.

For *ROSAT*, the design choices have been to use:
♦ Large diameter, short focal length, exact Wolter I optics providing high angular resolution and a wide field of view at low energies.
♦ Imaging, gas filled proportional counters providing modest imaging and spectral resolution for low energy X-rays.
♦ A high resolution, channel plate camera providing images at low energy at the limit of telescope resolution.

3.2 *ASCA*

ASCA (Fig. 3.2) (Inoue 1992) was launched from Japan on 20 February 1993. Its goal is to obtain broad band spectra with moderate resolution in an energy range that includes the Fe spectral features. *ASCA* (Tanaka *et al.* 1994) carries four thin foil conical telescopes provided by NASA/GSFC and Nagoya University (Serlemitsos *et al.* 1984). Two of them

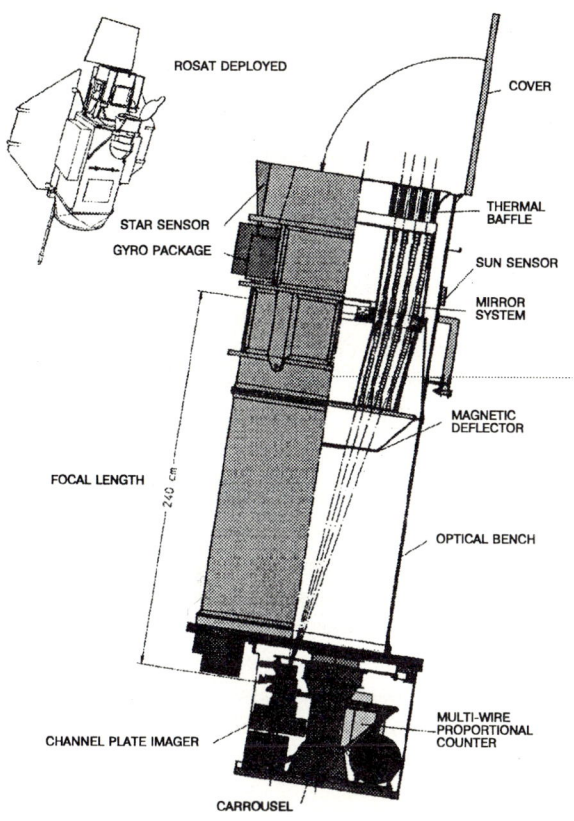

Fig. 3.1 The *ROSAT* Observatory.

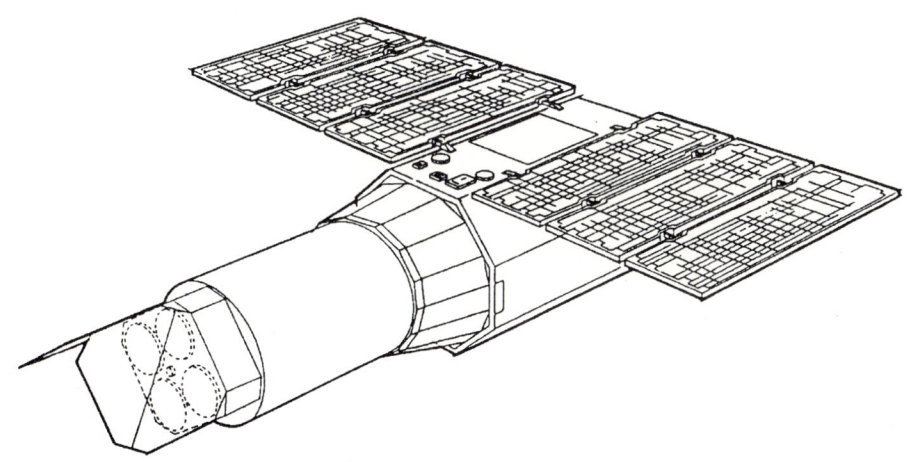

Fig 3.2 The *ASCA* Observatory.

are equipped with *CCD* cameras provided by MIT (Burke *et al.* 1993) and the other two are equipped with gas scintillating proportional counters provided by Tokyo University and ISAS (Kohmura *et al.* 1993) . *ASCA* has provided the first in orbit test of a CCD spectrometer and has demonstrated the enormous scientific potential of the combination of thin foil telescopes with semiconductor spectrometers.

For *ASCA*, the design choices have been to use.

♦   A compact array of 4, short focal length, thin foil optics providing maximum collecting area over the energy range including Fe emission while preserving reasonable energy resolution.

♦   Broad band, spectroscopic imaging *CCD* detectors providing a wide energy range and uniform gain.

3.3 Spectrum X-Gamma *(SRG)*

Work on the *SRG* mission (Fig. 3.3) (Schnopper 1994) began in 1987. It is to be the first of a series of new astronomical observatories being developed under the sponsorship of the Russian Academy of Sciences with financial support from the Russian Space Agency, scientific support from the Space Research Institute of the Academy and technical support from the Babakin Institute of the Lavochkin Association. *SRG*, NASA's *AXAF* and ESA's *XMM* share an overlapping set of scientific priorities. It is the emphasis within that set that serves to define the unique contribution that each of these observatory class missions can make in the study of cosmic X-ray sources.

The expected launch window opens in late-1996 and the observatory is intended to be operated for an initial period of 3 years. The satellite will be launched into a deep, highly eccentric orbit with a period of about four days. This choice of orbit enables long duration (up to about 80 h) observations to be made. Ground station operations which include up- and down-links for data and command transfer will take place approximately once every 24 hours. As experience grows, it should be possible to plan for 10 reorientations of the spacecraft per 24 hour period. With all the operational and pointing constraints taken into account, there will be access to approximately 80 per cent of the celestial sphere at any time.

Two telescope facilities, the Soviet Danish Röntgen Telescope (*SODART*, see Fig. 3.3) Schnopper, H. 1994) and the Joint European X-ray Telescope (*JET-X*, see Fig.3.4) (Wells, *et al.* 1994) will provide a broad range of plasma diagnostic capability. In addition to *SODART* and *JET-X*, other instruments on *SRG* will cover the energy range from EUV to gamma rays and are not discussed.

3.3.1 *SODART*

Two *SODART* (Fig.3.4) (Schnopper 1994) thin foil X-ray telescopes will fly on *SRG*. In the focal plane of one telescope there are: the high- and low energy imaging proportional counters (*HEPC/LEPC*), provided by a Danish led consortium (Budtz-Jørgensen *et al.* 1994); a solid state spectrometer array (*SIXA*), provided by a Finnish led consortium (Vilhu

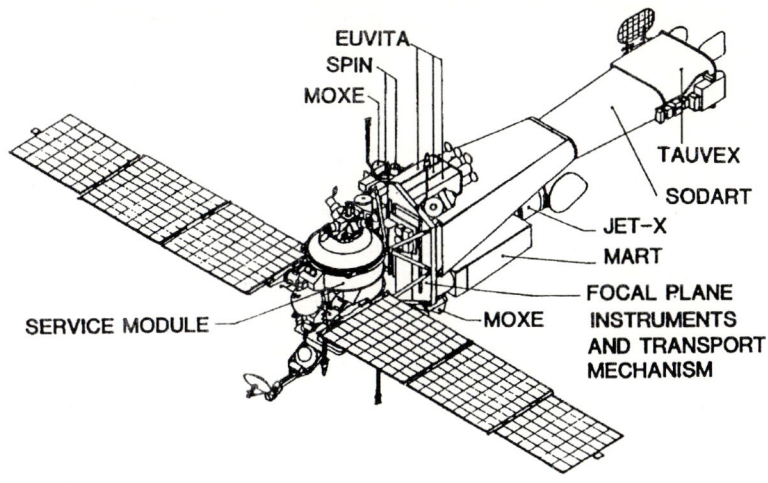

Fig. 3.3 The *SPECTRUM X-GAMMA* Observatory

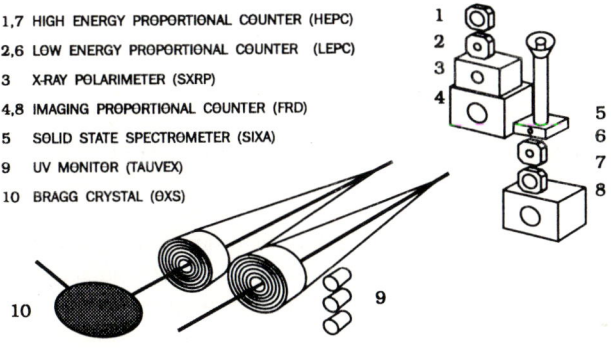

1,7 HIGH ENERGY PROPORTIONAL COUNTER (HEPC)

2,6 LOW ENERGY PROPORTIONAL COUNTER (LEPC)

3 X-RAY POLARIMETER (SXRP)

4,8 IMAGING PROPORTIONAL COUNTER (FRD)

5 SOLID STATE SPECTROMETER (SIXA)

9 UV MONITOR (TAUVEX)

10 BRAGG CRYSTAL (ΘXS)

Fig. 3.4 The elements of *SODART*.

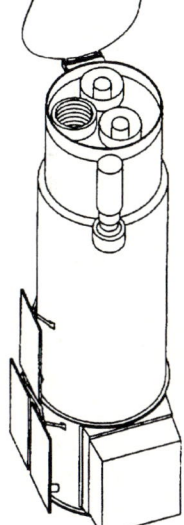

Fig. 3.5 The *JET-X* telescope system.

et al. 1994); and a focal plane X-ray detector *(KFRD)*, provided by a Russian led consortium. In the other, there are: another *HEPC/LEPC* pair; a stellar X-ray polarimeter *(SXRP)*, provided by NASA (Kaaret *et al.* 1990); and another *KFRD*. An objective Bragg crystal spectrometer *(OXS)*, provided by a Danish led consortium (Christensen *et al.* 1994), is mounted in front of one of the telescopes. Mounted alongside and co-aligned with the *SODART* telescopes is a unit consisting of three UV telescopes *(TAUVEX)*, provided by Israel (Brosch *et al.* 1994). In addition to its scientific use, *TAUVEX* can be linked to the spacecraft attitude control system to provide data to the station keeping control units that stabilized the orientation of the spacecraft. These instruments and their scientific goals will be described briefly.

For *SODART*, the design choices have been to use:
♦ long focal length, thin foil optics providing the maximum collecting area over the widest possible energy range while preserving a reasonable angular resolution.
♦ Broad band, large area, imaging detectors providing the widest possible energy range; uniform gain and imaging response; and high count rate capability.
♦ A high resolution Bragg spectrometer capable of providing spectral imaging of extended objects without severe loss of spectral resolution.
♦ A polarimeter capable of using the large collecting area and wide energy range provided by the *SODART* telescope.

3.3.2 *JET-X*

Two replicated, exact Wolter I telescopes, provided by Italy (Citterio *et al.* 1994) as part of *JET-X* (Fig. 3.5) (Wells *et al.* 1994), will fly on *SRG*. Focal plane imaging on each is provided by a cooled *CCD* detector, provided by the United Kingdom (Owens *et al.* 1994), that combines high spatial resolution with good spectral resolution. In addition, a star sensor is co-aligned with the telescopes and will provide the aspect information that is required to form accurate X-ray images from data acquired over a long integration time.

For *JET-X*, the design choices have been to use:
♦ Replicated Wolter I optics providing reasonable collecting area over the widest possible energy range while preserving good angular resolution.
♦ Broad band, spectroscopic imaging, *CCD* detectors providing a wide energy range, uniform gain, and high count rate capability.

3.4 Advanced X-ray Astrophysics Facility *(AXAF)*

AXAF (Fig. 3.6) (Zombeck 1994; Weisskopf *et al.* 1995), one of NASA's series of Great Observatories, is scheduled for launch in late 1998. *AXAF's* single telescope is a high resolution mirror assembly *(HRMA)* that has four nested pairs of ground and polished, exact Wolter I, grazing incidence mirrors that are provided by NASA. Low- and a medium/high-energy transmission grating spectrometers *(LETGS)* and *(HETGS)*, are provided by the Netherlands (Brinkman *et al.* 1987) and NASA (Markert *et al.* 1994), respectively. The two focal plane instruments, both provided by NASA, are a *CCD* imaging spectrometer *(ACIS)* and a high resolution, channel plate camera *(HRC)*. *ACIS* (Lumb *et al.* 1993, Burke *et al.* 1993) consists of two components: *ACIS-I* is a square array to be used as a broad band, imaging spectrometer and *ACIS-S* is a linear array to be used to read the image of the spectrum cast by the *HETGS* gratings. A channel plate detector, *HRC-S*, reads out the *LETGS* spectrum. Both detectors are mounted on curved surfaces to avoid aberration.

For *AXAF*, the design choices have been to use:

◆   Ground and polished, exact Wolter I, optics providing sub-arcsec angular resolution that is suitable for deep surveys and the study of fine extended structure.
◆   Broad band, spectroscopic imaging, *CCD* detectors providing a wide energy range, uniform gain, and high count rate capability.
◆   A high resolution, channel plate camera providing images at the limit of telescope resolution.
◆   Transmission grating spectrometers providing the highest possible low energy spectral resolution.

3.5 X-ray Multi-Mirror *(XMM)*

XMM (Fig. 3.7) is the third of ESA's Cornerstone Missions within the Horizon 2000 long term plan. It is scheduled for launch at the end of 1999. Three replicated, exact Wolter I telescopes, provided by ESA, will fly on *XMM*. Each telescope will have a *CCD* camera (*EPIC*), provided by a consortium led by Italy, with two of the cameras provided by the UK (Holland *et al.* 1993) and the other by Germany (Bräuninger *et al.* 1992). Two of the telescopes will be shared by reflection grating spectrometers (*RGS*), provided by the Netherlands (Brinkman *et al.* 1989; Kahn and Hettrick 1985) . Each will be read out with a linear *CCD* (Verhoeve *et al.* 1992). An optical monitor will be provided by the UK.

For *XMM*, the design choices have been to use:

◆   Replicated Wolter I optics providing reasonably high collecting area over the widest possible energy range with good (sub-arcmin) angular resolution.
◆   Broad band, spectroscopic imaging, *CCD* detectors providing a wide energy range, uniform gain, and high count rate capability.
◆   A reflection grating spectrometer providing medium to high energy resolution and good sensitivity, particularly at low energies.
◆   A co-aligned optical monitor.

4 X-ray Telescopes

There are a set of exclusion laws that apply to X-ray telescopes and their instrumentation. X-ray astronomers, with their twin goals of finding ever increasing numbers of sources further and further back in red shift time and of imaging the fine details in extended sources, put the requirement for angular resolution ahead of that for collecting area and energy resolution. Faint sources must be seen with statistical significance against the contribution from the ever present diffuse X-ray background. Poor angular resolution leads to point source confusion and the loss of structural detail.

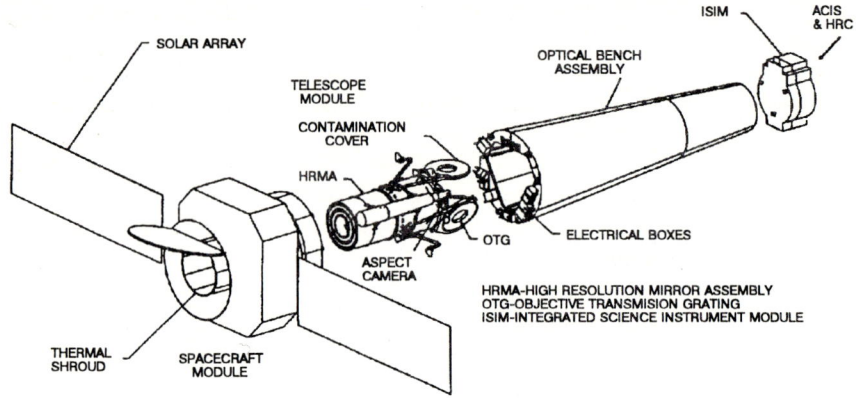

Fig. 3.6 The *AXAF* Observatory.

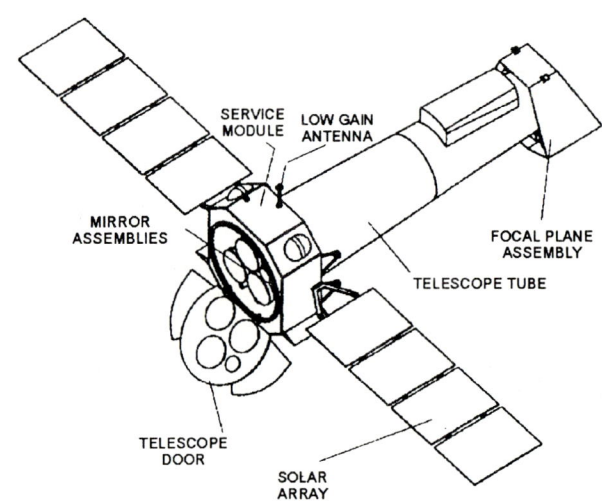

Fig. 3.7 The *XMM* Observatory.

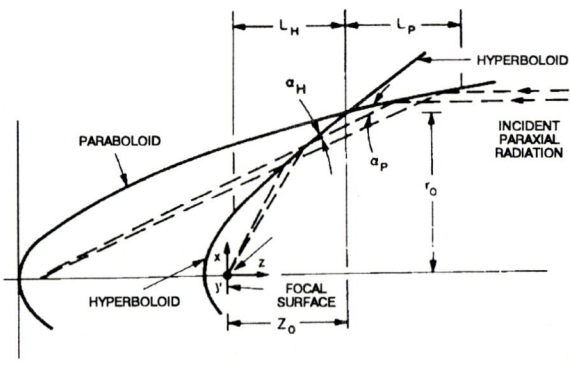

Fig. 4.1 Wolter I optics.

X-ray astrophysicists, with their goal of understanding the nature of the sources, put the requirements for energy resolution and collecting area ahead of angular resolution. High energy resolution spectra require photons to be detected in numbers sufficient to give statistical significance in each energy pixel of interest. Objects at the edge of the universe are not candidates for these investigations and source confusion is generally not a problem.

For most materials, the real part of the index of refraction for X-rays is generally less than 1. Thus, for a given energy, there is a limiting glancing (critical) angle angle above which X-rays are either absorbed or scattered. Below it, they are totally reflected (except for absorption in some cases). For a given energy, the value of the critical angle depends upon the atomic number of the reflecting material, the higher the better. For the diagnostic energy range, the critical angle is typically of the order of 1 deg or less. Therefore, except under special circumstances, it is not possible to fashion a reflecting X-ray telescope with the geometry normally associated with telescopes that operate at other wavelength bands. Grazing incidence reflection is the only choice for a broad band X-ray telescope.

Telescopes designed for the study of X-ray spectra should be capable of achieving significant sensitivity for energies up to 10 keV. This will allow them to explore the full range of Fe emission and absorption features that provide valuable diagnostic information. Given the limitations imposed by grazing incidence geometry, this means telescopes of rather large f-number (typically ~3-15). Sufficient collecting area can be obtained from one or more long focal length telescope *(SRG, AXAF, XMM)* or several shorter focal length telescopes *(ASCA, ASTRO-E)*.

Wolter (1952a,b) described a series of telescopes based upon grazing incidence reflection. The most common one used for x-ray astronomy is called Wolter I optics (Fig. 4.1). It consists of a nested set of confocal hyperboloid-paraboloid shells each with a slightly different range of grazing angles. The nesting is required to improve the filling factor of the circle formed by the outermost shell. This ranges from less than 10 percent to greater than 65 percent and depends upon the technology used to make the telescope.

There are various figures of merit *(FOM)* for X-ray telescopes, but the one most commonly quoted is called the full width at half power *(FWHP, HPW)*. It is theangular diameter of the circle that encloses half the photons collected from a point source. Equally interesting, but less quoted, is the full width at 90 percent power *(FW90)*. Another important *FOM* is the full width at half maximum *(FWHM)* of the point spread function *(PSF)*. This parameter is important in determining the ability of the telescope to detect closely separated sources. If the *PSF* has large "tails", then weak sources will be masked by nearby bright ones.

The figure of the shell i.e., how well the mean surface of the reflecting shell conforms to the geometric ideal, and the surface roughness usually determine the value of each *FOM*. The diffraction limit, which is of great significance in the performance of optical telescopes, is not likely to be attained by any of the present generation of X-ray telescopes. Contributions to the *FOM* from telescope surface figure errors are usually independent of X-ray energy. The surface roughness contribution, in particular the part arising from the distribution of the correlation lengths of surface irregularities, can be strongly dependent upon energy. All useful designs require a surface roughness of not more than several Å.

4.1 Polished Optics

The telescopes flown on *EINSTEIN* and *ROSAT* have provided images with the highest angular resolution yet to be obtained. *AXAF* (Ried 1995) will carry that record one step further. Telescope shells can be made from an optical quality, low expansion, fused quartz or Zerodur. The raw material is obtained in blocks from which the individual shells are sawn. Each shell is then machined, ground and finally polished to the exact figured required by the Wolter I geometry. The desired result is a telescope with figure errors that are less than a few μm and surfaces that are smoother than 5-10 Å. The design of the structure that holds the shell is equally important. The goal is that all shells shall be confocal with their optic axes parallel. Extensive testing takes place at all stages of production. This process can take years to complete. Excellent test results on a verification model of the AXAF telescope have been reported by Freeman *et al.* (1993) and Hughes *et al.* (1993).

The structural requirements for stability during thermal excursions and launch vibrations add to those already imposed by the stringent requirements for angular resolution. The result is that rather thick shells are necessary and only a few (four for *EINSTEIN, ROSAT* and *AXAF*) can be accommodated in the telescope cell. The thickness of the innermost shell and the difficult nature of the manufacturing process mean that there is not much collecting area close to the telescope axis. Telescopes with polished optics, therefore, will typically have an effective area *vs.* energy response that falls more rapidly than that for telescopes produced using the technologies described below.

4.2 Replicated Optics

Telescopes now being developed for *JET-X* (Wells *et al.* 1994) and *XMM* (Gondoin *et al.* 1994) consist of a nest of thin shell replicas of exact Wolter I optics. The shells replicate the figure obtained by grinding and polishing a suitable mandrel. One mandrel is required for each shell in the telescope. For both *JET-X* and *XMM*, each mandrel is ground and polished to have the appropriate hyperboloid and the paraboloid element.

If some compromise with angular resolution is allowed, the shell thickness can be as small as several mm. This allows for a more densely filled aperture and that leads to more collecting area *vs.* energy, especially at higher energies. A separate mandrel is required for each shell and, although costly, the net cost per telescope is less that for the glass version. Each mandrel can, however, be used more than once. Thus, there are additional scientific and economic benefits for missions such as *XMM* that have several telescopes.

Once the mandrel is prepared it is coated with an evaporated layer of the material that will ultimately become the reflecting surface of the telescope. The shell material is electroformed Ni for *JET-X* and *XMM*, but other materials such as carbon fiber reinforced epoxy can be used. After it is glued to the mandrel, a thermal shock is used to separate the shell from the mandrel. If all goes well, the reflecting material, originally on the mandrel, adheres preferentially to the glue on the inner surface of the shell and forms a perfect replica of the mandrel. Test results of verification models of JET-X (Wells *et al.* 1994; Citterio *et al.* 1995) and XMM (Gondoin *et al.* 1994) telescopes agree well with specifications.

4.3 Approximate Optics

A considerable portion of the cost of either ground and polished or replicated optics goes into the grinding and polishing of shells or mandrels in an attempt to make the surfaces as smooth as possible to prevent X-ray scattering. Although here are other ways of obtaining the required smooth surfaces, the simplest cannot be easily formed into Wolter I optics.

If a further loss of angular resolution can be tolerated, then approximations to Wolter I optics can be considered. The typical shell length of a Wolter I X-ray telescope is usually small when compared with its focal length. If it is reduced even further, then sections of conical surfaces will be good approximations to the paraboloidal and hyperboloidal surfaces required by the exact Wolter I geometry (Serlemitsos *et al.* 1984). Ray tracing studies have shown that the contribution to the *FWHP* from the conical geometry can be < 20 arcsec. This contribution decreases slightly with energy and does not increase significantly for a source located at the edge of the field of view. The object of this approach, in addition to cost benefits, is to make the most efficient use of the aperture by producing a very dense nest of thin shells. In addition, shells can be placed very close to the telescope axis. For a given telescope aperture, this design produces the largest collecting area and the highest maximum useful energy. The present designs call for the telescopes to be made from identical quadrants. Each is assembled from two subunits, one approximating the paraboloid the other the hyperboloid. This approach has been tested with *BBXRT* (Serlemitsos *et al.* 1984) and flown on *ASCA* (Tanaka *et al. 1994)*. It is also the basis of the *SODART* (Westergaard *et al.* 1990; Schnopper *et al.* 1994) and *ASTRO-E* (Serlemitsos and Soong 1995) designs.

Commercially available aluminum foils with thicknesses between 0.1 and 0.4 mm can be obtained with very smooth surfaces. After initial inspection and cleaning, they are cut to the appropriate shell length (usually a constant) and a width somewhat greater than one-fourth the circumference at the appropriate location in the nest. Next, they are rolled to an approximate cylindrical shape. Because of the straight cuts that define the shell's width, it does not define a precise frustum when properly mounted. The deviations are negligible for a large *f*-number. The surface of the rolled foil is then coated with a thin layer of acrylic lacquer in a batch process. The surface tension in the dried lacquer provides the extremely smooth surface required for minimum X-ray scattering. The lacquer does not, however, remove figure errors. Another batch process is used to gold coat the reflecting surface.

Before mounting in the telescope structure, each foil is cut to a precise length and tested for surface smoothness by making either laser or X-ray scattering measurements. Before being glued in place, the shells assembled in individual quadrants are checked once more for proper alignment and focal length with narrow beam laser and wide optical beam measurements. Excellent test results for the first of 3 flight quality *SODART* telescopes (one will become the flight spare) are reported by Cristensen *et al. (*1995). They were obtained at the Expanded Beam Facility at the Daresbury Synchrotron (Christensen *et al.*1993).

There are alternatives to using a curved foil as the figure of the reflector. Replication of highly polished mandrels formed with both conical sections is one. A hybrid form of replication, where Al foils are cast against a Au or Ir coated master form with an acrylic layer, is being developed for *ASTRO-E* (Serlemitsos and Soong 1995; Soong *et al. 1995)*.

4.4 X-ray Telescopes Compared

The most accurate X-ray telescopes are made from highly polished glass elements. The Ir coated telescope on *AXAF* will have a *FWHP* of < 1 arcsec. The Au coated, high-throughput, *SODART* telescopes on *SRG* are thin foil, conical approximations to Wolter I optics. They will have a *FWHP* of ~2 arcmin. The Au coated Ni, replicated Wolter I optics on *XMM* will have a *FWHP* of <30 arcsec.

The outermost of the four shells of the *AXAF* telescope has a diameter of 1.2 m. Its walls must be very thick to achieve the stability that is required for very high angular resolution. *AXAF*, therefore, makes very poor use of its geometrical aperture, sacrificing cost and collecting area to gain angular resolution. Each of the two *SODART* telescopes has an outer shell diameter of 0.6 m and is filled with 143 Al shells 0.4 mm thick. *SODART* makes good use of its geometrical aperture, sacrificing angular resolution to gain collecting area and low cost. In between lies *XMM*. Each of its three telescopes has an outer diameter of 0.7 m and is filled with 58 Ni shells. *XMM* sacrifices cost to achieve angular resolution and, with multiplicity, gains collecting area.

The various parameters that define the X-ray telescopes that are currently in orbit and those that are likely to be launched before the end of the decade are compared in Table 4.4.1 and in Figure 4.2.

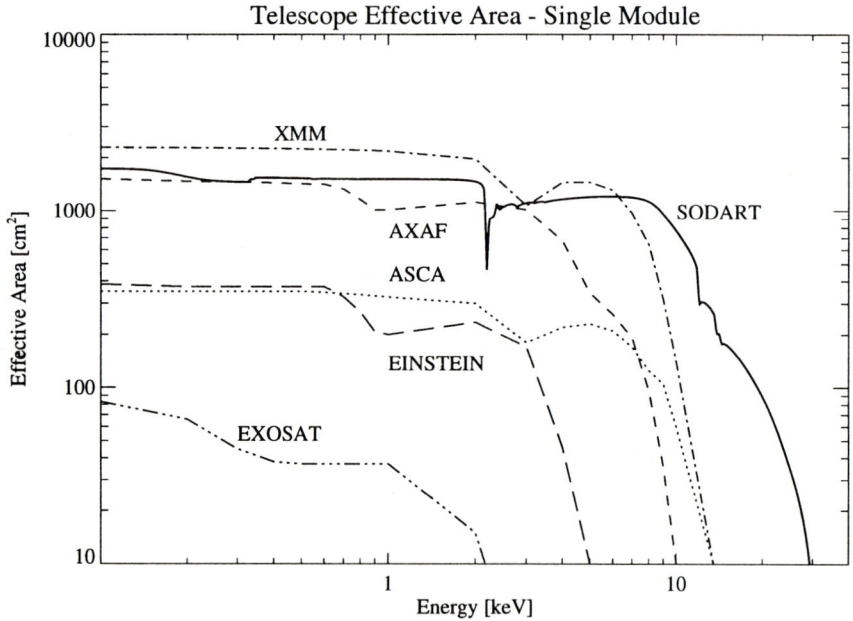

Fig. 4.2 Effective area for a single module of various telescopes.

Table 4.4.1 X-ray telescopes compared (single module)

		ROSAT	ASCA	SRG SODART	SRG JET-X	AXAF	XMM
Field of view	arcmin	120	16-24	60	30	40	30
FWHP	arcsec	3.1	N/A	N/A	12-20	0.7	<30
	arcmin	N/A	4.0	~2.0	N/A	N/A	N/A
Focal length	m	2.4	3.5	8.0	3.5	10.0	7.5
Shell length	cm	50.0 (x2)	10.0 (x2)	20.0 (x2)	30.0 (x2)	83.8 (x2)	30.0 (x2)
Reflecting surface		Au	Au	Au	Au	Ir	Au
Maximum energy	keV	2.4	10	25	10	10	10
Diameter: inner	cm	37	12.0	16	19	63	30.6
outer	cm	84	34.5	60	30	120	70
Number of telescopes		1	4	2	2	1	3
Geometric area/telescope	cm²	1140	558	1752	400	1100	1930
Eff. area @ (keV)	cm²	540 (0.25)	300 (1.5)	1460 (2)	320 (1.5)	900 (0.25)	1330 (2)
@ (keV)	cm²	400 (1.0)	150 (7)	1120 (8)	140 (8)	500 (5)	850 (3)
@ (keV)	cm²	260 (1.5)	60 (20)	90 (20)		120 (8)	620 (8)
Shell: material		Zerodur	Al	Al	Ni	Zerodur	Ni
thickness	mm	15-22	0.125	0.4	0.64-11	15-23	0.47-1.1
coating		Au	Au	Au	Au	Ir	Au
Number of shells		4	120	143	12	4	58
Mass per telescope	kg	760	10	101	62.5	1680	430

5 X-ray Detectors

As the scientific goals have become more directed towards the interpretation of spectra, a need for detectors with ever increasing energy resolution has arisen. As more distant and, therefore, faint sources became interesting, there was also a need for focal plane detectors serving as broad band spectrometers with both good imaging and energy resolution. They should be capable of covering the entire field of view and be sensitive over the energy range of interest. The capability for broad band energy resolution is a desirable feature although, until recently, the highest angular resolution has been obtained with detectors with little or no energy resolution. In those cases, a series of filters was used to isolate broad X-ray energy bands.

Since many interesting X-ray sources are variable on timescales below 1 ms, time resolution becomes an important factor for their study. There is also a kind of exclusion law for X-ray detectors. It is generally not possible to achieve high position-, energy- and time-resolution in one detector. There are also practical limitations. Since each of the measured parameters (position, energy and time) must be digitized, the effective downlink communications bit rate can easily be exceeded by event rates from sources with only a modest flux. There are three detector types commonly in use. Imaging proportional counters and microchannel plates (not discussed here) have flown on *EINSTEIN, EXOSAT* and *ROSAT*. *ASCA* has introduced charge coupled solid state detectors and *SRG* and *XMM* will follow suit. Decher *et al.* (1994) have produced a very thorough review of X- and gamma-ray detectors.

5.1 Gas Filled Proportional Counters

The traditional proportional counter consists of an outer conducting cylinder at ground potential and a fine central wire at a suitable high voltage. The cylinder is filled with a an appropriate gas mixture and, depending upon the energy range, may or may not have a thin entrance window to admit X-rays. Ar and Xe are the primary gases used in proportional counters. Additional "quench" gases are usually added improve stability and reduce the rate of detection of non-X-ray events. The process of detection proceeds as follows. An atom in the gas absorbs the incoming X-ray and emits a photoelectron. The resulting excited state decays in a process that can emit Auger electrons and/or other photons. The total energy of the incoming electron is shared, according to the appropriate laws of physics, among the photoelectron and the decay products.

The photons and electrons cause further ionization until a cloud of primary electrons and ions is produced. The exception to this rule is the escape from the active volume of the detector of a fluorescent photon characteristic of one of the outer shell to inner shell transitions in the counting gas. The escape event steals a precisely known energy from the initial charge cloud. The probability for an escape event to occur depends upon both the value of the fluorescent yield for the inner shell vacancy and the probability that the detector gas can capture the emitted photon. The energy resolution of a proportional counter depends upon the statistics of both the number of primary electrons created in the absorption event and in the subsequent multiplication process for a single electron.

The electrons and ions drift in opposite directions along the electric field lines between the wire (anode) and shell (cathode). The accelerated electrons moving towards the wire enrich the cloud through collisions with the gas in the counter. The high field near the anode causes a final avalanche where most of the charge is produced. This process, called multiplication, can produce as many as 10^4 electrons from each initial X-ray event. The electrons, with their greater mobility, reach the wire in a time much shorter than it takes the ions reach the shell. The event can be sensed on the wire as a current pulse with a short rise time characteristic of the arrival of the electrons and a longer decay time characteristic of the arrival of the ions at the wall. The pulse can be integrated in a charge-sensing preamplifier and the resulting signal, called the pulse-height, is proportional to the energy of the incoming X-ray minus any energy carried away by escaping photons. Care must be taken to avoid situations where successive X-rays arrive at time intervals that are short compared with the ion drift time. In such cases, events will "pile up", become indistinguishable, and give an erroneous energy signal.

This simple proportional counter concept together with a suitable mechanical collimator has served well as the basic detector in early balloon, rocket and satellite missions. Its transformation to modern imaging devices is discussed below.

5.1.1 Multiwire Proportional Counter *(MWPC)*

An array of micro-proportional counter cells can be created by sandwiching a parallel array of finely spaced anode wires between two similar arrays of cathode wires with the wires in one arranged to be orthogonal to wires in the other. Figure 5.1 shows the design of the *SODART KFRD* multiwire detector (unpublished document). Pfefferman *et al.* (1986) describe the *ROSAT* multiwire detector. The separation between the cathode planes and the anode plane should be on the order of the individual wire separation. It is, however, difficult to make fine wire grids with spacings much smaller that 1 mm. The anode wires must be extremely fine to provide the necessary field strengths to cause multiplication. The high voltage, however, causes repulsive forces between the anode wires and attractive forces between the anode and cathode wires that ultimately distort the gain and position uniformity.

The region around each cathode vertex defines a counter cell (pixel). An initial absorption and drift region is established in the space above the grid sandwich. The electrons drift past the cathode grid and the high voltage between the anode and cathode cells provides multiplication in the usual way. The signal from the electron avalanche at the anode is used to derive the X-ray energy. The cathode signals provide the position information. The lateral diffusion within the drift region causes the electron cloud to broaden and by the time it reached the grids it typically expands to extend over several wires. This feature is used to advantage since it allows a position determination to be made from the centroid of an event with an accuracy that can be much smaller that the wire spacing.

5.1.2 Microstrip Gas Counter *(MSGC)*

Conventional *MWPC* designs usually lead to gain and position non-linearities, count rate restrictions and high operating voltages. The *MSGC* approach adopted for the *SODART HEPC/LEPC* detectors, shown in Fig. 5.2, is a novel approach towards removing these unwanted characteristics (Budtz-Jørgensen 1992; Budtz-Jørgensen *et al.* 1993,1994). The wire grids are replaced by narrowly spaced, conducting microstrips which are accurately deposited (±0.1 µm) on an insulating substrate. The normal procedure would be to deposit a set of anode and cathode wires on one side of the substrate to obtain energy (anode) and one dimensional position (cathode) information. An additional deposition of cathode wires, with an orientation orthogonal to the wires on the front side, is made on the rear side of the substrate. They provide the other dimension to complete the position measurement. For very low energy X-rays, a very thin, and therefore, fragile substrate would be required to obtain sufficient charge induction. In this case, a fine wire grid suspended above the frontside of the substrate can be used instead of the rear deposition. The wire grid is operated at a relatively low voltage and does not distort.

The manufacturing precision insures that the gain will be constant over the entire area of the array. The close spacing of the anode and cathode wires yields the required field strength for the multiplication process with a relatively low applied voltage when compared with conventional *MWPC* wire grid designs. Uniform gain and imaging properties over the whole array are insured by the lack of distortion when the high voltage is applied. Budtz-Jørgensen *et al.* (1994) report the results of *HEPC/LEPC* performance tests.

5.1.3 Gas Scintillating Proportional Counter *(GSPC)*

The normal drift process in a proportional counter produces additional electrons through ionizing collisions. By adjusting the electric field to lower values, however, it is possible to cause only excitation rather than ionization. If Xe is used as the counter gas, then the primary emission when the atom de-excites is in the UV wavelength band.

There are two versions of the *GSPC*. In one, there is a drift region followed by a high-field scintillation region. In this case, the total light output is proportional to the energy of the X-ray. In the second, being developed for *SAX* (Fig 5.3) (Erd and Bavdas 1992), the drift and scintillation regions are combined into a driftless counter with only a high-field region. Since the incoming X-ray can be absorbed at any depth in the counter, the number of UV photons is variable. The time during which UV photons are produced, called the burst length, is related to the depth at which the X-ray is absorbed. A combination of the number of photons and the burst length is used to determine the X-ray energy. It is the second type that also appears on *ASCA* (Kohmura *et al.* 1993).

For a *GSPC*, the energy resolution is determined by the statistics of the total number of UV photons created. This leads, in general, to a higher resolution than for the other varieties of proportional counters discussed above. Imaging is accomplished by using a multi-anode phototube that is coupled to the scintillation region by a gas tight, UV transparent window (typically quartz). The centroid of the initial X-ray event can be determined from the position of and the number of photons detected in each measurement chain.

5.2 Semiconductor Detectors

Gas filled proportional counters have served the needs of X-ray astronomers very well for

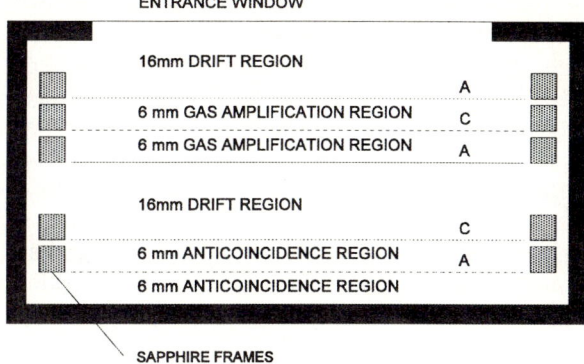

Fig. 5.1 The multiwire proportional counter for *KFRD* on *SODART*.

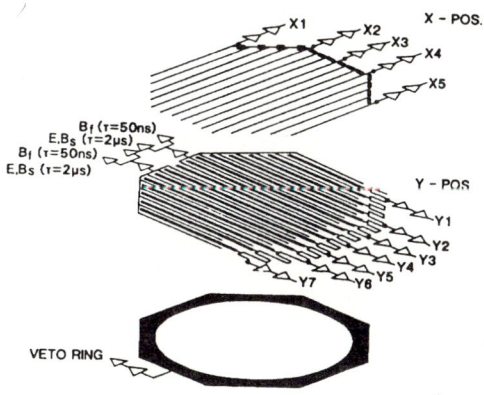

Fig. 5.2 The microstrip gas counter concept for *HEPC/LEPC* on *SODART*.

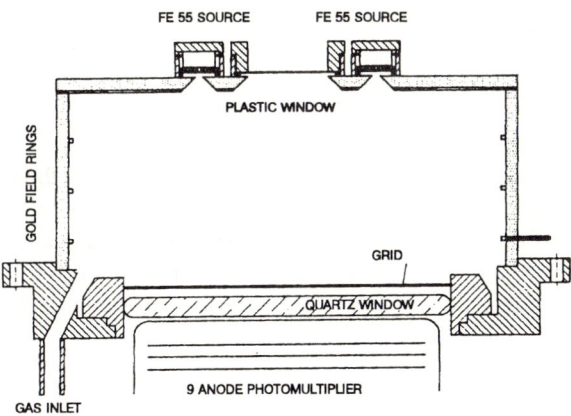

Fig. 5.3 The gas scintillation proportional counter concept. This is the configuration for *SAX*.

over three decades. The most refined gas filled detector designs, however, could not meet the expandind requirements for energy and position resolution. X-ray astronomers followed developments in ground based laboratories and developed the new technology required for space applications. The technology was based upon the properties of semiconductors although the devices are commonly called solid state detectors.

The goal, as with gas filled counters, is to convert the energy of an incoming X-ray photon into a current pulse that can be integrated into a signal that is proportional to the X-ray energy. The electronic structure of a semiconductor is the key to its successful use as an X-ray detector.

5.2.1 Lithium Drifted Silicon Detector *Si(Li)*

The defining element for a semiconductor is its band gap. It is the energy that an electron that normally resides in the valence band must acquire to be able to reach the conduction band. This process creates an electron-hole pair. Once in the conduction band, the electron can be collected as part of a current pulse if a suitable bias voltage is applied across the semiconductor. The energy that an absorbed X-ray deposits in a semiconductor goes partly into heat and partly into creating electron-hole pairs. Typically used materials, Si and Ge, have gaps between the valence and conduction bands on the order of 1 eV. Since not all of the absorbed energy goes into the creation of electron-hole pairs, the average energy for the creation of an electron-hole pair is ~3-4 eV. This number is about a factor of 10 less than the energy required to create and electron-ion pair in a gas filled detector. Thus, for a given X-ray energy, more primary electrons are created in a semiconductor detector and, as expected, the energy resolution is much better.

Semiconductor detectors are not without their problems. At room temperature, electrons in the valence band can, with some small probability, be thermally excited and reach the conduction band. This process contributes an unwanted noise that reduces the resolution. Semiconductors with a very small band gap should, therefore, be avoided. Room temperature conduction can never be avoided and most *Si(Li)* detectors are operated at liquid nitrogen temperature (77 °K) in the laboratory and at ~100 °K in space where cooling is difficult.

An always present, minute amount of impurity material leads, as well, to unwanted conductivity. Impurities in Si usually provide an excess of positive charge carriers and the material is then called p-type. Their effect can be neutralized by the addition of small amounts of Li into the detector. The process called drifting places the detector wafer against heated Li which diffuses (drifts) into the Si. With excess charge no longer available, the material is called intrinsic. Under ordinary circumstances, the Li is mobile at room temperatures, but modern techniques, including ion implantation have made it possible to avoid this problem and detectors can be stored at room temperature for long periods.

Unlike the case for gas filled detectors, there is no multiplication process. The only charge collected is the primary charge. This puts severe demands on the amplifying electronics attached to the detector. In present designs, despite being cooled, the input capacitance seen by the preamplifier contributes noise that becomes the dominant factor in the energy resolution of semiconductor devices below ~ 100 eV. *Si(Li)* detectors can tolerate fairly high counting rates before pulse pile-up occurs.

The most common form is called a *PIN* detector. The X-rays enter the device through a thin, hole enriched, p-type (undrifted) layer that is also used as one of the electrical contacts. Absorption takes place in a thick, lithium drifted layer (intrinsic) and the other face is an electron enriched, n-type layer. The n-type layer forms as part of the Li diffusion/implantation process. In principle, this technology could be extended into an imaging device, but there are practical limitations. High resolution telescopes require a detector an overall size of about 10 cm with pixel sizes of about 15 μm. Such large arrays made of tiny detectors are, however, achievable through another device called the *CCD*.

A single *Si(Li)* detector has been flown on *EINSTEIN*, a segmented array of five on *BBXRT* (on the Space Shuttle *COLUMBIA*) and an array of nineteen individual detectors will be flown on *SRG* (Fig. 5.4) (Vilhu *et al.* 1994).

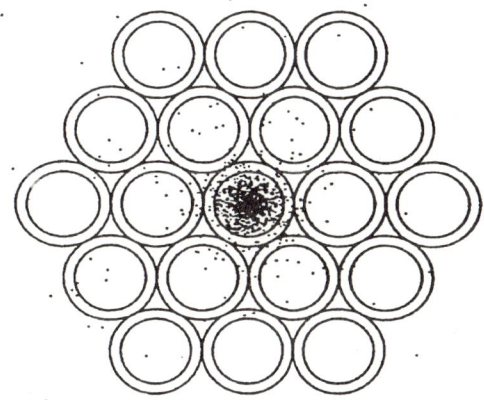

Fig. 5.4 The lithium drifted silicon array, *SIXA*, to be flown in the focal plane of the *SODART* telescope on *SRG*. The array is superimposed on a simulated *SODART* image of a point source.

5.2.2 Charge Coupled Device *(CCD)*

CCDs have become widely known because they are the recording device in the video camera. They are sensitive over a wide range of wavelengths from infrared through UV. Under appropriate conditions, they will also detect X-rays.

Modern microlithographic technology is used to grow arrays of capacitors in a piece of p-type Si. The capacitors are made by forming an insulating layer of silicon dioxide on one face of the silicon and then depositing an array of metal electrodes on the surface of the insulator. The technology is called metal-oxide-semiconductor *(MOS)*. The thickness of the silicon is adjusted to insure that X-rays in the desired energy range are absorbed with reasonable efficiency. The metal layer must be thin enough to let them through. All the metal electrodes are biased with a positive voltage, but in a way that creates a series of cells that are completely surrounded by cells with a smaller positive voltage. The cells with the higher positive voltage then become traps for any electrons that are liberated when X-rays are absorbed in their neighborhood.

The *CCD* is read out row by row. Increasing the voltage on the row next to the trap forms a deeper trap into which the electrons are "bumped". *CCDs* typically have two arrays. One detects events which are then shifted out to the other which stores them. The first then collects a new "frame", while the second is being read out. Each frame is translated into a serial string of current pulses. Each pulse corresponds to a particular pixel in the detector according to its position in the string. The current pulses can be integrated and, as with the *Si(Li)* detector, each signal is proportional to the energy of the absorbed X-ray.

There are, however, three differences with respect to the *Si(Li)* detector. For the *CCD*, the frame is analyzed at a time later that the arrival of the X-ray. For the *Si(Li)* detector the analysis is nearly instantaneous. If two X-rays are absorbed in the same *CCD* pixel during the integration time, the trapped charges will add and the event will be interpreted as a single event with an energy equal to the sum of both events. Pile-up causes the same problem in *Si(Li)* detectors, but their characteristic time constant is much faster than the framing time for a *CCD*. "Fast" *CCDs* are being developed to avoid these problems. The second difference is that, in some cases, where the X-ray event occurs deep in the semiconductor, it is possible to have an event shared by several adjacent pixels. This problem is dealt with by appropriate software for the *CCD*. Since individual pixels in *Si(Li)* array detectors are usually large, these events occur relatively less often. Finally, the volume of the pixel and, therefore, its capacitance is much smaller for the *CCD* than for the *Si(Li)* detector. This means a much smaller contribution to the limiting energy resolution, especially important at low energies where the electronic noise contribution dominates for the case of the *Si(Li)*.

CCD detectors are flying on *ASCA* and will fly on *SRG, AXAF* and *XMM*.

5.3 X-Ray Detectors Compared

The various parameters that define the X-ray detectors that are currently in orbit and those that are likely to be launched before the end of the decade are compared in Table 5.3.1.

6 X-ray Spectrometers

Spectral measurements with broad band detectors having poor energy resolution are most effective when the shape of a featureless continuum is to be studied. When many emission lines and/or absorption features are present, the diagnosis is reduced to finding the best fit to a model that includes all of the plasma parameters plus an additional set of parameters that describes the response function of the telescope/detector system as a function of X-ray energy and position in the field of view. Until now, this has been the case for almost all of the astrophysics that has been derived from X-ray observations to date. The primary exceptions have been the Sun, for which high energy resolution data have been routine, and for the few objects observed with wavelength dispersive spectrometers: The transmission grating instruments on *EINSTEIN* and *EXOSAT* and the Bragg Spectrometer on *EINSTEIN*.

Table 5.3.1 X-ray detectors on various missions compared

		ROSAT	ASCA	SODART	SRG	JET-X	AXAF	XMM	
		PSPC (MWPC)	CIS (CCD)	MSGC (H/L)*	SIXA	CCD	CCD	CCD (MOS)	CCD (PIN)
Energy res., ΔE	keV	$0.43(E/0.93)^{0.5}$		$0.33E^{0.5}$	0.17-0.26	0.05-0.15	0.05-0.15	0.05-15	0.05-15
Time resolution	μs	130	16×10^3	<5	30	24	60	20	
Maximum rate	cts s⁻¹	1×10^3	5×10^2	5×10^3	6×10^3	6×10^4	N/A	N/A	N/A
Gas: Ar	%	65	N/A	N/A	N/A	N/A	N/A	N/A	N/A
Xe	%	20	N/A	90	N/A	N/A	N/A	N/A	N/A
CH₄	%	15	N/A	10	N/A	N/A	N/A	N/A	N/A
Semiconductor		N/A	MOS	N/A	Si(Li)	MOS	N/A	MOS	PIN
Pixel array		N/A	840×848	N/A	19	1024×768	1024□	600□	400□
Number of detectors		1	4	1	1	2	4	7	1
Field of view:									
Total	arcmin	120	22×22	60/30	25	40	16×16	30	53×53
Per pixel	arcmin	N/A	0.026	N/A	3.9	0.027	0.0083	0.019	0.12
Active dia.	cm	8	2.2×2.2	15/7.5	5	2.07×2.76	2.46×2.46	6	6×6
Energy range	keV	0.1-2.4	0.4-12	2-25/0.2-8	0.5-20	0.3-10	0.4-10	0.1-15	0.2-10
Efficiency >(%)	keV	0.13-2.4(10)		2-15/1-8 (70)	0.5-20 (10)	0.35-10 (10)	0.4-8 (30)	0.3-10 (50)	
Thickness									
Gas:	cm	N/A	N/A	4/4	N/A	N/A	N/A	N/A	N/A
Si(Li):	mm	N/A	N/A	N/A	3	N/A	N/A	N/A	N/A
CCD	μm	N/A	40	N/A	N/A	38	N/A	65	N/A
Gas pressure	atm	1.5	N/A	1/0.5	N/A	N/A	N/A	N/A	N/A
Window mat.									
Polyimide	μm	N/A	N/A	7.5/0.85	0.25	N/A	N/A		
Aluminum	nm	N/A	N/A	40/40	120	40/80	N/A		
Au	nm	N/A	80	N/A	10	N/A	N/A		
Lexan	nm	40 (μg cm⁻²)	100	N/A	N/A	200	N/A		
Carbon	μg cm⁻²	50	N/A	N/A	N/A	N/A	N/A		
Polypropylene	μm	1	N/A	N/A	N/A	N/A	N/A		

*H/L=HEPC/LEPC; N/A=not applicable

Practical difficulties limit the number of cosmic X-ray sources for which good high resolution studies can be obtained. The sources should be bright enough to provide a statistically significant signal in each energy and spatial pixel within an observing time that is likely to be awarded by a time allocation committee. The achievable sensitivity of planned missions limits the possible number of high resolution studies to only a small portion of a potential list of known sources numbering over 100,000. It should, however, be possible to find archetypes from each classification of X-ray sources that are sufficiently bright to observe at high energy resolution. Although the number is small, the quality of the data obtained from them and the uniqueness of it interpretation will serve as a benchmark against which the data from broad band measurements can be interpreted more reliably.

6.1 Wavelength Dispersive Spectrometers

6.1.1 Transmission Grating Spectrometer (TGS)

Modern grazing incidence spectrographic technology with its complex ruled grating surfaces has made the transmission grating spectrometer all but obsolete, except for X-ray observations in space. The ruling of lines with spacings that are comparable with X-ray wavelengths is not easily accomplished. Transmission grating spectrometers for X-rays have, therefore, a rather low dispersion. Modern microlithographic technology has made it possible to manufacture accurately reproducible, free standing transmission gratings with line densities up to 5,000 lines mm^{-1}. Individual grating facets can be assembled into circular arrays that can fill the collecting apertures of an X-ray telescope. Although the dispersion angles are quite small for ordinary X-ray wavelengths, the combination of a long focal length telescope with high angular resolution and an imaging detector with high spatial resolution, is a good recipe for a medium to high resolution spectrometer. With dispersion angles typically greater than the field of view of the telescope, the best position for the grating is behind the telescope. It should be mounted on a toroidal surface (Fig. 6.1) (Brinkman *et al.* 1987) that compensates for the shape of the wave front emerging from the telescope. This geometry will eliminate most of the comatic aberrations.

The grating forms a spectrum on a curved surface (Fig. 6.2) (Markert *et al.* 1994). For a point source, the spectrum is a curved line. For an extended source, each spectral feature forms a image. The dispersion is usually insufficient to resolve the images and the overlap of spectral and image features is difficult to interpret.

The focal plane dispersion is $x \sim (\lambda/d)f$ where λ is the X-ray wavelength, d is the grating spacing and f is the focal length of the telescope. The angular resolution of the telescope, $\Delta\theta_H$, causes blurring in the focal plan of constant size $\Delta x \sim \Delta\theta_H f$ that limits the spectral resolution to $\Delta\lambda/\lambda = \Delta E/E = \Delta x/x \sim \Delta\theta_H (d/\lambda)$ which implies a constant $\Delta\lambda$. Grating limited resolution is very difficult to obtain. Thus, transmission gratings work best for long wavelength X-rays. Maximum resolution is obtained when the position resolution of the detector is at least $\Delta x/2$.

Transmission grating spectrometers have flown on *EINSTEIN* and *EXOSAT* and will fly on *AXAF*.

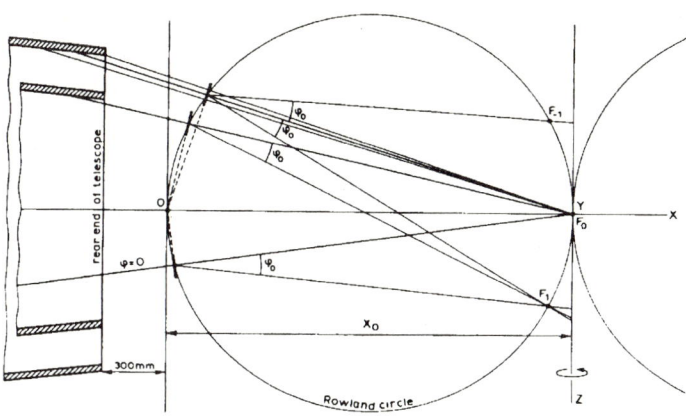

Fig. 6.1 The grating mounting surface is a toroid formed by rotating the Rowland circle about a line in the plane of dispersion that passes through the focus of the telescope. The center of each flat grating facet is mounted perpendicular to andat the intersection of the central ray in each reflected bundle with the toroidal surface.

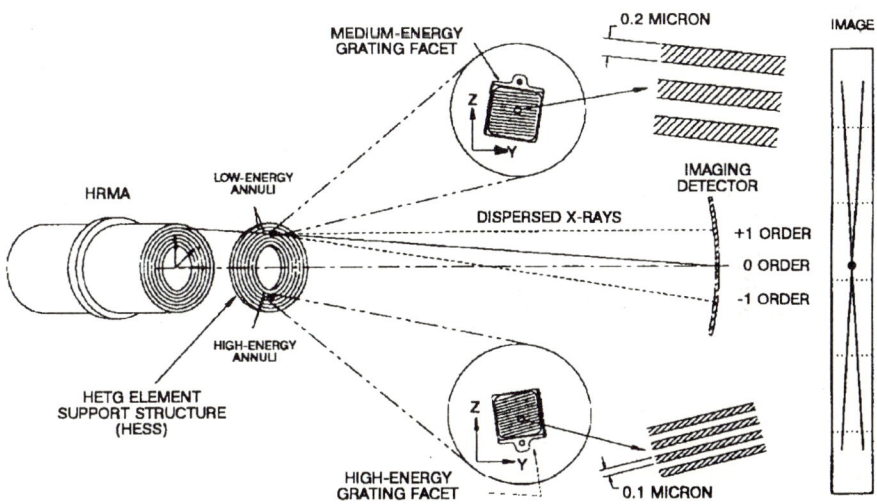

Fig. 6.2 The high and low energy transmission grating arrays that will fly on *AXAF*. The *HETG* and *LETG* grating facets are mounted with a 10 deg angle between their dispersion planes. Spectra from each grating can, therefore, be read independently by the segmented array of 6 *CCD* detectors that conforms to the curved focal surface.

6.1.2 Reflection Grating Spectrometer *(RGS)*

Optimal results will not be obtained when a *TGS*, with low dispersion, is coupled to a telescope with a *FWHP* much greater than several arcsec. The telescope blur can only be overcome with gratings having a very large dispersion. This implies d-spacings that are much smaller than present day technology can manufacture. It is well known, however, that the effective grating spacing can bedecreased when the rays are incident at a grazing angle rather a right angle. The effective line spacing is given by $d_e \sim d\sin\theta_i$ where θ_i is the angle of incidence. This design could be adopted for transmission gratings, but they must have a thickness that is sufficient to have opaque bars and, therefore, when tilted to grazing angles, only a very small transmission aperture would be left. This loss of efficiency is remedied by the use of reflection gratings. The same optical complications that preclude the placement of a *TGS* in front of the telescope also apply to the *RGS*. Again, the solution is to place the *RGS* behind the telescope and then deal with the problem of reducing the comatic aberration (Fig. 6.3) (Kahn and Hettrick 1985).

For a *TGS*, the comatic aberrations are largely removed by choosing the appropriate mounting surface for the multitude of individual facets of identical d-spacing that make up the whole grating. The manufacturing process for reflection gratings is completely different from that for transmission gratings. The *RGS* must be ruled and there is little or no flexibility in choosing the geometry of the substrate. It must be flat to insure that the ruling process will achieve the required accuracy. The grating facets must be tilted to an appropriate blaze angle, γ, that insures maximum reflection efficiency for a specific wavelength range of interest (Fig.6.4) (Kahn and Hettrick 1985). The design being developed for *XMM* is based on the in plane mounting geometry where parallel rulings are made in a direction perpendicular to the optical axis of the telescope. The only parameter left is the line spacing and, if it is varied approximately according to the square of the distance from the focus, the image will not contain primary coma. The use of one, very long, grating is impractical. Instead, a chevron array is used to cover one half the telescope aperture (Fig. 6.3). The X-rays that come through the other half are focused on a *CCD* detector that is used to monitor the broad band spectrum.

The *RGS* has the same difficulties with extended sources as does the *TGS*. Its greater dispersion makes the problem less severe. There is, however, an additional complication. Unlike the *TGS*, the *RGS* itself becomes a part of the imaging system and must not introduce significant new contributions to the *FWHP* of the grating image. This condition imposes strict requirements on both the figure of the *RGS* substrate and the roughness of the grating facets.

6.2 Bragg Crystal Spectrometers

The most successful, high resolution, non-solar, plasma diagnostics have been obtained with the Bragg crystal spectrometer flown on *EINSTEIN*. The diffraction angle, θ, for X-rays of wavelength, λ, by crystal planes with spacing, d, is given by Bragg's Law, $\lambda=2d\sin\theta$. Many efficient crystals can be found with d-spacings that are not much different in size from X-ray wavelengths. Bragg crystals can offer, therefore, a high wavelength dispersion when compared with gratings, a very high spectral resolution and a very high reflection efficiency.

The problem is that spectrometers based on these features can, in general, be used to study only a very limited wavelength range at one time. In contrast with grating spectrometers, all Bragg instruments must either be scanned or else must give up efficiency. Until they are ultimately replaced by microcalorimeter spectrometers (see section 6.3.1), they offer the capability of the highest resolution attainable over selected regions of a very wide range of energy that includes the Fe lines. Thus, the philosophy behind most Bragg spectrometers is that it is not absolutely necessary to have high resolution performance at every point in the spectral range from C through Fe. A few well chosen regions, rich in spectral features, will serve most of the needs of plasma diagnosticians. Broad band, lower resolution results will provide complementary diagnostics. In addition, certain Bragg instruments can be designed to image extended sources without loss of spectral resolution.

As is the case for grating spectrometers, there are two possible locations for Bragg spectrometers with respect to an X-ray telescope: They go either in front or behind. The Focal Plane Crystal Spectrometer *(FPCS)* on *EINSTEIN* (Schnopper and Kalata 1969) is an example of the latter and the *OXS* on *SODART* (Schnopper and Byrnak 1987) is an example of the former. The lack of throughput of the *EINSTEIN* telescope at the Fe line range coupled with the inherent inefficiencies in the *FPCS* design limited the use of the instrument to the study, at low energy, of only a handful of the brightest objects.

6.2.1 Objective Crystal Spectrometer (OXS)

The only Bragg spectrometer currently under development for the study of cosmic X-ray sources is the *OXS* that will fly on *SRG* (Schnopper 1994). The Bragg spectrometer that was planned for *AXAF* (similar to the one flown on *EINSTEIN)* has fallen victim to the *AXAF* restructuring and will not be flown. *SRG* is a very favorable platform for a Bragg spectrometer. The *SODART* telescopes have a very large collecting area and only a modest angular resolution. The angular extent of the response of a Bragg crystal to a parallel beam of monochromatic X-rays is not much different from the angular resolution of one of the *SODART* telescopes. This means that contributions from the telescope blur will not greatly degrade the *OXS* energy resolution performance.

A number of the deficiencies attached to Bragg spectroscopy, in general, will be removed in the *SODART/OXS* combination (Fig. 6.5). Placing the spectrometer in front of the telescope separates the processes of wavelength dispersion and imaging. The Bragg panel, consisting of a mosaic of many crystals, acts as both a narrow pass filter and a mirror. Each pixel in the field of view reflected by the panel and imaged by the telescope satisfies a unique Bragg angle on the panel. Each pixel in the detector can, therefore, be identified with a particular energy. The Bragg angle (and, therefore, the energy) is defined by the angle between the surface of the panel and the line that points from the panel to a specific location in the field of view. For a point source, only one wavelength satisfies a Bragg condition on the panel. An extended source defines a range of wavelengths incident on the crystal panel. The range is determined by either the source extent or the field of view, whichever is smaller. Each wavelength satisfies a locus of constant Bragg angle in the field of view. Scanning the panel about an axis perpendicular to the dispersion plane records either an extended portion of the spectrum for a point source or energy resolved images for an extended source. For practical reasons, the range of Bragg angles is limited to 45±15 deg.

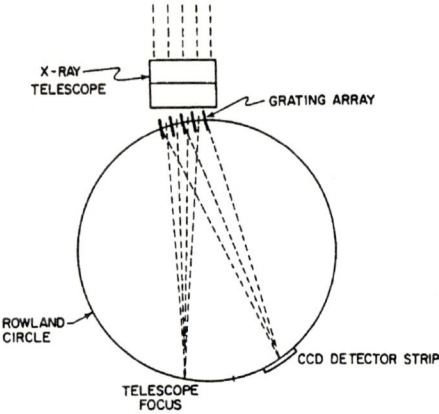

Fig. 6.3 The *XMM* reflection grating concept. The grating plates have variably spaced rulings that intersect the Rowland circle at 90 deg. Ideally, the rulings should be circles with a radius equal to the length of the cord that joins the intersection of the ruling with the Rowland circle with the grating focus. Each ruling has its own Roland circle.

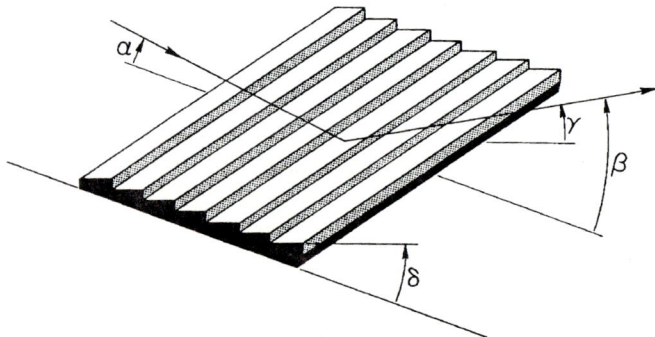

Fig. 6.4 An example of a grazing incidence , reflection grating plate.

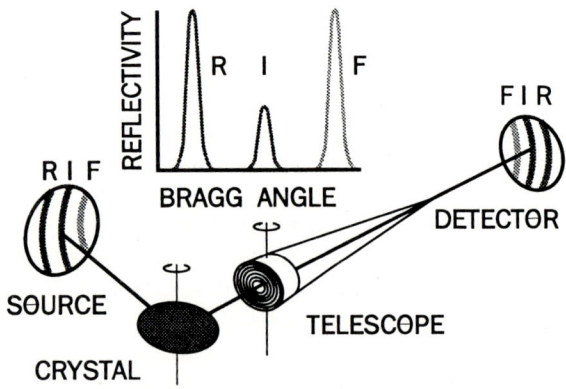

Fig. 6.5 The objective crystal spectrometer concept.

The scanning process is somewhat complicated. If the angular range of the spectrum exceeds the telescope field of view, then the telescope will have to be repositioned as necessary to avoid serious vignetting. If, in addition, the source is extended, further repositioning will be necessary. The typical observation will have the source located at ~90 deg from the spacecraft pointing axis. This may or may not inconvenience other instruments on board.

The *OXS* design for *SRG* has crystals mounted on both sides of the panel; Si and RAP share one side and LiF on the other. One feature unique to *OXS* is that the Si crystals (Si and S spectra) are coated with a multilayer structure that allows simultaneous measurements in the energy window below the C K-edge (Christensen *et al.* 1994; Louis *et al.* 1995).

6.3 Energy Dispersive Spectrometers

Silicon, or something similar, makes best semiconductor spectrometer i.e., a device that converts the energy of an X-ray into electrons. It is unlikely that other semiconductor technologies will produce a device with a spectral resolution that is $>\Delta E/E \sim 10^{-2}$. Various grating and Bragg crystal spectrometers can, in some cases, achieve a $\Delta E/E$ from 10^{-2} to nearly 10^{-4}. Grating spectrometers have difficulty imaging extended objects such as supernova remnants and clusters of galaxies, two important candidates for plasma diagnostic studies at high spectral resolution. Bragg spectrometers avoid that problem, but are very inefficient and are limited to studies of only the brightest sources. What is needed is a device that creates many more primary events that a solid state spectrometer, operates over a wide spectral band and images extended sources. It does exist and is called a X-ray microcalorimeter.

6.3.1 Low Temperature Microcalorimeters

At some time after an X-ray of energy E interacts within a detector of heat capacity C, when all of the electronic excitation has decayed, the net result is the conversion of X-ray energy into thermal energy. The temperature of the absorber, T, will rise by a small amount, $\Delta T = E/C$, depending upon its thermal properties. The temperature rise is proportional to the energy of the absorbed X-ray. It can be converted into an electrical signal if the absorber is connected to a temperature sensitive resistor called a thermistor (Fig. 6.6) (Moseley *et al* 1985). At room temperature, the small change is not noticeable against the thermal noise in the system. If the absorber is connected to a very low temperature heat sink through a weak thermal link then things improve. There is a limit on the energy resolution that is set by thermal fluctuations and is given by $\Delta E(FWHM)=2.35\,\xi(kT^2C)^{\frac{1}{2}}$, where ξ is a variable that is weakly dependent upon temperature and the electrical properties of the detector and k_B is the Boltzmann constant. $\Delta E(FWHM)$ is independent of energy. Candidate materials with appropriate values of C operating at temperatures less than 100 m°K have formal values of $\Delta E(FWHM)< 1$ eV. Measured values of $\Delta E(FWHM) \sim 7$ eV have been obtained for 6 keV X-rays (Fig. 6.7) (Silver and LeGros 1995).

The thermal bath cools the absorber with a thermal time constant given by $\tau=C/G$, where G is the thermal conductance of the link. In general, the absorber cools rather slowly and presently developed laboratory systems are limited to a counting rate of $\lesssim 100$ hz. These have sizes of about 0.5×0.5 to 1×1 mm^2. Small arrays of 3×3 or 4×4 units could be put at the center

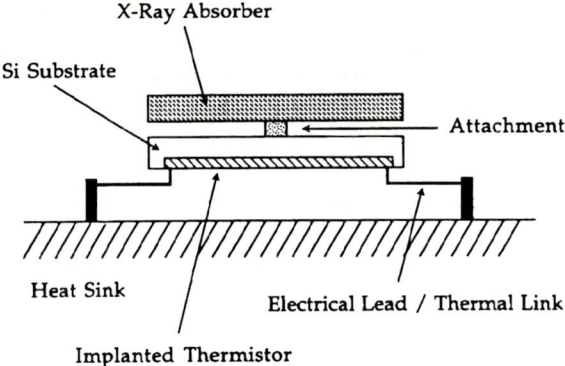

Fig. 6.6 The microcalorimeter spectrometer concept. X-rays depsit their energy and heat the absorber. Heat is transferred to the thermistor changing its resistance which is detected electronically. The heat sink is in contact with the cold environment that restores the detector to its working temperature after each event.

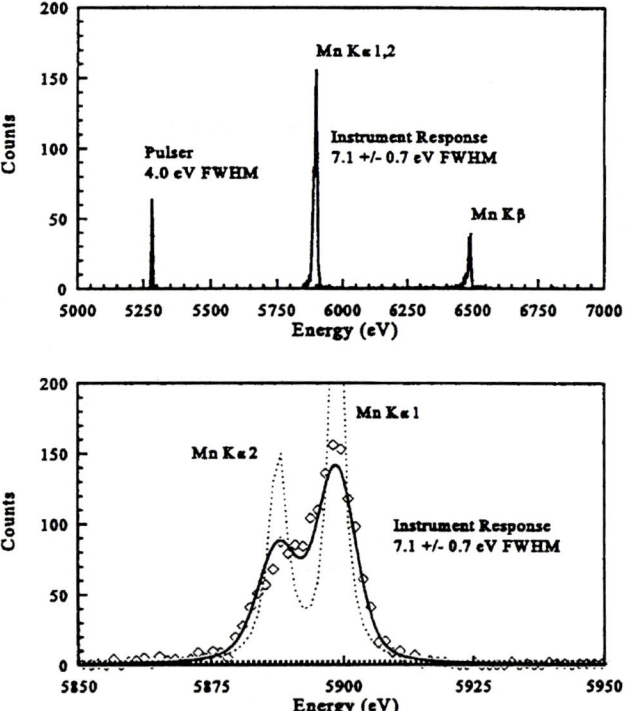

Fig. 6.7 The Fe55 X-ray spectrum obtained with a microcalorimeter. The ◇ symbols are the experimental data, the solid line is the best fit model and the dashed like is the comparison spectrum obtained from a Mn target X-ray tube with a two crystal spectrometer with a $\Delta E \sim 0.6$ eV.

of the field of view of an X-ray telescope. Since *AXAF* will probably be the last of the large, high angular resolution telescopes for some time to come, it is more likely that microcalorimeters will be used at the focus of telescopes similar to those being developed for *SRG* and *XMM*. Care must be taken to keep most of the X-rays received on the array otherwise observing efficiency will be lost. One way would be to use telescopes with smaller focal lengths that would have a proportionately smaller focal plane footprint of the *FWHP*. For telescopes with the same outer diameter, however, the one with the longer focal length has the greater collecting area at high energy. The high energy efficiency for small focal length telescopes can be improved by either applying graded multilayers to the reflecting surfaces or changing to a 4 reflection design that leads to smaller grazing incidence angles. The latter approach is being developed for *ASTRO-E* (Serlemitsos and Soong 1995).

6.4 X-ray Spectroscopic Capabilities Compared

The various parameters that define the X-ray spectrometers that are currently in orbit and those that are likely to be launched before the end of the decade are compared in Table 6.4.1. and in Figs. 6.8 (Schnopper 1994), 6.9 (adapted from Zombeck 1994 and Markert *et al.* 1994), 6.10 (Brinkman *et al.* 1989) , 6.11 (Markert *et al.* 1994) and 6.12 (Brinkman *et al.* 1989).

7 Line Detection Sensitivities Compared and Conclusions

This comparison is intended to give an overview of the performance of the various instruments described above. A series of simulated spectra are made for each spectrometer as it views the same source. The least complicated case of a coronal plasma in thermal equilibrium is assumed. A point source has also been assumed. Four energy bands have been chosen that are representative of the emission from cosmically abundant ions. Thermal spectra were derived from the work of Mewe and Kaastra (1994). These were then folded together with the detection properties of each of the instruments. The observing time was, in each case, scaled to provide the same count total per energy bin, where the bin size is determined by the energy resolution. Cosmic and instrumental background contributions add to the continuum and serve only to decrease the statistical significance i. e., signal-to-noise ratio, of the line detection. They are not included in the calculations. The simulations made by Westergard (1995) are shown in Figs. 7.1.

In most cases, it takes a $\Delta E \sim$ 1-5 eV to resolve the spectral detail. This is particularly true in the Fe and S spectral regions shown in Fig. 7.1. Except at the very lowest energy, the best spectra can be obtained with the *SRG OXS*, but the price is a very long observing time. Results that are nearly as good can be obtained with much greater economy of time with grating spectrometers on *AXAF* and *XMM* and with the microcalorimeter on *ASTRO-E*. It can also be seen that microcalorimeter spectrometers with their constant energy resolution perform best at high energy, while grating spectrometers with their constant wavelength resolution, $\Delta \lambda$, work best at low energy. Calorimeter spectrometers, with their high quantum efficiency, make the best use of observing time. They can, therefore, remain competitive when coupled with small diameter telescopes such as *ASTRO-E*.

Table 6.4.1 Spectrometers compared

Table 6.4.1a Grating spectrometers

		AXAF			*XMM*
		LEG	*MEG*	*HEG*	*RGS*
Type		trans.	trans.	trans	refl.
Dimensions	cm^2	1.6 (dia.)	2.62×2.63	2.39×2.39	10.0×20.0
No. required		600	192	144	200
Period	μm	0.991	0.4	0.2	1.5
Material		Au	Au	Au	Au
Bar thickness	μm	0.5	0.4	0.7	N/A
Support film:					
Polyimide	μm	N/A	1	0.5	N/A
SiC	mm	N/A	N/A	N/A	1
Energy range	keV	0.3-5.0	0.4-5.0	0.8-10	0.35.2.5
ΔE (point)	eV	0.3-62.5	0.4-42	0.7-83	1.25*

N/A=not applicable; *at 500eV

Table 6.4.2b Bragg crystal and micro calorimeter spectrometers

		SRG/SODART				*ASTRO-E*
		Ni/C(001)*	RAP(001)	Si(111)	LiF(220)	
Type		multiayer	crystal	crystal	crystal	calorimeter
Dimensions	cm^2	6×6		6×6	2.3×6.3	
No. required		36	140	36	212	
2d-spacing	Å	78	26.121	6.271	2.848	N/A
Ni: d-spacing	Å	15	N/A	N/A	N/A	N/A
Thickness:						
Crystal	mm	3	0.5	3	4.5-6.0	
Ni/C	μm	0.3	N/A	N/A	N/A	N/A
Energy range	keV	0.17-0.28	0.55-0.81	2.3-4.6	5.0-7.4	
H-and He-like		**	O	S, Ar	Fe	N/A
ΔE (point)	eV	3	1	1	5	

*Ni/C is deposited on the Si(111) crystals; **Various L-lines; N/A=not applicable

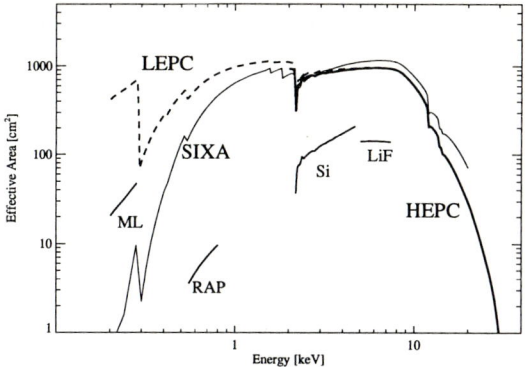

Fig. 6.8 The effective area *vs.* energy for the *SODART* detectors and Bragg spectrometer. The Bragg spectrometer curves include a model for the telescope.

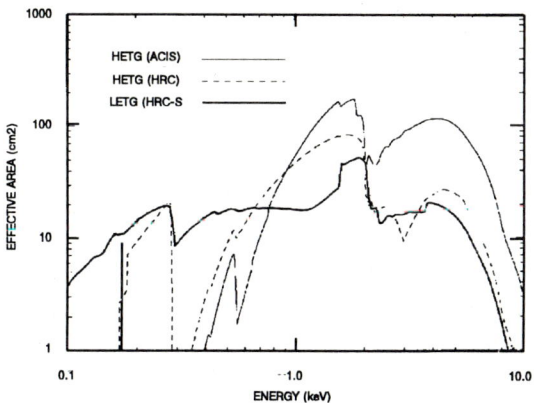

Fig. 6.9 The effective area *vs.* energy for the *AXAF* transmission grating spectrometers. The curves include models for the efficiency *vs.* area for the telescope and the *CCD* and channel plate cameras.

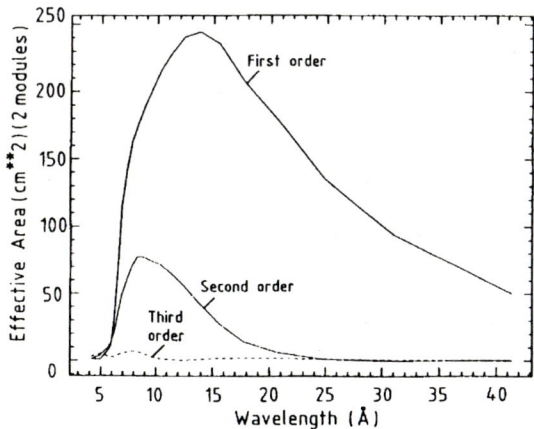

Fig. 6.10 The effective area *vs.* energy for both of the *XMM* reflection gratings. The curves include models for the telescope and the *CCD* camera.

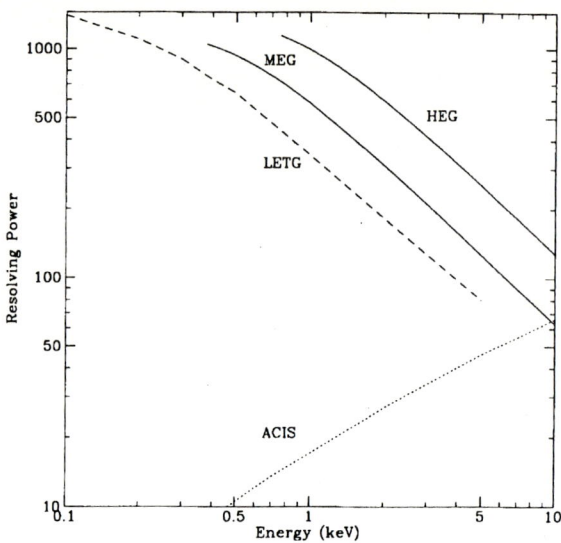

Fig. 6.11 The resolving power, $E/\Delta E$, *vs.* energy for the *AXAF* transmission grating and *CCD* spectrometers. The curve for *ACIS* is typical for all *CCD*s. The curves for the grating spectrometers are valid only for point sources.

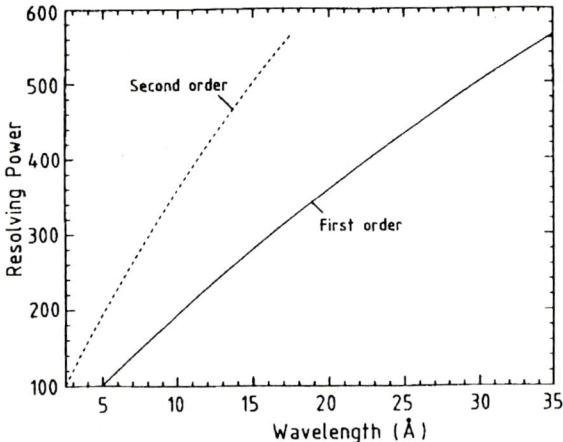

Fig. 6.12 The resolving powerr, $\lambda/\Delta\lambda$ (=$E/\Delta E$), *vs.* wavelength (~1/E) for the *XMM* reflection grating spectrometers. Because of their higher wavelength dispersion, the resolving power of these spectrometers are less sensitive to source extent and telescope blur than the *AXAF* transmission gratings.

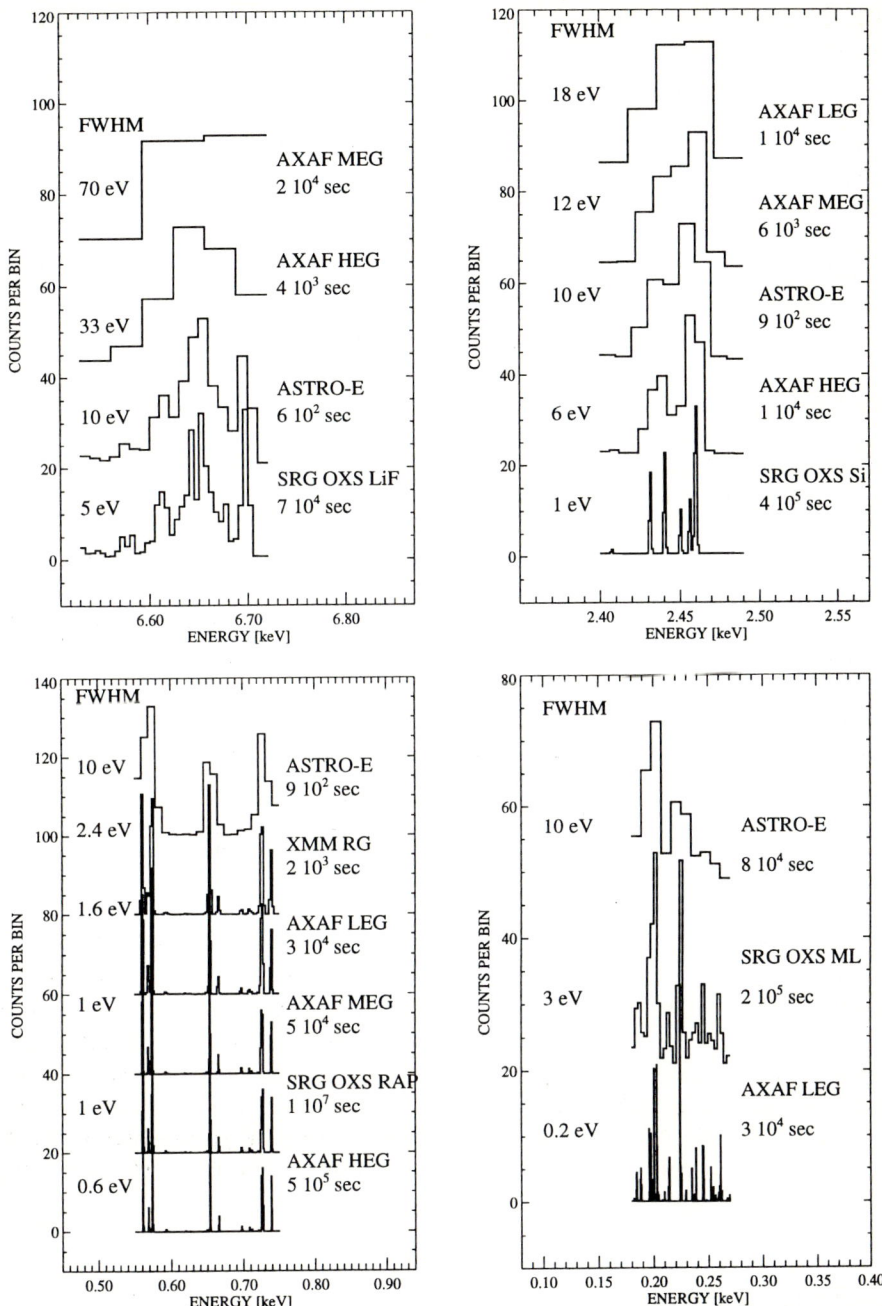

Fig. 7.1 Simulated spectra for the various instruments described above. The observing timetime required to produce the same counting rate in the peak channel is obtained from a plasma model that uses the relevant parameters for the performance of each instrument. A point source is assumed in all cases. Top left: He-like Fe; top right, He-like S; bottom left, He-like O and Ne, Fe L; bottom right, various lines in the C window. Note how energy resolution affects the quality of the spectrum.

The significance of the *FWHP* of the telescope is different role for each instrument. For a point source, the effect of the *FWHP* on energy resolution: is absent for microcalorimeters since they do not depend on wavelength dispersion; is modestly important for Bragg spectrometers since the have high wavelength dispersion; is very important for grating spectrometers since they have low wavelength dispersion. The *XMM RGS* has higher dispersion than the *XMM TGS*. Thus, the *RGS* would work quite well on *AXAF* while that would not be the case for the *TGS* on *XMM*. The influence of the background is ignored here.

Extended sources pose other problems. The *FWHP* defines the angular size of the element on an extended source that is viewed by imaging spectrometers such as the *ASTRO-E* microcalorimeter (provided that the angular size of the microcalorimeter chip is smaller than the *FWHP*). For the *SRG OXS* the angular width of the crystal response adds to the size of the element on the source. If *OXS* were to be installed on a higher resolution telescope such as *AXAF* or even *XMM*, the *FWHP* would become unimportant for spectral resolution. Overlapping spectral images will be the result when grating spectrometers observe extended sources with their low wavelength dispersion.

Another problem with extended sources is the requirement to detect the bright lines with a statistical significance of at least 5σ. This means that the surface brightness of the source pixel must be high enough to provide ~25-30 detected photons in the peak of each line feature within a reasonable observing time.

Many spectroscopic instruments will be flown before the end of the decade. They represent a powerful capability to make comprehensive plasma diagnostics of cosmic X-ray sources. For some of them, the results will be as significant as those presently obtained for the Sun.

8 Acknowledgements

Niels Jørgen Westergaard prepared the figures illustrating the performance of the *XSPECT* instruments and the line sensitivities for the various instruments. Ana M. Ulla contributed many thoughtful suggestions that improved the organization of the paper. Joachim Trümper and Alan Wells provided up-to-date information for the tables. Their efforts are much appreciated. Part of this work was prepared as a National Research Council Senior Research Associate while on a sabbatical leave from DSRI at the Laboratory for High Energy Astrophysics, NASA/GSFC. Financial support from the National Research Council, the Danish Space Research Institute and the Danish Natural Science Research Council is gratefully acknowledged.

10 List of Acronyms

ACIS	*AXAF CCD* Imaging Spectrometer *(NASA)*
ASCA	Advanced Satellite For Cosmology and Astrophysics (means "flying bird" Japan)
ASTRO-E	The fifth in a series of Japanese astronomical satellites
BBXRT	Broad Band (thin foil) X-ray Telescpe on Space Shuttle Columbia *(NASA)*

CCD	Charged Coupled Device
CFA	Center for Astrophysics
DSRI	Danish SpaceRresearch Institute
EINSTEIN	The first cosmic X-ray observatory with a grazing incidence telescope *(NASA)*
EPIC	European Photon Imaging *(CCD)*Camera on *XMM* (Italy,United Kingdom and Germany)
ESA	Europen Space Agency
EUV	extreme ultraviolet
EXOSAT	European X-ray Observatory Satellite *(ESA)*
FOM	figure of merit
FPCS	Focal Plane Crystal Spectrometer on *EINSTEIN (NASA)*
FWHP	full width at half power (see *HEW*)
FWHM	full width at half maximum
FW90	full width at 90 percent power
GSFC	Goddard Space Flight Center *(NASA)*
GSPC	gas scintillating proportional counter
HEPC	High Energy Proportional Counter on *SODART* (Denmark)
HETGS	High Energy Transmission Grating Spectrometer on *AXAF (NASA)*
HEW	Half energy width (see *FWHP*)
HRC	High Resolution (channel plate) Camera on AXAF (NASA)
HRMA	High Resolutiuon Mirror Assembly on *AXAF (NASA)*
ISAS	Institute of Space and Astronautical Science (Japan)
JET-X	Joint European Telescope for X-ray s (United Kingdom, Italy, Germany)
KFRD	Focal Plane Röntgen Detector on *SODART* (Russia)
LEPC	Low Energy Proportional Counter on *SODART* (Denmark)
LETGS	Low Energy Transmission Grating Spectrometer on *AXAF (NASA)*
MOS	metal-oxide semiconductor
MPE	Max-Planck-Institue für Physik und Astrophysik: Institut für Extraterrestriche Physik
MSGC	microstrip gas counter
MWPC	multiwire proportional counter
NASA	National Aeronautics and Space Administration *(USA)*
OXS	Objective Crystal Spectrometer on *SODART* (Denmark)
PIN	p-type-intrinsic-n-type semiconductor
PSF	point spread function
RAP	rubidium acid pthalate (a crystal on *OXS*)
RGS	Reflection grating spectrometer on *XMM* (Netherlands)
ROSAT	Röntgen Satellite (Germany)
SAO	Smithsonian Astrophysical Observatory
SAX	X-ray Astronomical Satellite (Italy)
Si(Li)	lithium drifted silicon
SIXA	Silicon X-ray Array on *SODART* (Finland)
SMM	Solar Maximum Mission *(NASA)*
SNR	Supernova remnant
SODART	Soviet Danish Röntgen Telescope
SRG	Spectrum Röntgen Gamma (Russia)
SXRP	Stellar X-ray Polarimeter on *SODART (NASA)*

TAUVEX	Tel Aviv University *UV* Experiment on *SODART* (Israel)
TGS	Transmission Grating Spectrometer on *AXAF (NASA)*
UHURU	The first X-ray satellite (means "peace" in Kenya) *(NASA)*
UV	ultraviolet
XMM	X-ray Multi-Mirror satellite *(ESA)*
XUV	extreme ultraviolet
YOHKOH	Japanese Solar X-ray Observatory

11 References

Bräuninger, H., Danner, R., Findeis, N., Hauff, D., Holl, P., Kemmer, J., Kendziorra, E., Krämer, J., Lechner, P., Lutz, G., Meidinger, N., Mohan, M., Pinotti, E., Reppin, C., Staubert, R., Strüder, L., Trümper, J.,and von Zanthier, C. 1992, *ESA SP-356*, 69-73.

Brinkman, A. C., van Rooijen, J. J., Bleeker, J. A. M., Dijkstra, J. H., Heise, J., de Korte, P. A. J., Mewe, R., and Pearels, F. 1987, *Astro. Lett. and Commun.*, *26*, 87.

Brinkman, A. C., Aarts, H. J. M., Branduardi-Raymont, G., Hailey, C. J., Jasen, F. A., Kahn, S. M., de Korte, P. A. J., and Zender, A. 1989, *Proc. SPIE.* **1159**, 495-509.

Brosch, N., Shemi, A., and Netzer, H. 1994, *Proc. SPIE*, **2279**, 469-479.

Burke, B. E., Mountain, R. W., Daniels, P. J., Cooper, M. J., and Dolat, V. S. 1993, *Proc. SPIE*, **2006**, 272-285.

Budtz-Jørgensen, C. 1992, *Rev. Sci. Instrum.*, **63**, 648-654.

Budtz-Jørgensen, C., Bahnsen, A., Madsen, M. M., Olesen, C., and Schnopper, H. W. 1993, *Proc. SPIE*, **1748**, 162-173.

Budtz-Jørgensen, C., Bahnsen, A., Madsen, M. M., Olesen, C., Jonasson, P., and H. W. Schnopper, H. W., 1994, *Proc. SPIE*, **2279**, 517-525.

Christensen, F. E., Hornstrup, A., Frederiksen, P., Abdali, S., Grundsøe, P., Schnopper, H. W., Lewis, R., Hall, C., and Borozdin, K. 1993, *Proc SPIE*, **2011**, 540-548.

Christensen, F. E., Westergaard, N. J., Rassmussen, I., Schnopper, H. W., Wiebicke, H.-J., Halm, I.,and Geppert, U. 1994, *Proc. SPIE*, **2279**, 511-516.

Christensen, F. E., Hornstrup, A., Frederiksen, P., Addali., S., Grundsøe, P., Polny, J., Westergaard, N. J., Nørgaard-Nielsen, H.-U., Schnopper, H. W., Hall, C., and Lewis, R. 1995, *Proc. SPIE*, **2515**, 458-467.

Citterio, O., Conconi, P., Ghigo, M., Loi, R., Mazzoleni, F., Conti, G., Mineo, T., Sacco, B., and Bräuninger, H. 1994, *Proc. SPIE*, **2279**, 480-492.

Citterio, O., Conconi, P., Ghigo, M., Mazzoleni, F., Poretti, E., Conti, G., Cusumano, G., Saco, B., Bräuninger, H., and Burket, W. 1995, *Proc. SPIE*, **2215**, 44-54.

Culhane, J. L., Gabriel, A. H., Acton, L. W., Rapley, C. G.., Phillips, K. J., Wolfson, C. J., Antonucci, E., Bentley, R. D., Catura, R. C., Jordan, C., Kayat, M. A., Leibacher, J. W., Parmar, A. N., Sherman, J. C., Springer, L. A., Strong, K. T., and Veck, N. J.. 1981, *Ap. J. (Letters)*, **244**, L141-L146.

Decher, R., Ramsey, B. D., and Austin, R. 1994, *NASA SP-517*.

Erd, C., and Bavdas, M. 1992, *Proc. SPIE*, **1743**, 133-138.

Freeman, M., Hughes, J., VanSpeybroeck, L., Bilbro, J., and Weisskopf, M. 1993, *Proc. SPIE*, **1742**, 136-151.

Gondoin, Ph., De Chambure, D., van Katwijk, K., Aschenbach, B., Citterio, O., and Willingale, R. 1994, *Proc. SPIE*, **2279**, 86-100.

Harra-Murnion, L. K. *et al.*, *Ap. J.*, in press.

Holland, A. D., Turner, M. J. L., Burt, D. J., and Pool, P. 1993, *Proc. SPIE*, **2006**, 2-10.

Hughes, J., Schwartz, D., Szentgyogyi, A., VanSpeybroeck, L., and Zhao, P. 1993, *Proc. SPIE*, **1742**, 152-161.

Inoue, H. 1992, *ISAS Symposium on Astrophysics*, 149-153.

Kahn, S. M., and Hettrick, M. C. 1985, *ESA SP-239*, 237-244.

Kaaret, P., Novick, R., Martin, C., Shaw, P., Fleischman, J. R., Hamilton, T., Sunyaev, R., Lapshov, I., Silver, E., Ziock, K., Weisskopf, M., Elsner, R., Ramsey, B., Chanan, G., Manzo, G., Giarruso, S., Santangelo, A., Costa, E., Piro, L., Fraser, G., Pearson, J. F., Lees, J. E., Perola, G. C., Massaro, E., and Matt, G. 1990, in *Observatories in Earth Orbit and Beyond*, Y. Kondo, ed., *Kluwer Academic Publishers*, 443-449.

Keenan, F. P., McKenzie, D. L., Phillips, K. J. H., and Conlon, E. S. 1994, *Ap. J.*,**426**, 454-458.

Kohmura, Y., Fukazawa, Y., Ikebe, Y., Ishisaki, Y., Kamijo, S., Kaneda, H., Makishima, K., Matsushita, K., Nakagawa, K., Tashiro, M., Ohashi, T., Inoue, H., Ishida, M., Makino, F., Murakami, T., Ogawara, Y., Tanaka, Y., Ueda, Y., Ebisawa, K., Mihara, T., Takeshima, T., Hiyoshi, K., Horii., Shoumura, R., And Taguchi, K. 1993, *Proc. SPIE*, **2006**, 78-89.

Louis, E., Spiller, E., Abdali, S., Christensen, F. E., Voorma, H.-J., Koster, N. B., Frederiksen, P., Tarrio, C., Gullikson, E. M., and Bijkerk, F. 1995, *Proc. SPIE*, **2215**, 194-203.

Lumb, D. H., Bautz, M. W., Burrows, D. N., Doty, J. P., Garmire, G. P., Gray, P., Nousek, J. K. A., and Ricker, G. R. 1993, *Proc. SPIE*, **2006**, 265-271.

Markert, T. H., Canizares, C. R., Dewey, D., McGuirk, M., Pak, C., and Schattenburg, M. L.. 1994, *Proc. SPIE*, **2280**, 168-180.

McKenzie, D. L., Broussard, R. M., Landecker, P. B., Rugge, H. R., Young, R. M.., Doschek, G. A., and Feldman, U. 1980, *Ap. J. (Letters)*, **238**, L43.

Mewe, R. and Kaastra, J. S. 1994, *European Astronomical Society Newsletter*, **Issue 8**, 3.

Moseley, S. H., Juda, M., Kelly, R. L., McCammon, D., Stahle, C. K., Szymkowiak, A. E., and Zhang, J. 1985, *ESA SP-356*, 13-19.

Owens, A., McCarthy, K. J., and Wells, A. A. 1994, *Proc. SPIE*, **2279**, 493-503.

Pallavicini, R. 1995, *Lecture Notes in Physics*, this volume.

Pfeffermann, E., Briel, U. G., Hippmann, H., Kettenring, G., Metzner, G., Predhel, P., Reger, G., Stephan, K.-H., Zombeck, M.. V., Chappel, J., and Murray, S. S. 1986, *Proc. SPIE*, **733**, 519.

Ried, P. B. 1995, *Proc. SPIE*, **2215**, 361-374.

Serlemitsos, P. J., Petri, R., Glasser, C., and Birsa, F. 1984, *IEEE Trans. Nucl.Sci.*, **Vol. NS-31**, 786-790.

Serlemitsos, P. J., and Soong, Y. 1995, *NASA/GSFC LHEA Preprint 95-08*.

Schnopper, H. W. 1994, *Proc. SPIE*, **2279**, 412-423.

Schnopper, H. W., and Kalata, K. 1969, *Appl. Phys. Lett.*, **15**, 134-136.

Schnopper, H. W., and Byrnak, B. P. 1987, *Appl. Opt.*, **26**, 2871-2876.

Silver, E., and LeGros, M. 1995, *X-ray Spectrometry*, in press.

Soong, Y., Jalota, L., and Serlemitsos, P. J., 1995, *Proc. SPIE*, **2215**, 64-69.

Tanaka, Y., Inoue, H., and Holt, S. S. 1994, *Publ. Astron. Soc. Japan*, **46**, L37-L41.

Trümper, J. 1984, *Physica Scripta*, **T7**, 209-215.

Verhoeve, P. W. A. M., Jansen, F. A., De Korte, P. A. J., den Boggende, A. F. J., Brinkman, A. C., Aarts, H. J. M., Burt, D., and Pool, P. 1992, *ESA SP-356*, 75-80.

Vilhu, O., Huovelin, J., Tikkanen, T., Hakala, P., Muhli, P., Kämäräinen, V. J., Sipilä, H.,

Taylor, I., Pohjonen, J., Päivike, H., Toivanen, J., Sunyaev, R., Kuznetzov, A., and Abrosimov, A. 1994, *Proc. SPIE*, **2279**, 532-543.

Weisskopf, M. C., O'Dell, S. L., Elsner, R. F., and VanSpeybroeck, L. P. 1995, *Proc. SPIE*, **2215**, 312-329.

Wells, A. A., Owens, A., and Sims, M. R.1994, *Proc. SPIE*, **2279**, 424-445.

Westergaard, N. J. 1995, *Private communication.*

Westergaard, N. J., Byrnak, B. P., Christensen, F. E., Grundsøe, P., Hornstrup, A., Henrichsen, S., Henriksen, U., Jespersen, E., Nørgaard-Nielsen, H. U., Polny, J., Schnopper, H. W., and Ørup, P 1990, *Optical Eng.*, **26**, 658-664.

Winkler, P. F., Canizares, C. R., Clark, G. W., Markert, T. H., Kalatta, K., and Schnopper, H. W. 1981, *Ap. J. (Letters)*, **246**, L27-L31.

Wolter, H. 1952a, *Ann. Phys.*, **NY 10**, 94.

Wolter, H. 1952b, *Ann. Phys.*, **NY 10**, 286.

Zombeck, M. V. 1994, *Harvard/SAO Center for Astrophysics, Preprint Series No. 4003.*

Lecture Notes in Physics

For information about Vols. 1–434
please contact your bookseller or Springer-Verlag

New Series m: Monographs

EADN Proceedings in Springer's Lecture Notes in Physics

EADN I

I. Appenzeller, H. J. Habing, P. Léna (Eds.)

Evolution of Galaxies.
Astronomical Observations

Proceedings of the Astrophysics School I, Organized by the European Astrophysics Doctoral Network at Les Houches, France, 5 – 16 September 1988
1989. X, 391 pages. Hardcover. ISBN 3-540-51315-9

These eight lectures have been written up in a clear and pedagogical style in order to serve as an introduction for students to fields of modern astrophysical and astronomical research where otherwise textbooks are not available. The first four lectures cover topics in galactic astronomy (formation, structure and evolution of galaxies) and the remaining four are devoted to observational methods and astronomical instrumentation. The lectures in the European Astrophysical Doctoral Network rank among the most highly respected specialists, and their lectures have been carefully edited and updated before publication.

EADN II

B. C. de Loore (Ed.)

Late Stages of Stellar Evolution

Computational Methods in Astrophysical Hydrodynamics
Proceedings of the Astrophysics School II, Organized by the European Astrophysics Doctoral Network at Ponte de Lima, Portugal,
11 – 23 September 1989
1991. VIII, 390 pages. Hardcover. ISBN 3-540-53620-5

This collection of 7 lectures is aimed to be a textbook for graduate students who want to learn about modern developments in astronomy and astrophysics. The first part surveys various aspects of the late stages of stellar - evolution, including observation and theory.
B. C. de Loore's long article on stellar structure is followed by reviews on supernovae, on circumstellar envelopes, and on the evolution of binaries. The seond part deals with the important problem of modeling stellar evolution based on the computational hydrodynamics.

Springer

Please order by
Fax: +49 30 8207 301
e-mail: orders@springer.de
or through your bookseller

Springer-Verlag, P. O. Box 31 13 40, D-10643 Berlin, Germany.

EADN Proceedings in Springer's Lecture Notes in Physics

EADN III
A. Sandqvist, T. P. Ray (Eds.)
Central Activity in Galaxies
From Observational Data to Astrophysical Diagnostics

Proceedings of the Predoctoral Astrophysics School III,
Organized by the European Astrophysics Doctoral Network
(EADN) in Dublin, Ireland, 10 – 22 September 1990
1993. XIII, 235 pages. Hardcover. ISBN 3-540-56371-7

This outstanding collection of surveys addresses graduate and
predoctoral students. It reports on theoretical research and
observational data on active galactic nuclei: The enigma of the
nuclei of galaxies with their central "monster" driving the vast
range of activity observed in quasars, own Galaxy are explored
in this volume.

Topics covered include: the impact of recent measurements in
the infrared and radio region on our knowledge of the nucleus
of our Galaxy; the spectra and classification of active galactic
nuclei, the properties of their host galaxies, their cosmological
distribution and evolution, the role of stars and the hydrodynamics of the interstellar medium in the nuclei; the
description of the inner parsec of a standard active galactic
nucleus based on direct interpretation of the observations; the
infrared activity of galaxies; the physics of radio galaxies and
their jets, emphasizing the physics of gas flow and high-energy
particle interactions as well as shock acceleration. These are all
discussed in considerable depth and presented in self-contained
chapters with exhaustive reference lists of the scientific
literature.

■ ■ ■ ■ ■ ■ ■ ■ ■ ■ ■

Please order by
Fax: +49 30 8207 301
e-mail: orders@springer.de
or through your bookseller

Springer

Springer-Verlag, P. O. Box 31 13 40, D-10643 Berlin, Germany.

EADN Proceedings in Springer's Lecture Notes in Physics

EADN IV
J. van Paradijs, H. M. Maitzen (Eds.)
Galactic High-Energy Astrophysics
High-Accuracy Timing and Positional Astronomy
Lectures Held at the Astrophysics School IV, Organized by the European
Astrophysics Doctoral Network (EADN) in Graz, Austria, 19 – 31 August 1991
1993. XIII, 293 pages. Hardcover. ISBN 3-540-56874-3

This book addresses graduate students in astronomy and astrophysics. The
first part is devoted to galactic high-energy astrophysics. It treats particle
accelerations (including shocks), the interstellar medium and supernova
remnants, high-energy emissions from normal stars and accretion in close
binaries. The second part deals with observations, such as pulsar timing, and
its measurement with radioastronomical tools, and astrometry as performed
in the HIPPARCOS satellite program.

EADN V
T. P. Ray, S. V. W. Beckwith (Eds.)
Star Formation and Techniques in Infrared and mm–Wave Astronomy
Lectures Held at the Predoctoral Astrophysics School V, Organized by the
European Astrophysics Doctoral Network (EADN) in Berlin, Germany, 21
September – 2 October 1992
1994. XIII, 314 pages. Hardcover. ISBN 3-540-58196-0

The rapid growth in our understanding of how stars form owes a lot to recent
developments in techniques for carrying out infrared and millimeter-wave
astronomy. Thus **Star Formation and Techniques in mm-Wave Astronomy**
were natural joint themes for the Fifth EADN Predoctoral Astrophysics
School held at the Technische Universität Berlin.
The lecture courses by six world-class experts
are aimed at postgraduate students and
scientists with a non-specialist interest
in the field. Topics include molecular
clouds, T Tauri stars, OB stars,
observation methods in infrared
and mm astronomy, as well as high
resolution techniques.

Springer

Please order by
Fax: +49 30 8207 301
e-mail: orders@springer.de
or through your bookseller

Springer-Verlag, P. O. Box 31 13 40, D-10643 Berlin, Germany.

EADN Proceedings in Springer's Lecture Notes in Physics

EADN VI

G. Contopoulos, N. K. Spyrou, L. Vlahos (Eds.)

Galactic Dynamics and N-Body Simulations

Lectures Held at the Astrophysics School VI, Organized by the
European Astrophysics Doctoral Network (EADN) in
Thessaloniki, Greece, 13 – 23 July 1993
1994. XIV, 417 pages. Hardcover. ISBN 3-540-57983-4

This book provides an in-depth coverage of modern research
on dynamical systems. The first part discusses stellar dynamics,
integrable systems, the transition to chaos and instabilities in
stellar dynamics as well as the dynamics of spiral galaxies.
Models are given and compared with observations. The second
part is devoted to the direct method of N-body simulations, to
gas dynamics simulations and to galaxy formation. Special care
is taken to give a pedagogical presentation of the material making this a unique text well suited for graduate courses in astrophysics.

In preparation:

EADN VIII

V. Icke (Ed.):

The Structure of the Universe

Lectures Held at the European Astrophysics Doctoral Network
Predoctoral School VIII, Organized by the European
Astrophysics Doctoral Network (EADN) in Leiden,
The Netherlands, 13 – 22 July 1995
Due July 1996

Topics:
Cosmological models – Statistical analysis
of galaxy clustering – Tools of trade –
The cosmic microwave background
radiation – Observational cosmology –
Observable signatures of inflation –
Point processes and random fields.

Springer

Please order by
Fax: +49 30 8207 301
e-mail: orders@springer.de
or through your bookseller

Springer-Verlag, P. O. Box 31 13 40, D-10643 Berlin, Germany.